Air Conditioning

A practical introduction

Third edition

David V. Chadderton

LONDON AND NEW YORK

First edition published 1993
by Routledge
Second edition 1997

This third edition published 2014
by Routledge
2 Park Square, Milton Park, Abingdon, Oxon, OX14 4RN

and by Routledge
711 Third Avenue, New York, NY 10017

Routledge is an imprint of the Taylor & Francis Group, an informa business

British Library Cataloguing in Publication Data
A catalogue record for this book is available from the British Library

Library of Congress Cataloging-in-Publication Data
Chadderton, David V. (David Vincent), 1944-
Air conditioning : a practical introduction / David V. Chadderton. – Third edition.
pages cm
Includes bibliographical references and index.
1. Air conditioning. I. Title.
TH7687.C43 2014
697.9'3–dc23
2013040197

ISBN13: 978-0-415-70338-3 (pbk)
ISBN13: 978-1-315-79406-8 (ebk)

Typeset in Frutiger Light by
Cenveo Publisher Services

Printed and bound in Great Britain by
TJ International Ltd, Padstow, Cornwall

Contents

Figures

Tables

Preface

Air Conditioning: A practical introduction, third edition, is a textbook for undergraduate courses in Building Services and Environmental Engineering, Mechanical Engineering, BTEC Continuing Education Diploma, Higher National Diploma and Certificate courses in Building Services Engineering, and will be of considerable help on National Certificate and Diploma programmes. Heating, ventilating and air conditioning is studied on undergraduate, CED, HND and HNC courses in Architecture, Building, Engineering, Building Management and Building Surveying and is part of all courses relevant to the design, construction and use of buildings.

The design of air conditioning systems involves considerable calculation work which is now mainly carried out with dedicated computer software; however, the engineering principles need to be fully comprehended in the first instance, as are the basic formulae and calculation techniques utilised. The reader is actively involved in the use of such data by the use of worked examples and copious exercises and design assignments.

Downloadable spreadsheets are used extensively throughout the book for assessing many cases of peak summertime temperature in buildings including The Shard at London Bridge, White Tower, London Olympic Velodrome, Solent University, Queens Building and housing.

Workbook cases are provided for assessment of peak design cooling load and annual energy for locations around the world. A downloadable file of climate data for many world locations demonstrates plant loads relative to London for quick assessment; these are mainly tropical and Middle East climates as these produce multiples of the London load for the same building, for example The Shard, if it were transplanted elsewhere and among other similar towers. The Bourke Street case study is the result of an energy audit and discusses annual energy use. User data can easily be added to any workbook.

Downloadable workbooks are provided for air duct sizing, fan and system integration, air duct acoustic design, plus other assignment applications. All formulae used are explained with copious examples. The reader is encouraged to make full use of spreadsheets as a valuable aid to understanding without the need to be taught how to use dedicated software. The spreadsheets provided can be edited and easily enlarged or applied to other cases with the sample data provided or with the user's own data.

Each chapter is introduced with lists of learning objectives and the key terms and concepts employed.

Approximate samples of data from the Chartered Institution of Building Services Engineers (CIBSE) Guides are given for educational purposes in order to demonstrate how the reader can utilise the reference data when undertaking professional contracts. Sample data alone will be insufficient for anything other than the set exercises and only the most recent CIBSE Guide edition is to be used more widely.

The stages of an air conditioning design process are clearly stated and are often given in the form of numbered lists. This approach may assist the testing of competences.

The vitally important tasks of commissioning and maintaining systems are explained and provided with extensive checklists so the reader understands that design calculation alone will not make a mechanical and electrical system function or keep running by itself; pressing the on switch is not enough. Testing air leakage is explained and this is applied to measure building air tightness in the CIBSE PROBE reports mentioned.

Mass cooling has been used widely in recent years to reduce plant cooling load and these opportunities are discussed. Standard topics of system types, psychrometrics, load calculation, air duct and pipe sizing, fan and system interaction, control methodology and thermal storage are extensively explained with many worked examples and assignments.

A control system worked example explains how a building management system is integrated with the HVAC system in very practical terms.

Air duct acoustics are adequately explained with worked examples and a workbook to enable the HVAC designer to make an assessment of plant sounds transferring to the occupied rooms – a difficult subject made easy.

Finding the project sale price for an air conditioning system is made easier with the aid of a downloadable workbook with sample prices and an established method of costing. The user can input current data from *Spon's Mechanical and Electrical Services Price Book*, or other source of pricing information, and calculate costs, change margins and discounts and find out what to charge the main contractor or final customer for the project.

A question bank is provided in addition to extensive questions in each chapter to provide self-learning material and resources for assignments and tests. In my experience in-class quizzes always proved popular and educational and this question bank makes the task of creating a University Challenge match very easy and enjoyable. Happy quizzing!

Acknowledgements

I am particularly grateful to the publishers for their investment in much of my life's work. A production like this only becomes possible through the efforts of a team of highly professional people. An enthusiastic, harmonious and efficient working relationship has always existed in my experience with Taylor & Francis. All those involved are sincerely thanked for their efforts and the result. My wife Maureen is thanked for her encouragement and understanding while I have been engrossed in keyboard work, writing, drawing, making spreadsheet workbooks and checking proofs. I would specifically like to thank those who have refereed this work. Their efforts to ensure that the book has comprehensive coverage, introductory work, adequate depth of study, valid examples of design, good structured worked examples and exercises are all appreciated. Many thanks also to the Chartered Institution of Building Services Engineers (CIBSE) and The Australian Institute of Refrigeration, Air-Conditioning and Heating (AIRAH) which inform the industry so efficiently and regularly through their excellent publications. Users and recommenders of the book are all thanked for their support; without them, it would not exist.

Introduction

This third edition of *Air Conditioning: A practical introduction* expands this established textbook into new areas. Air duct acoustics, peak summertime temperature in low energy buildings, case studies, mass cooling and energy demand around the world, extensive use of downloadable workbooks, building management services integration and air conditioning project selling price all add to the previously established essential topics of study for all students of air conditioning.

Units and constants

Système International units are used throughout. Table 0.1 shows the basic and derived units with their symbols and common equalities.

Table 0.1 SI units

Quantity	*Unit*	*Symbol*	*Equality*
Mass	kilogramme	kg	
	tonne	tonne	1 tonne = 10^3 kg
Length	metre	m	
Area	square metre	m^2	
Volume	litre	l	
	cubic metre	m^3	1 m^3 = 10^3 l
Time	second	s	
	hour	h	1 h = 3600 s
Energy	joule	J	1 J = 1 Nm
Force	newton	N	$1\ N = 1 \frac{kg\ m}{s^2}$
			1 kg = 9.807 N
Power	watt	W	$1\ W = 1\ \frac{J}{s}$
			$1\ W = 1\ \frac{N\ m}{s}$
			1 W = 1 VA
Pressure	pascal	Pa	$1\ Pa = 1\ \frac{N}{m^2}$
			$1\ bar\ (b) = 10^5 \frac{N}{m^2}$
			1 b = 10^3 mb
			1 b = 100 kPa
Frequency	hertz	Hz	$1\ Hz = 1\ \frac{cycle}{s}$
Temperature	Celsius	°C	
	Kelvin	K	K = °C + 273

Electrical units

Quantity	*Unit*	*Symbol*	*Equality*
Resistance	ohm, Ω	R	
Potential	volt	V	
Current	ampere	I	$I = \frac{V}{R}$

Table 0.2 Multiples and sub-multiples

Quantity	*Name*	*Symbol*
10^{12}	tera	T
10^{9}	giga	G
10^{6}	mega	M
10^{3}	kilo	k
10^{-3}	milli	m
10^{-6}	micro	μ

Table 0.3 Physical constants

Quantity	*Symbol*	*Equality*
gravitational acceleration	g	$9.807\ \frac{m}{s^2}$
specific heat capacity of air	*SHC*	$1.012\ \frac{kJ}{kg\ K}$
specific heat capacity of water	*SHC*	$4.186\ \frac{kJ}{kg\ K}$
density of air at 20°C, 1013.25 mb	ρ	$1.205\ \frac{kg}{m^3}$
density of water at 4°C	ρ	$10^3\ \frac{kg}{m^3}$
exponential	e	2.718

Symbols

Symbol	*Description*	*Units*
A	area	m^2
	solar altitude angle	°
	absorptivity of glass	
A_f	area of opaque fabric	m^2
A_g	area of window	m^2
B	solar azimuth angle	°
b	barometric pressure	bar
°	angle	degree
C	wall azimuth angle	°
	flow coefficient	
D	wall-solar azimuth angle	°
	direct irradiance	$\frac{W}{m^2}$
d	diameter	m or mm
	suffix for design and diffuse irradiance	
d.b.	dry bulb temperature	°C
Δp	pressure drop	Pa
Δp_{12}	pressure change from node 1 to node 2	Pa
E	emissivity	
	slope angle	°
EA	exhaust air	$\frac{m^3}{s}$ or $\frac{l}{s}$
EL	equivalent length	m
F	force	N
	incidence angle	°
	surface factor	
FSP	fan static pressure	Pa
FTP	fan total pressure	Pa
FVP	fan velocity pressure	Pa
F_1, F_2	heat loss factors	
F_u	thermal transmittance factor	
F_v	ventilation factor	
F_y	admittance factor	

Symbol	Description	Units
f	decrement factor friction factor	
g	gravitational acceleration	$\frac{m}{s^2}$
	air moisture content	$\frac{kg\ H_2O}{kg\ dry\ air}$
g_s	saturation air moisture content	$\frac{kg\ H_2O}{kg\ dry\ air}$
H	height head overcome by pump or fan horizontal surface	m m
h	time	hours
h	specific enthalpy	$\frac{kJ}{kg}$
h_{fg}	latent heat of vapourisation	$\frac{kJ}{kg}$
h_o	outdoor air specific enthalpy	$\frac{kJ}{kg}$
h_r	room air specific enthalpy	$\frac{kJ}{kg}$
h_{so}	outside surface heat transfer coefficient	$\frac{W}{m^2\ K}$
h_{si}	inside surface heat transfer coefficient	$\frac{W}{m^2\ K}$
I	electrical current	ampere
	solar irradiance	$\frac{W}{m^2}$
I_{DH}	direct solar irradiance on horizontal surface	$\frac{W}{m^2}$
I_{DV}	direct solar irradiance on vertical surface	$\frac{W}{m^2}$
I_{TV}	total solar irradiance on a vertical surface	$\frac{W}{m^2}$
I_{TH}	total solar irradiance on a horizontal surface	$\frac{W}{m^2}$
I_{dH}	diffuse solar irradiance on a horizontal surface	$\frac{W}{m^2}$
I_{DS}	direct solar irradiance on a sloping surface	$\frac{W}{m^2}$
I_{TS}	total solar irradiance on a sloping surface	$\frac{W}{m^2}$
I_{NH}	solar irradiance normal to a horizontal surface	$\frac{W}{m^2}$
I_{NV}	solar irradiance normal to a vertical surface	$\frac{W}{m^2}$

(continued)

Symbol	*Description*	*Units*
I_l	long wave radiation from a surface	$\frac{W}{m^2}$
J	quantity of energy, joule	J
K	constant	K
k	thermal conductivity	$\frac{W}{m\ K}$
kg	mass	kilogramme
kJ	quantity of energy	kilojoules
kVA	kilovolt ampere	
kW	power	kilowatt
kWh	energy consumed	kilowatt-hour
L	length	m
LH	latent heat of evaporation	kW
l	length	m
	volume, litre	l
LEL	lower explosive limit	ppm or %
LOG	logarithm to base 10	
$LTHW$	low temperature hot water	
m	mass flow rate	$\frac{kg}{s}$
	mass	kg
MJ	quantity of energy, megajoule	MJ
$MTHW$	medium temperature hot water	
MW	power, megawatt	MW
m	length	metre
mm	length, millimetre	mm
N	air change rate	$\frac{\text{air changes}}{\text{hour}}$
	number of occupants	number
	rotational speed	Hz
N	force, newton	N
OA	outside air intake	$\frac{m^3}{s}$ or $\frac{l}{s}$
OEL	occupational exposure limit	ppm or %
P	power, watt	W
p	pressure, pascal	Pa
p_a	atmospheric pressure	Pa
π	ratio of circumference to diameter of a circle	
ppm	concentration	parts per million
PS	percentage saturation	%
p_s	saturated vapour pressure	Pa
	static pressure	Pa
p_t	total pressure	Pa
p_v	vapour pressure	Pa
	velocity pressure	Pa
p_{sl}	saturated vapour pressure at sling wet bulb air temperature, t_{sl}	Pa
Q	volume flow rate	$\frac{m^3}{s}$ or $\frac{l}{s}$
	heating or cooling power	kW

Symbol	*Description*	*Units*
Q_D	direct transmitted solar irradiance	W
Q_d	diffuse transmitted solar irradiance	W
$\tilde{Q}_t$	swing in total heat exchange	W
Q_u	steady state fabric heat loss	W
$\tilde{Q}_u$	cyclic variation in fabric heat loss	W
Q_v	ventilation heat exchange	W
R	resistance, electrical, ohm	Ω
	thermal resistance	$\frac{m^2K}{W}$
	specific gas constant	$\frac{kJ}{kg\ K}$
	reflectivity of glass	
	overall motor drive power ratio	
	recess dimension	m
RA	return or recycled air	$\frac{m^3}{s}$ or $\frac{l}{s}$
RH	relative humidity	%
ρ	density	$\frac{kg}{m^3}$
RPM	rotational speed	$\frac{\text{revolutions}}{\text{minute}}$
R_a	air space thermal resistance	$\frac{m^2K}{W}$
R_{si}	internal surface thermal resistance	$\frac{m^2K}{W}$
R_{so}	external surface thermal resistance	$\frac{m^2K}{W}$
SA	supply air	$\frac{m^3}{s}$ or $\frac{l}{s}$
SH	sensible heat transfer	kW
SHC	specific heat capacity	$\frac{kJ}{kg\ K}$
SMR	square of the mean of the square roots	
SR	regain of static pressure	Pa
$\frac{S}{T}$	sensible to total heat ratio	
s	time	seconds
T	absolute temperature	Kelvin, K
	torque	N m
	total irradiance	$\frac{W}{m^2}$
	transmissivity of glass	

(continued)

Continued

Symbol	Description	Units
t	temperature	°C
	dry bulb air temperature	°C d.b.
t_a	air temperature	°C
t_{ai}	inside air temperature	°C
t_{ao}	outside air temperature	°C
t_c	operative temperature	°C
	resultant temperature at centre of room	°C
t_{dp}	dew point temperature	°C
t_e	environmental temperature	°C
t_{ei}	indoor environmental temperature	°C
$\tilde{t}_{ei}$	swing of indoor environmental temperature	°C
$\tilde{t}_{eo}$	swing of outdoor environmental temperature	°C
t_f	flow temperature	°C
t_g	glass temperature	°C
t_m	mean temperature	°C
	mixed air temperature	°C
t_r	mean radiant temperature	°C
	return temperature	°C
	room temperature	°C
t_s	supply air temperature	°C
	surface temperature	°C
t_{sl}	sling wet bulb air temperature	°C w.b.
U	thermal transmittance	$\frac{W}{m^2K}$
UEL	upper explosive limit	ppm or %
U_f	thermal transmittance of opaque fabric	$\frac{W}{m^2K}$
U_g	thermal transmittance of window	$\frac{W}{m^2K}$
V	volume	m^3 or litre
	vertical surface	
V	electrical potential	volt
v	velocity	$\frac{m}{s}$
	specific volume	$\frac{m^3}{kg}$
W	width	m
w.b.	wet bulb temperature	°C
Y	admittance	$\frac{W}{m^2K}$
	length	m
10^6	10 raised to the power of 6	

1 Uncooled low energy design

Learning objectives

Study of this chapter will enable the reader to:

1. calculate peak summertime temperature in uncooled buildings;
2. analyse case study buildings;
3. assess a floor of The Shard if not cooled;
4. use CIBSE Post Occupancy Review of Building Engineering (PROBE) case studies;
5. assess the London 2012 Olympic Velodrome;
6. compare White Tower with modern buildings;
7. assess lightweight habitable spaces;
8. understand conditions in traditional old and new UK houses.

Key terms and concepts

Air conditioning 2; air temperature 2; building air leakage 2; discomfort 2; environmental temperature 2; evaporative cooling 2; internal heat gains 2; low energy building 2; mean 24 h heat gain 5; mean radiant temperature 8; overheating 8; Passivhaus 2; peak summertime temperature 2; PROBE 3; Simple (cyclic) Model 2; swing in heat gain 2; thermal analysis software 2; ventilation 2; windows 2.

Introduction

This chapter calculates the hourly predicted peak summertime temperature in uncooled buildings with a workbook using the Simple (cyclic) Model. Case studies are calculated and discussed. Examples analysed include The Shard, London 2012 Olympic Velodrome, a university office and a small home. It uses CIBSE Post Occupancy Review of Building Engineering (PROBE) reports as case studies. It also helps in the decision on the need for air conditioning.

Peak summertime temperature

The internal environmental temperature created in a room or building that does not have refrigeration or an evaporative cooling system, is calculated using a workbook and checked against measurements taken in a sample office. Low energy buildings are likely to have natural ventilation with operable windows and ventilators. These rely on architecture to limit solar radiation, convection and conduction heat gains from the warmer external environment. Ventilation air may be from any combination of natural, mixed mode or entirely mechanical systems. A really low energy building design might rely on natural, or assisted natural ventilation, make use of solar heat gains, have controllable shading devices, maximise the use of internal heat gains from people, lighting, computers and machinery, and have a minimal system providing top-up space heating; for example, the Passivhaus design. What quality of comfort conditions are likely to be found in UK buildings of this type? Most of us know the answer because we live in such a building, travel by car, train or bus in a mobile equivalent of such a building, and work in one as well. Motor cyclists and cyclists are more tolerant of discomfort while travelling.

When we have experimented with all the possible combinations of solar shading, operable windows and ventilators, some summertime overheating is experienced by most people. Then we resort to cool drinks, adjusting clothing, switching on portable fans, taking breaks from work and perspiring a lot. Thunderstorms and heavy rain invariably follow a series of uncomfortably hot days, unavoidably raising humidity and discomfort. We tend to adapt to a series of uncomfortable days, knowing it will not last long (CIBSE TM52, 2013). Designers should have access to a simple method of predicting whether a building, typical module or a room, is expected to overheat unless it has air conditioning. Thermal analysis software models real-time conditions and calculates what internal air temperatures will be on an hourly basis. These cost thousands of pounds, require extensive training to use them and have ongoing maintenance and upgrade costs for the design office. The spreadsheet file provided here, gives a suitably accurate hourly assessment using the Simple (cyclic) Model (CIBSE Guide A, 2006, Example 5.2, pages 5–19 to 5–21). We know how airtight constructed buildings really are, as distinct from design load calculations and computer modelling, from the PROBE reports. Surprisingly perhaps, measured building air leakage rates are higher than some might expect. Air leakage standards for the buildings are: leaky 36 $m^3/h\ m^2$, meaning typically 12 air changes/h; average 18 $m^3/h\ m^2$, meaning 6 air changes/h; and tight 9 $m^3/h\ m^2$, meaning 3 air changes/h for a 3 m high ceiling height (Chadderton, 2013, page 77, Chapter 5). We will use these standards of measured infiltration rates as a starting point for assessment of peak summertime temperature. We know these are real values from audited buildings that were constructed and maintained to standards of good practice. There may be other ways of establishing air flow rates through a building, such as when it has a mechanical ventilation system running, but an idealised zero leakage does not happen in the real world when a building is closed up.

During warm sunny weather, buildings without any method of lowering the internal temperature may become overheated and uncomfortable for normal work or habitation. The upper limit of acceptability may be as high as an internal environmental temperature of 27°C. Such a choice is arbitrary and does not take account of the glaring effect of direct solar irradiance upon the person, their activity level, their clothing's thermal insulation or the temperature and speed of the air around them. Increased air velocity aimed at a sedentary person reduces discomfort in excessively hot conditions. Athletes on running or cycling machines in a gymnasium extend their performance time in the presence of high air flows similar to being outdoors. An assessment of the peak summertime temperature within a building ought to be made before the decision to design a cooling system is made. The provision of low cost cooling systems can be investigated.

Internal environmental temperature will be combinations of the 24 hour mean heat gains, cyclic gains producing temperature swings about the mean, and heat loss from the internal environment mainly due to external air ventilation. Some of this heat loss might be accomplished using some form of mechanical cooling. The final temperature reached is a balance between gains and losses, some of which are potentially under the control of the occupier or engineer. Painting the exterior of the roof with white paint or spraying water onto the roof can reduce the heat gains. Some large areas of glass that were built during the 1950s and 1960s when there was little thought given to energy economy are known to have been painted white. Additional mechanical ventilation with outdoor air that is already at 25°C to 30°C or more, may not produce human comfort conditions. It may be sufficient to avoid the overheating of hardware such as computer servers, stored goods and operational electric motors when personnel are not at risk. The increased air velocity around personnel will alleviate discomfort and may produce tolerable conditions. Ideally there needs to be manual control over the direction and velocity of increased air circulation as weather conditions vary quickly.

The method of assessing the peak environmental temperature is to calculate the:

1. 24 hour mean solar heat gains;
2. 24 hour mean internal heat gains;
3. Mean internal environmental temperature from the known gains and the 24 hour mean external air temperature;
4. Peak swing in heat gains above the 24 hour mean;
5. Swing in environmental temperature due to the swing in heat gains;
6. Peak environmental temperature (the mean plus the swing values).

Mean heat gain from people, lights and equipment is found by multiplying their power by the hours of usage and dividing by 24 hours. The 24 hour mean solar gains come from the daily mean total irradiance from CIBSE Guide A (2006), Table 2.30 and correction factors for shading, Table 5.7.

$$\text{Mean gain } Q = \sum (A_g U_g)(t_{ei} - t_{ao}) + 0.33NV(t_{ei} - t_{ao}) + \sum (A_f U_f)(t_{ei} - t_{eo})$$

A_f = area of opaque fabric, m^2
A_g = area of window, m^2
U_f = thermal transmittance of opaque fabric, $\frac{W}{m^2 K}$

U_g = thermal transmittance of window, $\frac{W}{m^2 K}$

t_{ei} = mean internal environmental temperature, °C
t_{eo} = mean external environmental temperature, °C
t_{ao} = mean internal air temperature, °C

The mean internal environmental temperature t_{ei} is found by rearranging the equation.

$$Q = \sum (A_g U_g) t_{ei} - \sum (A_g U_g) t_{ao} + 0.33NVt_{ei} - 0.33NVt_{ao} + \sum (A_f U_f) t_{ei} - \sum (A_f U_f) t_{eo}$$

$$Q + \sum (A_g U_g) t_{ao} + 0.33NVt_{ao} + \sum (A_f U_f) t_{eo} = \sum (A_g U_g) t_{ei} + 0.33NVt_{ei} + \sum (A_f U_f) t_{ei}$$

$$Q + \sum (A_g U_g) t_{ao} + 0.33NVt_{ao} + \sum (A_f U_f) t_{eo} = t_{ei} \left(\sum (A_g U_g) + 0.33NV + \sum (A_f U_f) \right)$$

$$t_{ei} = \frac{Q + \sum (A_g U_g) t_{ao} + 0.33NVt_{ao} + \sum (A_f U_f) t_{eo}}{\sum (A_g U_g) + 0.33NV + \sum (A_f U_f)}$$

Table 1.1 Heat transfer data for example 1.1

Surface	*A, m^2*	*$U \frac{W}{m^2K}$*	*AU*	*$Y \frac{W}{m^2K}$*	*AY*	*f*	*Lag, h*
Glass	10	5.7	57	5.7	57	1	0
External wall	8	0.4	3.2	4.3	34.4	0.18	10
Internal wall	54	1.7	0	3.5	189	0.72	1
Floor	36	1.7	0	5.2	187.2	0.72	3
Ceiling	36	1.7	0	2.2	79.2	0.86	1

EXAMPLE 1.1

A south facing Brighton office 6 m x 6 m and 3 m high has single glazed clear float window openings of 10 m^2. The surrounding rooms, and also above and below, are all similar. There are three occupants emitting 90 W and four electrical items of 150 W each. The office is used for 8 hours in each 24 hours. Windows and the door are shut at almost all times and the ventilation rate is one air change per hour. Use the data provided to estimate the 24 hour mean internal environmental temperature. The peak solar irradiance on a south facing vertical window is 710 W/m^2 at noon on 22 September in south east England and the daily mean is 200 W/m^2. The solar gain correction factor for the glazing without blinds is 0.76. Daily mean t_{ao} is 15.5°C, t_{ao} at noon is 18.5°C and mean t_{eo} is 25°C. Thermal data are shown in Tables 1.1 and 1.2.

$$\sum (A_g U_g) = 57 \frac{W}{K}$$

$$\sum (A_f U_f) = 3.2 \frac{W}{K}$$

$$\sum (AY) = 546.8 \frac{W}{K}$$

$$\text{Mean solar gain} = 200 \frac{W}{m^2} \times 10\ m^2 \times 0.76$$

$$\text{Mean solar gain} = 1520\ W$$

$$\text{Mean internal gain} = \frac{(3 \times 90\ W \times 8\ h) + (4 \times 150\ W \times 8\ h)}{24}$$

$$\text{Mean internal gain} = 290\ W$$

$$\text{Total mean gain } Q = (1520 + 290) W$$

$$\text{Total mean gain } Q = 1810\ W$$

$$\text{Mean gain } Q = \sum (A_g U_g)(t_{ei} - t_{ao}) + 0.33 NV (t_{ei} - t_{ao}) + \sum (A_f U_f)(t_{ei} - t_{eo})$$

$$N = 1 \frac{\text{air change}}{h}$$

$$\text{Room volume } V = 6m \times 6\ m \times 3\ m$$

$V = 108 \text{ m}^3$

$$t_{ei} = \frac{Q + \sum(A_g U_g)\, t_{ao} + 0.33NVt_{ao} + \sum(A_f U_f)\, t_{eo}}{\sum(A_g U_g) + 0.33NV + \sum(A_f U_f)}$$

$$t_{ei} = \frac{1810 + 57 \times 15.5 + 0.33 \times 1 \times 108 \times 15.5 + 3.2 \times 25}{57 + 0.33 \times 1 \times 108 + 3.2}$$

$$t_{ei} = 34.7^\circ\text{C}$$

The mean 24 hour environmental temperature could reach 34.7°C under these conditions and of course would be entirely unsatisfactory. It is a south facing office with single glazing and minimal outdoor air ventilation. Occupants would very quickly learn to open windows and the internal door to promote air flow as well as insisting on hiding the glass with blinds or external shading. This is an example of what can happen within a horticultural glass house, and, unfortunately, some uncooled commercial buildings. Mean environmental temperature exceeds the recommended 27°C and unless the swing in heat gains is a large negative quantity, discomfort will result, also evacuation to a cooler part of the building. A positive swing is expected due to the additional irradiation through the glazing at noon. Temperature swings from the mean 24 hour value of the environmental and air temperatures are shown in Table 1.2. An alternating solar gain factor is applied to the increase of the noon irradiance above the 24 hour mean value from CIBSE Guide A (2006), Table 5.7, page 5–16. This is 0.66 for clear single glazing in a lightweight building. The use of heat absorbing or reflective glass, double glazing, exterior shading or internal blinds produces lower factors. Greatly increased ventilation outdoor air flow rate will lower room temperature. The reader may like to experiment with such alternatives to discover whether suitable design improvements can be made.

Example 1.2 calculates the total swing in the solar heat gains through the glazing, from the ventilation air, internal sources and the structure, between the 24 hour mean and the peak, $\tilde{Q}_t$. The swing, mean to peak, in the internal environmental temperature, $\tilde{t}_{ei}$, is then found from:

$$\tilde{Q}_t = \left(\sum AY + 0.33NV\right) \tilde{t}_{ei}$$

$$\tilde{t}_{ei} = \frac{\tilde{Q}_t}{\sum AY + 0.33NV}$$

The final room environmental temperature is the sum of the mean and swing figures.

Note that a solid concrete floor in contact with the ground does not have a separately calculated conduction heat gain swing. Its time lag is very long, days rather than hours. Heat flow is not directly related to the rapid changes in solar radiation, air or sol-air temperatures on such a short time lag scale as are exposed surfaces. The effect of the thermal storage in the floor is included in the $\sum AY$ calculation.

EXAMPLE 1.2

Continue example 1.1 to calculate the office peak environmental temperature at noon, 12.00 h. The alternating solar gain factor for the glass is 0.66. Data are provided in Table 1.2.

Table 1.2 Sol-air data for example 1.2

Surface	*Lag h*	*24 h* t_{eo}	*24 h* t_{ao}	*Time, h*	t_{eo}	*Swing* $\tilde{t}_{eo}$
External wall	10	25	15.5	02.00	11	−14
					t_{ao}	$\tilde{t}_{ao}$
Window	0	—	15.5	12.00	18.5	3

The external wall has a thermal time lag of 10 h, so subtract 10 hours from noon and look up the external sol-air temperature at 02.00 h, 11°C. That is when the solar heat gain or loss occurred at the outside surface of the wall.

$$\text{Swing } \tilde{t}_{eo} = t_{eo} \text{ at time of heat gain} - 24 \text{ h mean } t_{eo}$$

$$\tilde{t}_{eo} = (11 - 25)°\text{C}$$

$$\tilde{t}_{eo} = -14°\text{C}$$

Glass has no appreciable thermal lag. Heat gains through glass occur at the air temperatures.

$$t_{ao} = 18.5°\text{C at noon}$$

$$\text{Swing } \tilde{t}_{ao} = t_{ao} \text{ at time of heat gain} - 24 \text{ h mean } t_{ao}$$

$$\tilde{t}_{ao} = (18.5 - 15.5)°\text{C}$$

$$\tilde{t}_{ao} = 3°\text{C}$$

$$\text{Swing in solar radiation gain} = 0.66 \times (710 - 200)\frac{\text{W}}{\text{m}^2} \times 10 \text{ m}^2$$

$$\text{Swing in solar radiation gain} = 3366 \text{ W}$$

$$\text{Swing in south wall heat gain} = fAU\tilde{t}_{eo}$$

$$\text{Swing in south wall heat gain} = 0.18 \times 8 \text{ m}^2 \times 0.4\frac{\text{W}}{\text{m}^2\text{ K}} \times (-14)\text{K}$$

$$\text{Swing in south wall heat gain} = -8 \text{ W}$$

$$\text{Swing in glass conduction} = fAU\tilde{t}_{ao}$$

$$\text{Swing in glass conduction} = 1 \times 10 \text{ m}^2 \times 5.7\frac{\text{W}}{\text{m}^2\text{ K}} \times 3 \text{ K}$$

$$\text{Swing in glass conduction} = 171 \text{ W}$$

$$\text{Swing in ventilation gain} = 0.33NV\tilde{t}_{ao}$$

$$\text{Swing in ventilation gain} = 0.33 \times 1 \times 108 \times 3$$

$$\text{Swing in ventilation gain} = 107 \text{ W}$$

$$\text{Swing in internal heat gains} = \text{peak gain} - \text{mean gain}$$

$$\text{Swing in internal heat gains} = (3 \times 90) + (4 \times 150) - 290 \text{ W}$$

$$\text{Swing in internal heat gains} = 580 \text{ W}$$

Total swing in heat gains $\tilde{Q}_t = 3366 - 8 + 171 + 107 + 580$ W

Total swing in heat gains $\tilde{Q}_t = 4216$ W

From Table 1.1, $\sum(AY) = 546.8\ \frac{W}{K}$

$$\tilde{t}_{ei} = \frac{\tilde{Q}_t}{\sum AY + 0.33NV}$$

$$\tilde{t}_{ei} = \frac{4216}{546.8 + 0.33 \times 1 \times 108}$$

$$\tilde{t}_{ei} = 7.2°C$$

Peak environmental temperature $t_{ei} = 34.7 + 7.2$

Peak environmental temperature $t_{ei} = 41.9°C$

The result confirms the earlier conclusion and shows that a south facing conservatory is unsuitable for work. Increased ventilation, heat absorbing or reflecting glass and shading devices can be investigated.

Solent case study

The author once lectured in what is now Southampton Solent University. A typical staff office was designed for 1–4 lecturers and was constructed in a new teaching block in 1970 to the same design and standard as the remainder of the campus that had commenced building in 1960 in the city centre. Visit the campus on Google Earth at East Park Terrace, Southampton and take the street view of the three-storey green tiled block from the Nicholstown bus stop at 2 New Road. This is the smallest building on the campus, it faces south onto New Road and had class and staff rooms along the south and north sides connected by an east–west central corridor. There is no mechanical ventilation, air conditioning or cooling for the purpose of this case study, as was the case up to 1993. Low temperature hot water radiators and operable windows maintained comfort conditions. Venetian blinds were lowered and the windows remained closed for the weekend.

Heavyweight reinforced concrete floors and structural frame were topped with an asphalted concrete roof – a heavyweight building. Take thermal data as: concrete block tiled cavity walls thermal transmittance U 1.7 W/m^2 K , admittance Y 4.3 W/m^2 K, decrement factor f 0.18, time lag 10 h; glazing U 3.5 W/m^2 K, Y 3.5 W/m^2 K, f 1.0, lag 0 h; internal walls and door U 0 W/m^2 K as there is no calculated heat flow through them, Y 2.3 W/m^2 K because heat flows into thermal storage in the walls, f 0, lag 0 h; second floor roof U 1.6 W/m^2 K, Y 2.6 W/m^2 K, f 0.5, lag 10 h; concrete floor U 0 W/m^2 K, Y 2.6 W/m^2 K, f 0, lag 0 h. Single glazed steel framed external hinged operable windows had aluminium framed sliding double glazing internally with venetian blinds between the panes. Three internal walls and identical rooms the other side, so there was no heat transmission between rooms but admittance of heat gain into and from the solid surfaces did matter. Ceiling height 2.8 m, external wall length 2.5 m, room length 4 m and window height 1.8 m.

According to PROBE measurements of tested buildings, airtightness could be considered to be 'tight', as the building was closed for the whole weekend, at maybe 9 m^3/h m^2, meaning 3.2 air changes/h due to natural ventilation (Chadderton, 2013, page 77, Chapter 5). Occupiers controlled window and door openings to minimise cross draughts in cold weather and maintain workroom privacy. In summer, opening the hinged and sliding glazing and lowering blinds allowed cool air from the internal corridor and class rooms on the north side to maximise ventilation in hot weather. Fire doors at entrances to stairways on each floor and in long corridors between major sections of the other buildings restricted through flow of ventilation.

This same modular design applied to most rooms on the campus, including class rooms, administrative offices, the board room, main library, staff lounge, student refectory and many laboratories.

The workbook file *Solent1south.xls* is provided for analysis of the peak summertime temperature likely in this case study. Exactly the same calculations are performed in the workbook as we just did for the first two examples, with the additional benefit of hourly analysis, chart presentations and checking the result against a preferred air condition. Data are provided in the case file for use only within this book. Users may enter their own data for commercial project uses.

Observe the calculated and measured results for the south facing office and analyse what this means for the staff. Identical staff offices also faced east, west and north in the same campus, as can be seen by moving around the perimeter roads. Save a file for each orientation and discuss what you find from predictions for these other sides. State what actions the occupants needed to undertake during their working days and evenings. Recommend improvements that might be taken to reduce summer overheating, if occurring, by increasing the natural ventilation rate, opening the blinds, installing tinted exterior glass, or ideas of your own. The following observations are provided for consideration.

The resulting calculations are highly dependent on the airtightness standard selected for the prediction.

Glazing area is a large proportion of the external wall; good for daylight penetration. No external shading.

Environmental temperature, $t_{ei} = 0.5t_{ai} + 0.5t_r$

Mean radiant temperature t_r of the room is expected to be 2°C or more above t_{ai} due to solar radiation through the glazing, so, t_{ei} is expected to be at least 1°C above the measured t_{ai}. Predicted peak summertime environmental temperature is 1.1°C higher than measured indoor air temperature, as expected.

CIBSE Guide A outdoor air temperature for design purposes is much lower than what occurred on a series of hot days.

Windows and doors remained closed outside of occupation hours for security and fire safety. During occupied hours, leaving office doors open to the corridor and class rooms facing north, coupled with many open doors and windows, provided significant cross ventilation from cooler rooms and much lower office temperatures. Students and staff simply had to cope with occasional overheating. Some days in June and July caused overheating but few members of staff were in occupation in August.

The calculated air temperature profile for this office is comparable with that measured and published (Chadderton, 2013, Figure 5.21, page 108, Chapter 5), and shown in the workbook chart. This is considered to be a reasonably satisfactory validation of the assessment and a pointer to what a comprehensive software analysis should produce.

The south office needs cross ventilation for an average building value of 18 $m^3/h\ m^2$, 6.4 air changes/h, so that the peak indoor environmental temperature reduces to the incoming outdoor air temperature of 30.4°C. Replacing the exterior window with tinted or reflecting glass could lower the peak indoor environmental temperature to 30°C; nothing lower than these conditions could be achieved with this basic module design.

A north facing office needs to be a leaky room with 12.9 air changes/h to reduce peak indoor environmental temperature to 26.5°C, and is only just acceptable.

The east facing office needs to be a leaky room with 12.9 air changes/h but the environmental temperature remains above 26°C from 09.00 h to 18.00 h and would be too hot without tinting the glass.

A west facing office needs to be a leaky room with 12.9 air changes/h, but the environmental temperature remains above 26°C from 12.00 h to 18.30 h and would be too hot without tinting the glass.

It is obvious today that a 1960s simple modular design was never going to avoid some summer overheating on all facades. Occupants simply had to put up with the conditions. Natural ventilation with tall, openable double glazing that flooded the teaching and administrative rooms with daylight, and low temperature hot water central heating radiators, was the design standard for the era. There would never have been any public finance for refrigerated air conditioning, mechanical ventilation or mechanized ventilators for any of the rooms. Modern architectural appearance would have been the controlling sentiment at the time of design in the late 1950s.

This university developed from the old brick buildings of City College, 41 Saint Mary Street, Southampton. Locate the street view; compare the old and modern building designs from the point of view of avoiding summertime overheating. Note the heavyweight thermal mass brick walls, shading from plan layout and eaves, small operable windows, pitched slate roof and low building height. Look around at nearby buildings to note similar characteristics. Older buildings are less likely to overheat as the interior remains shaded and cool for most of the day. However, that may never have been a design consideration as small windows were always used.

Queens Building case study

Queens Building, De Montfort University, Leicester, built in 1993 featured in The Government's Energy Efficiency Best Practice Programme (1997). There was a full performance study (DUALL Project, 1996)). It won the HVCA Green Building of the Year award in 1995 (HVCA, 1995). Natural ventilation with automatically controlled inlet and outlet dampers, massive brick thermal storage walls, daylighting, passive cooling and prominent ventilation stacks were the features of Queens Building. Conventional low temperature hot water heating was used for all buildings while the Machine Hall had high level radiant panels. The building is occupied by few staff and researchers during the hottest summer weeks. Queens Building has a striking and almost a gothic look that reflected the appearance of nearby older buildings, suggestive of a prison or Victorian era factory.

Visit the campus on Google Earth along Mill Lane and Grasmere Street and take the street view of the brick buildings. Locate 28 Grasmere Street, look north east across the gravel car park at the two-storey industrial building stretching to the right and disappearing from view behind houses in Grasmere Street; this is the Mechanical Laboratories Machine Hall building. It has eight pitched roofs, each with a pair of wing shaped ventilators on top. There are few small windows on this side facing south west, the long side facing Grasmere Street. Move over to the other side of the Machine Hall and view it from the car park barrier in Havelock Street; there are few windows on that side facing north east and also few on the end of the building that faces south east. The other short side facing north west attaches to a taller building which includes the plant rooms and combined heat and power room on the corner of Grasmere Street and Mill Lane.

This building is of a similar design to the rest of Queens Building and will be used as an example of the natural ventilation passive cooling design. Approximate dimensions were calculated from the views: Machine Hall length 55 m, width 27 m and brick wall height 6 m. Average hall height to the roof including gables is 7.45 m. Net roof area is estimated to be 1400 m^2 including gables and ventilator stacks. The south west wall has 70 m^2 glazing, the south east wall 6 m^2 glazing and the north east wall has 10 m^2 glazing. The roof has eight pitched sections with a total of 130 m^2 of horizontal glazing, 40 m^2 of south west glazing and openable louvres. Internal volume of the Machine Hall is 11000 m^3. Enter formulae into the workbook cells when calculating areas to leave a record of where data came from.

Take thermal data as: brick and block walls thermal transmittance U 0.3 W/m^2 K, admittance Y 5 W/m^2 K, decrement factor f 0.25, time lag 10 h; all glazing U 3 W/m^2 K, Y 3 W/m^2 K, f 1.0, lag 0 h; internal walls and door U 2 W/m^2 K, Y 5 W/m^2 K, f 0, lag 0 h; roof U 0.25 W/m^2 K, Y 0.3 W/m^2 K, f 0.5, lag 1 h as it had a lightweight construction; concrete floor U 0.4 W/m^2 K, Y 1.5 W/m^2 K, f 0.9, lag 10 h. Take all glazing as single clear glass, solar correction factor 0.76 and alternating solar correction factor 0.5 for a heavyweight building. Gable windows facing south west were triple glazed, but we will ignore that in the first instance. No window shading was indicated.

Airtightness was not measured in the PROBE report as there was never a design intention to make the building sealed from the external air. As a start, take the airtightness as the average at 18 m^3/hm^2, creating 2.4 air changes/h due to natural ventilation (Chadderton, 2013, page 77, Chapter 5). The building energy management system (BEMS) controlled low level and roof ventilator dampers, similarly to most rooms on the campus, including class rooms, offices and laboratories.

Internal heat gains will be taken as five occupants emitting 90 W each, no lighting needed, 3 kW machinery heat output during the period 08.00 to 18.00 h.

Formulae in the roof conduction swing calculation column have been edited so that there is only 1 h time lag, that is, the horizontal sol-air temperature only 1 h earlier is used and not the 10 h figures set previously for the 10 h time lag.

The file *Queens.xls* is provided for analysis of the likely peak summertime temperature. Data are provided in the case file. Observe the calculated results and analyse what this means for the staff. State what actions the occupants needed to undertake during their working days and evenings. Recommend improvements that might be taken to reduce summer overheating if occurring.

The PROBE report included these observations: An innovative building aiming in the right direction for the HM Government Carbon Plan (2011). Some summer overheating in a naturally ventilated building in the UK is bound to happen; most people treat it as normal and do not demand air conditioning. Passive design maintained generally satisfactory conditions.

Combination of thermal mass, controllable natural ventilation, central heating, daylighting and minimum solar intrusion, were an endorsement of the passive design principle. Does this sound familiar? Of course it does, it is exactly the design principle for almost every home, commercial building, educational building, university, church, stately home and stone castle built prior to 1960 in the UK (CIBSE, 1997).

Predicted internal temperatures show that no amount of natural ventilation avoids summertime overheating, confirming PROBE expectations. Tinted glass and shading the extensive roof lights solves the problem but reduces daylight and may be impractical.

Case studies came from the CIBSE Post Occupancy Review of Building Engineering (PROBE), investigations conducted in 1995–2002 (*CIBSE Journal*). PROBE reports are available from https://www.cibse.org. Go to 'Freely download selected CIBSE publications, CIBSE Low Carbon Consultants Training Material', and download PROBE reports.

Users are rarely concerned whether their building is fully air conditioned, naturally ventilated, low or high technology, low energy or costs a fortune to run. Complex ventilation mechanisms relying on high user interaction and understanding may annoy users and not be used as intended by the designers. Users appreciate being able to make adjustments to their immediate environment. But there are limits; users do not expect to have to behave as a frequent control mechanism to maintain the air conditions.

Air leakage through a building is measured by removing a main door and connecting a ducted fan to the opening; you will see how this is done in the PROBE reports. Outdoor air is blown in to pressurise the building to a target static air pressure of 50 Pa. The air flow through the fan and duct is measured and this is the air leakage rate for the building. Reversing the direction of flow through the axial fan allows negative pressurisation so that inward air leaks can be found by holding smoke candles around their sources. Some buildings cannot maintain a 50 Pa pressure due to extensive openings, such as when the building has natural ventilation openings, openable windows, permanent ventilation louvres, swing doors or is excessively leaky. A lower static pressure is held for air flow measurement and the equivalent at 50 Pa is calculated. Air leakage flow is related to the floor area where a tight building passes around 8 m^3/hm^2 at 50 Pa while leaky buildings go up to 35 m^3/hm^2 or more.

Questions

1. What are the differences between the designers and users of a building?
2. Why would a very simple calculation of steady state heat gains and losses not provide an assessment of peak summertime temperature within a building?
3. How many people use the computer building energy management system (BEMS) in a large office building, university campus and hospital every day and week?

    1. Everyone in the building.
    2. Specialist maintenance contractor.
    3. One person has the expertise and time to use it.
    4. Nobody.
    5. Everyone in the property and facilities management department.

4. Who is most interested in a macroscopic appreciation of a building?

    1. Owner.
    2. Facilities manager.
    3. Building services engineer.
    4. Employees.
    5. Architect.

5. Who is most concerned with the microscopic scale aspects of a building?

    1. Architect.
    2. Employees.
    3. Owner.
    4. Facilities manager.
    5. Building services engineer.

6. What is your opinion of The Shard building at London Bridge station (4–6 London Bridge Street, London, SE1 9SG)? A one word answer was it? Prefer to pass on to the next question? Know the answer without calculation? A great deal of work went into it and the substantial building is expected to stand there for 50 plus years, so we cannot ignore it. In the context of peak summertime temperature, what would happen on the intermediate 68th floor viewing deck?

    Visit the-shard.com website, and also in Wikipedia, and look at the features. Locate the corner of Borough High Street, St. Thomas Street and Bedale Street in Google Earth and stand there to view The Shard at a distance. Move around The Shard and its surroundings. View the many photographs taken and relate the design to its surroundings. Many views show construction underway.

    Each level has floor to ceiling triple glazing with automatic internal blinds. Assume the glazing to be clear/clear/heat absorbing. There is no point in lowering internal blinds on the windows of this viewing deck as there would be no view out. The 68th floor dimensions are approximately 12 m × 12 m × 3 m high. The building faces approximately north, south, east and west. The inner core comprises lift shafts and various rooms. We will consider this to be a lightweight construction. Only a few lower floors will have any shading from surrounding low height buildings.

    Copy a workbook file, rename it as *shard* and enter the data to predict the peak summertime temperature without any air conditioning. As a simplification, ignore the inner core walls as they add thermal mass and make the internal conditions worse.

    Triple glazing U 3 W/m^2 K, Y 3 W/m^2 K, f 1.0, lag 0 h, solar correction factor 0.37 and alternating solar correction factor 0.35; ceiling U 0 W/m^2 K, Y 0.3 W/m^2 K, f 0.5, lag 10 h; floor U 0 W/m^2 K,

Y 1.5 W/m^2 K, *f* 0.9, lag 10 h. There are no walls, the perimeter is all glass, there are 20 occupants during 10.00–22.00 h and no lights are switched on.

7. An outstanding design using natural ventilation with no cooling is the London Olympic 2012 Velodrome in Stratford (*CIBSE Journal*, October 2012, Built for Speed, pp. 30–36). Locate Quartermile Lane alongside the Velodrome, look around the site and view the many photographs. Some show construction work underway. Note the extensive low and high level exterior ventilation louvres and strips of roof lights admitting fully diffused daylight. The entrance foyer is glazed but fully shaded with a large flat veranda. Concourse glazing provides daylight into the spectator area all around the building. The designer's intention was to maintain 28°C d.b. air at track level.

   Visit the london2012.com website, look at the features and take the virtual tour of the Velodrome and watch the construction videos. There are two entrance foyers on the short sides, one facing approximately south and the other approximately north.

   All dimensions are approximations for the purpose of this exercise; floor 100 m × 50 m; average internal height 12 m; perimeter double glazing, above the track, is 2.5 m high; each entrance foyer double glazing is 32 m × 3 m; eight double glazed roof lights of 80 m × 1 m. We will consider this to be a lightweight construction. Copy a workbook file, rename it as *Velodrome* and enter the data to predict the peak summertime temperature without any air conditioning. Enter data for the building as if it were rectangular with a flat roof.

   Triple glazing *U* 3 W/m^2 K, *Y* 3 W/m^2 K, *f* 1.0, lag 0 h, solar correction factor 0.6 and alternating solar correction factor 0.5; wall *U* 0.3 W/m^2 K, *Y* 0.8 W/m^2 K, *f* 0.6, lag 6 h; roof *U* 0.25 W/m^2 K, *Y* 1.5 W/m^2 K, *f* 0.9, lag 3 h; concrete floor *U* 0.1 W/m^2 K, *Y* 3 W/m^2 K, *f* 0.7, lag 10 h.

   Make sure to correct the cell references for the time lags given. Calculate the peak summertime environmental temperature for an empty building with no lighting switched on. Select an assessment for the building leakage rate, and test other values. Then save the file as *Velodrome2* and add 6000 spectators, 200 track personnel and 356 down lights of 1 kW each for television use. Consider occupancy time to be 10.00 h to 24.00 h.

8. Locate the Woodhouse Medical Centre at 5–7 Skelton Lane, Woodhouse, near Sheffield with Google Earth. Constructed in 1989, it was intended as a low energy green building, single-storey with brick/block walls and high thermal insulation and natural ventilation (PROBE 6, Woodhouse Medical Centre, *BSJ* August 1996). Look around to observe how the medical centre fits in with nearby architecture consisting of 1–2-storey brick and tile traditional and well established houses and public buildings. It blends very easily with its semi-rural surroundings. Windows are small, openable and are shaded with eaves. We will calculate summertime temperature for the front part of the left hand building facing onto Skelton Lane showing three small windows facing the lane. This square floor plan is 14 m × 14 m and room height is 3 m. 4 m^2 of single glazing face north onto Skelton Lane, 12 m^2 face west, none face east. The south end is an attached internal wall. Consider the 6 m^2 of openable single glazed roof lights to be on a horizontal roof. Window *U* 3 W/m^2 K *Y* 3 W/m^2 K, *f* 1.0, lag 0 h, solar correction factor 0.76 and alternating solar correction factor 0.5; walls *U* 0.4 W/m^2 K, *Y* 3 W/m^2 K, *f* 0.4, lag 10 h, ceiling *U* 0.3 W/m^2 K, *Y* 0.7 W/m^2 K, *f* 1, lag 10 h; floor *U* 0.25 W/m^2 K, *Y* 1.5 W/m^2 K, *f* 0.9, lag 10 h; internal wall *U* 1.5 W/m^2 K, *Y* 5 W/m^2 K, *f* 0.5, lag 10 h. There are 10 occupants during 09.00–20.00 h. Copy a workbook file, rename it as *Woodhouse* and enter the data to predict the peak summertime temperature.

9. What result do you expect from calculation of peak internal summertime temperature in a UK commercial building that was designed for air conditioning, windows not openable, every workstation having a computer, shaded glazing to avoid sun glare for all workstations, artificial lighting continuously on and occupancy around 8 m^2 per person?

10. Locate the Tower of London, central White Tower, with Google Earth. Constructed in 1078, it would have been the epitome of construction design and skill at the time. What sort of peak summertime

temperature might the original residents and guests' experience? Look around to observe how the White Tower fits in within the Tower of London architecture. Other surrounding buildings did not exist in the present form, so ignore them. Ignore the dungeons for calculation purposes. Windows are small, probably not openable and have no shading or eaves. The square floor plan is approximately 20 m × 20 m and a total of 13 m in height. Facades face virtually north, south, east and west. Natural ventilation from doors and chimneys from open log fireplaces would create a draughty environment. There is around 25 m^2 of lead light single glazing on each facade. Solid stone walls probably 1 m thick or more U 3 W/m^2 K, Y 5 W/m^2 K, f 0.5, lag 10 h, windows U 6 W/m^2 K, Y 6 W/m^2 K, f 1.0, lag 0 h, solar correction factor 0.76 and alternating solar correction factor 0.5; roof heavyweight construction U 3 W/m^2 K, Y 5 W/m^2 K, f 0.5, lag 10 h; stone floor U 0.4 W/m^2 K, Y 4 W/m^2 K, f 0.6, lag 10 h. Assume 120 occupants 24 hours a day for staff, soldiers and prisoners. Combustion of hydrocarbon fuel, wood and candles, for cooking, lighting and water heating, will not be counted in this calculation. Copy a workbook file, rename it as *White Tower* and enter the data to predict the peak summertime temperature. Overall time lag for the building would be measured in days and it probably felt cold indoors all summer. Visit a castle or cathedral to observe internal air conditions in summer, that is, on a warm summer day. Compare what you calculate with modern buildings known to you.

11. Calculate the peak summertime temperature in a caravan. These are a popular means of holidaying, preferable to frame tents for keeping rain out and for surviving a camping experience in muddy fields. Windows are small, openable, no eaves and have interior shading. A typical floor plan is 6 m × 2 m and 2 m in height. A camper selects a southern England coastal site and places the caravan so the long side faces south for the view. There are 2 m^2 of darkened single polycarbonate glazing on each facade U 3 W/m^2 K, Y 3 W/m^2 K, f 0.5, lag 0 h, solar correction factor 0.2 and alternating solar correction factor 0.2; walls and roof are of the same construction of galvanised steel frame, aluminium external skin, 50 mm glass fibre insulation and 5 mm board U 1 W/m^2 K, Y 1 W/m^2 K, f 1.0, lag 1 h; carpeted, uninsulated wood floor U 2 W/m^2 K, Y 2 W/m^2 K, f 1, lag 1 h. Two occupants remain in the caravan all day, resting. They keep all windows and doors fully open and the ventilation standard is leaky. Ignore lighting, cooking, refrigerator, TV and hot water heat gains. Compare the living conditions with that of a brick building. Copy a workbook file, rename it as *Caravan* and enter the data to predict the peak summertime temperature.
12. Corrugated galvanised, or colour bonded, sheet steel commenced being used as a building material in the 1840s and remains in extensive use today. It was easily transported to Australia and the Americas by early traders until local production took over. Some architects choose to use it as a feature of their designs. Calculate the peak summertime temperature in a building constructed of a steel frame and uninsulated corrugated steel on a cast concrete floor. Such buildings include a well-known van factory in Southampton, many industrial buildings and garden sheds. Single-storey colonial period houses in the 1800s were lined with painted hessian, and later insulated and lined with plasterboard. Copy a workbook file, rename it as *corrugated iron* and enter the data to predict the peak summertime temperature.

    Take a floor plan of 15 m × 15 m and 2.8 m in height facing north, south, east and west. There are 5 m^2 of clear single glazing on each facade U 6 W/m^2 K, Y 6 W/m^2 K, f 0.5, lag 0 h, solar correction factor 0.76 and alternating solar correction factor 0.66; walls and roof are of the same construction of galvanised steel frame, single sheet of corrugated galvanised sheet steel U 6 W/m^2 K, Y 6 W/m^2 K, f 1.0, lag 1 h; concrete floor U 0.4 W/m^2 K, Y 2 W/m^2 K, f 1, lag 10 h. There are no permanent occupants, lights and equipment are not operating, doors are closed and air leakage rate is taken as tight. What happens if the walls and roof are insulated to U 1 W/m^2 K and Y 1 W/m^2 K? Compare the living conditions with those within other buildings.
13. Holidaying or living for other reasons under canvas is an experience of living outdoors with minimal shelter that many of us have had. An aluminium frame tent 4 m × 4m and 2 m high covered in

thin waterproofed canvas with a built-in ground sheet, has a south facing front entrance and window amounting to clear glazing of 4 m^2. Window, door, walls and roof all have U 6 W/m^2 K, Y 6 W/m^2 K, f 1.0, lag 0 h. Glazing solar correction factor 0.76 and alternating solar correction factor 0.66; earth floor U 1 W/m^2 K, Y 4 W/m^2 K, f 0.6, lag 10 h. There are no occupants for the purpose of this calculation. Calculate summertime temperature in the tent when the door is closed and outside air is almost still. Comment on the living conditions provided. Copy a workbook file, rename it as *tent* and enter the data to predict the peak summertime temperature.

14. Many people in the UK live in a traditional unimproved terraced 2–3 bedroom house having brick cavity walls, a suspended timber ground floor and a pitched grey slate roof. Original houses had no such thing as thermal insulation or draught proofing. Significant natural ventilation was needed to provide draught for the chimneys and black coal burnt. Locate the corner of Hurst and Hilda Streets, Oldham, Lancashire with Google Earth. Many of these very small homes have been extended by building a bedroom into the roof space and/or by building out into the tiny back yard. We will calculate for an unimproved original design.

    Look west to the end of Hilda Street. What do you see? The chimney of a steam boiler plant that powered the adjoining cotton mill of the 1800s industrial revolution. Walk around the old four-storey industrial building and imagine working there, walking or cycling to work from the houses in nearby streets – no electricity, piped gas, cars or indoor plumbing for those occupants. Every room had, and still has the facility for, an open coal-burning fireplace and a closed coal-burning cooking stove in the main living/kitchen room. Larger homes had a coal cellar and access manhole in the front footpath for deliveries by manually carried sacks from a horse-drawn cart. Central heating? No such thing for mill workers. Such houses date back to the 1850s. Look around to observe nearby architecture consisting of small terraced shops, two-storey red Lancashire brick and tile traditional houses, modern houses looking the same as the 1800s design, larger houses and public buildings. New industry and housing is changing the landscape. What is the large green dome seen from the corner of Hurst and Hilda Streets? A mosque. This shows the changing demographics. Also cars fill any available parking space. Notice the modern UPVC framed windows, probably double glazed, rather than the original wood frames. Doors now are also either smart, glazed UPVC or decorative wood. A permanent air brick below the front window allows through draft for the open coal fires.

    Calculate summertime temperature for an inner terraced house in Hurst Street. Floor plan is 4.25 m width of house, 9 m length and room height is 2.7 m (bricks counted for dimensions). 9 m^2 of single glazing face east onto Hurst Street, 10 m^2 at the rear elevation face west, none face north or south. Window U 5.7 W/m^2 K, Y 5.7 W/m^2 K, f 1.0, lag 0 h, solar correction factor 0.76 and alternating solar correction factor 0.5; walls U 1.5 W/m^2 K, Y 4.5 W/m^2 K, f 0.4, lag 10 h, roof U 2.3 W/m^2 K, Y 2 W/m^2 K, f 1, lag 10 h; suspended timber floor with perimeter wall air bricks U 2 W/m^2 K, Y 3 W/m^2 K, f 0.6, lag 10 h; north and south party walls U 0 W/m^2 K, Y 6 W/m^2 K, f 1, lag 10 h. Ignore the intermediate timber first floor thermal data as it has negligible thermal storage and time lag. Use the average leakage standard. Adjoining houses are maintained at the same temperature. No solar shading was used as daylight penetration was paramount. There are four occupants (parents plus two working young people) when they are not at work during 18.00–07.00 h. Ignore heat gains from candle lighting, cooking and water heating; bathing was likely conducted in public baths or in front of the open coal fire in the living room.

15. Traditional style small terraced 2–3 bedroom homes formed the basic accommodation for working families in the UK after the industrial revolution of the 1700s. They still do. Many have become second homes in holiday regions. Many are larger, have gardens and a garage, but have a look at any street of newly constructed terraced homes; they are fundamentally the same as Hurst Street, Oldham, as in question 14. Hurst Street houses need to be modernised to work towards meeting the objectives of the HM Government Carbon Plan (2011).

Copy the *Oldham* file and save as *Oldham improved*. Floor plan is 4.25 m wide, 9 m long and room height is 2.7 m. UPVC framed double glazing 9 m^2, with internal blinds, face east onto Hurst Street. Heritage laws do not allow changes to the street frontage of these houses. Rear windows are 10 m^2 UPVC framed double glazing facing west. Window U 2.3 W/m^2 K, Y 2.3 W/m^2 K, f 1.0, lag 0 h, solar correction factor 0.15 and alternating solar correction factor 0.11; walls U 0.2 W/m^2 K, Y 2 W/m^2 K, f 0.4, lag 10 h, roof U 0.2 W/m^2 K, Y 2 W/m^2 K, f 1, lag 10 h; insulated suspended timber floor U 0.2 W/m^2 K, Y 2 W/m^2 K, f 0.6, lag 10 h; north and south party walls U 0 W/m^2 K, Y 6 W/m^2 K, f 1, lag 10 h. Ignore the intermediate timber first floor thermal data as it has negligible thermal storage and time lag. Use the tight air leakage standard, as mechanical ventilation with heat reclaim is installed. Adjoining houses are maintained at the same temperature.

There are two occupants in the house all day and night on a particular summer weekend day, or public holiday. There are six lights of 70 W each on during 17.00–24.00 h. Electrical equipment, refrigerators, dishwasher, laundry, computers, communications, entertainment, garden lighting, ornamental pond pump, exterior security lights, battery charging, cooking and water heating provide an average electrical demand of 500 W continuously.

What effect do raising thermal insulation values, solar and ventilation control, and today's use of electrical energy have on a Hurst Street house peak summertime temperature and achievement of the Government's emission targets?

16. Which is correct about what we have learnt from these peak summertime temperature predictions?

    1. Every UK building is comfortable all year;
    2. Air conditioning is a necessity in the UK;
    3. UK buildings become overheated occasionally;
    4. White Tower is a better design than The Shard;
    5. Nothing has been learned from history.

2 Air conditioning systems

Learning objectives

Study of this chapter will enable the reader to:

1. define the term air conditioning;
2. state the reasons for air conditioning;
3. categorise the available systems;
4. be aware of low cost cooling methods;
5. understand the scheduling of plant air dampers;
6. know why building zones are necessary;
7. know that zone peak loads are not simultaneous;
8. understand why some zones are air pressurised;
9. understand the working principles, applications and limitations of 16 categories of air conditioning system;
10. produce schematic diagrams of practical systems;
11. explain why a particular system is being proposed for an application by means of description and sketches;
12. apply air conditioning to the project building provided;
13. know the systems used in an airport terminal through a case study.

Key terms and concepts

Air conditioning 17; air handling zones 21; changeover system 28; chilled ceiling 40; Coanda effect 25; containment 22; district cooling 40; dual duct 30; dumping 25; evaporative cooling tower 20; fan 17; fan coil unit 29; fresh air intake 18; heat gains 17; heating and cooling coils 29; independent unit 35; induction unit 27; low cost cooling 32; mechanical ventilation 18; mixing box 30; motorised damper operation 20; multizone plant 24; orientation 21; peak solar gains 21; perimeter heating 27; recirculation 18; refrigeration compressor 19; reversible heat pump 30; single duct 23; split system 36;

stagnation 26; static pressure 22; terminal unit 25; throttling valve 26; variable air volume 24; variable frequency control 25.

Introduction

Air conditioning means the full mechanical control of the internal environment to maintain specified conditions for a certain purpose. The objective may be to provide a thermally comfortable temperature, humidity, air cleanliness and freshness for the users of the building or it may be to satisfy operational conditions for machinery or processes. The term air conditioning may be used to describe an air cooling system that reduces excessive temperatures but does not guarantee precise conditions, to minimise capital and operational costs. This is better described as comfort cooling.

The decision to adopt air conditioning is analysed and a list of the possible systems given. So called free cooling and how to achieve it, is explained. The choice of zones is important to air conditioning and the reasons for their selection are discussed.

The main systems of air conditioning are described. The list is not exhaustive. The large number of possible combinations of components makes each system at least partly unique, commensurate with the diversity of building services engineering.

Drawings of a project building are provided for assignment work, discussion of possible solutions, examination and testing purposes plus the assessment of competences. The drawings are not printed to scale and need to be drawn to an appropriate size and scale. They are referred to in further chapters and are sufficiently detailed for most purposes. Detailed dimensions will need to be decided by inference.

A case study of a recently completed airport demonstrates that multiple systems are used for complex buildings. *Building Services, The CIBSE Journal* is a regular source of practical cases and is recommended reading.

Assignment questions are provided to encourage wider use of references and discover more of the technical solutions for air conditioning.

The decision to air condition

Mechanical ventilation includes the moving of air by means of fans, air filtration, heating, humidification, heat reclaim for economy of operation and free cooling that can be obtained from the external atmosphere. Air conditioning differs from mechanical ventilation by the incorporation of refrigeration. Adding mechanical refrigeration equipment and a cooling coil to a mechanical ventilation system turns it into air conditioning. Low cost air conditioning may only cool inside the building by a specific number of degrees Celsius below the outside air temperature or limit the internal percentage saturation to 70%. This is the upper limit for comfort and for the risk of microbiological growth. The term air conditioning does not mean that specified comfort conditions can or will be maintained. Achieving comfort requires the provision of correctly designed, installed and maintained systems.

Air conditioning is included in a total building design for a variety of motives that range from industrial necessity to the prestige to be gained from marketing the building or its use. The overall technical and aesthetic solution includes the use of passive design features to minimise the use of mechanical plant. Good design is visually acceptable as well as being technically competent. The environmental and energy cost implications are analysed. Suitable reasons include the following.

1. Unacceptable high internal summer air temperature resulting from solar heat gains if it is not provided with refrigeration.
2. Heat gains occurring within the building, from people, lights, electrical, catering and mechanical equipment, which produce uncomfortable air temperatures for the occupants.

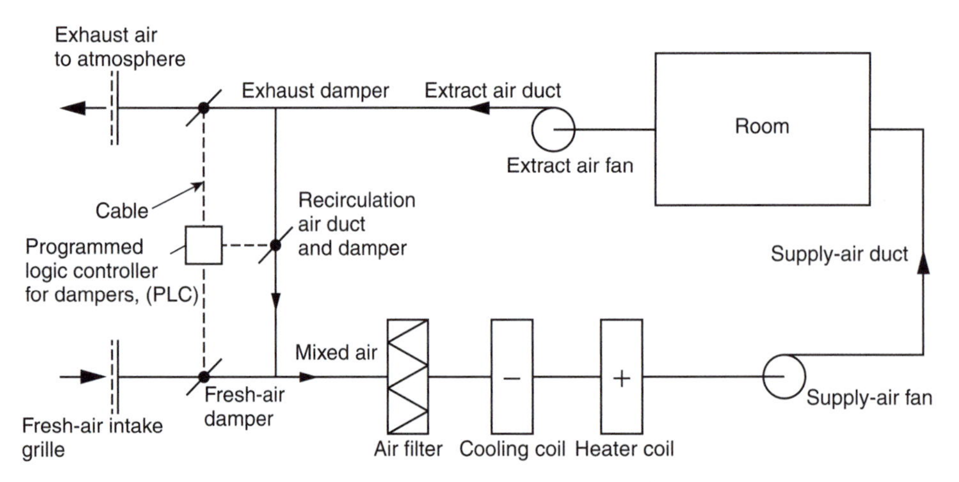

2.1 Schematic layout of a single duct variable air temperature air conditioning system, with recirculation of room air (SDVATR).

3. Occupied areas that cannot be satisfactorily supplied with enough fresh air by natural ventilation.
4. High rise buildings where prevailing wind pressure precludes the opening of windows to provide the necessary ventilation.
5. Road traffic, aircraft or train noise close to the building would cause too much disturbance if windows were opened. The building has to be air sealed from the external environment to limit noise penetration and consequently mechanical ventilation and possibly refrigeration.
6. External air pollution requiring the building to be sealed.
7. Close control of the internal atmosphere is required for manufacturing processes such as pharmaceuticals, electronics, nuclear components, paper and cotton production.
8. Secure containment of radioactive processes and materials requires that all possible leakages of contaminated air and dust are eliminated. Full mechanical control of the internal environment becomes necessary for both the process and the personnel.
9. The reliable operation of most microprocessors and electric motors depends upon maintaining a maximum surrounding ambient air temperature of up to 40°C and the plant room may need to be air conditioned.
10. Shops, hotels and commercial buildings where customers are admitted may be air conditioned for their comfort and to provide a marketing advantage over competitors.
11. Countries in the tropics have air conditioned buildings, homes, vehicles and cars through necessity.
12. Storage and display of works of art, antiques, furniture, fabrics, paintings and paper archives.
13. Sterile conditions needed for health care.

Methods of system operation

Air conditioning systems are categorised by their mode of operation:

1. single duct variable air temperature with 100% fresh air, SDVATF;
2. single duct variable air temperature with recirculation of room air, SDVATR. (Figure 2.1 is a schematic layout);
3. single duct variable air temperature for multiple zones, SDVATM;
4. single duct variable air volume, SDVAV;

5. single duct variable air volume and temperature, SDVAVT;
6. single duct variable air volume with separate perimeter heating system, SDVAVPH;
7. single duct with induction units, SDI;
8. single duct with fan coil units, SDFC;
9. single duct with reversible heat pumps, SDRHP;
10. dual duct with variable air temperature, DDVT;
11. dual duct with variable air volume, DDVAV;
12. independent unit through the wall, IU;
13. split system, SS;
14. reversible heat pump, HP;
15. chilled ceiling, CC;
16. district cooling, DC.

Low cost cooling

Cooling of the building can be obtained by passing outside air through it, overnight, or by circulating water that has been cooled in an evaporative cooling tower on the roof, through the chilled water coil in the air handling unit. Direct evaporative cooling from sprayed water within the air handling plant is used in hot climates where the outdoor air moisture content is low. Evaporative cooling is used widely in Australia in homes, retail premises and public houses. A desiccant wheel regenerative heat exchanger may be combined with evaporative cooling sprays and a heating coil to provide a cooling system. The desiccant wheel dehumidifies the supply air while the spray provides cooling. A heating coil in the exhaust air duct drives moisture from the desiccant. Such low cost cooling methods consume electrical energy in the operation of fans and pumps. Maintenance costs are incurred in keeping the system clean and operational. The capital cost of a ducted evaporative cooling system is not significantly less than that for a refrigeration system, but it provides inferior control of the indoor air condition. The automatic control system consumes energy with attendant cost and maintenance. The same plant is used as when the refrigeration system is in operation. The cash saving available from using an evaporative cooling system is from the power consumption of the refrigeration compressor or the heat supplied to the absorption chillers.

Cooling with outside air rather than by refrigeration, requires that up to 100% of the room air supply can be by cool outdoor air. This method can be acceptable when a wide fluctuation in the room percentage saturation can be allowed. Comfort air conditioning can tolerate such swings, but critical industrial production facilities, such as printing, cannot. The availability of this air depends on the following factors.

1. Time of year. Winter, spring and autumn seasons have low external air temperatures. Winter air below 10°C will have minimum usage.
2. Time of day. Sufficiently cool outdoor air will not be available during the warm months in the day-time occupancy hours of offices and shops. Outdoor air above 20°C will have minimum usage.
3. Fresh air intake location. The fresh air intake grille is positioned in a shaded location above street level where cool and uncontaminated air can be found. Availability of shade depends upon the orientation of the side of the building and the time of day. Underground tunnels or air ducts can be used to lower the temperature of the fresh air intake.
4. Space availability. The dimensions of the building's facade available for the fresh air intake and exhaust louvres must be sufficient for the passage of 100% of the supply air volume flow rate. Restricted plant room locations near ground level limit the amount of free cooling that is achievable. Extensively louvred plant room walls at roof or intermediate floor levels can be utilised.

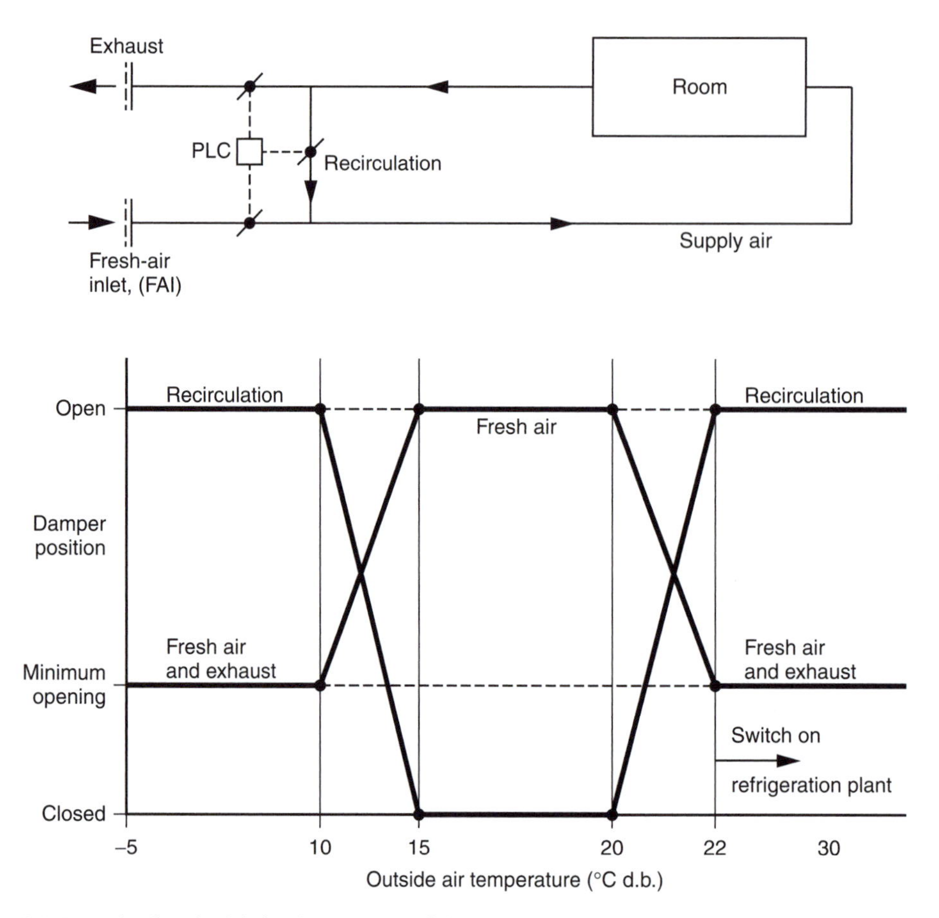

2.2 Example of a schedule for the operation of dampers in a SDVATR system.

An evaporative cooling tower lowers the water temperature circulated through it towards the outdoor wet bulb air temperature. Water evaporation increases the cooling effect of the outdoor air. When the outdoor wet bulb air temperature is below the supply air temperature needed to cool the room, usable free cooling is potentially available.

Internal rooms within the building that have no direct view or connection with the external walls or roof require continuous cooling through the year. Outdoor air will be suitably cool for much of the year in the UK.

The amount of free cooling provided is controlled by varying the proportion of fresh air in the room air supply system. The proportion moves from the minimum up to 100% by the operation of motorised dampers on the fresh air, exhaust and recirculation ducts. The minimum fresh air intake corresponds to the occupancy or process fresh air requirement at the peak summer or winter external design conditions. Variable amounts of fresh air are admitted between these extremes. Figure 2.2 shows how the three dampers are scheduled for a single duct system with recirculation according to outdoor air temperature. The dampers have a linear characteristic that relates flow to opening. The objective is to minimise the use of heating and cooling energy while maintaining the specified internal air conditions. The external air temperatures indicated are sample values that do not apply to all cases and are sensed at the fresh air intake grille.

Table 2.1 Peak solar heat gains through unprotected clear vertical glazing

Orientation	*Peak gain* $\frac{W}{m^2}$	*Sun time, h*	*Date*
S	510	1300	22 September
E	477	0900	21 June
W	477	1700	21 June
N	161	1300	23 July

Data in this table are approximate and only for use within the examples in this book.
Sun time is Greenwich Mean Time, GMT. British Summer Time, BST is GMT plus 1 hour and is used from the end of March to the end of October

The fresh and exhaust air dampers are interlinked to have identical openings and they remain at the minimum settings below 10°C and above 22°C. These dampers begin opening at 10°C and are fully open at 15°C. They remain fully open from 15°C to 20°C while cooling is available to remove solar and internal heat gains from the building. At 20°C the dampers begin to close to reduce admission of heat gains into the building from the warm ventilation air. They reach their minimum setting at 22°C and remain there during hot weather.

The recirculation air damper is fully open until the winter air reaches 10°C. Its opening is ramped down to be fully closed at 15°C. The fresh air and exhaust dampers are fully open. The recirculation damper remains closed to allow the building to be flushed with fresh air until 20°C is reached. The recirculation damper is ramped up to be fully open at 22°C where it remains static during the summer.

A programmable logic controller (PLC) microprocessor retains these instructions and operates the components. This program can be reset manually or through the wiring from the supervising computer of a building energy management system (BEMS). The refrigeration system is switched on at 22°C and is separately controlled. The shape of the damper movement graph shown is for illustration purposes only and each control system is designed according to its unique needs.

Air handling zones

Areas of the building that have a similar heating, cooling and humidity control plant load are grouped into zones. The south facing rooms of a modular office block are all exposed to an identical pattern of solar heat gains during the working day in that their peak gains occur simultaneously. Each office has the same occupancy usage. When the south facing glazing of rooms on the lower storeys is shaded from the direct solar radiation by other buildings, they have lesser heat gains than higher level rooms on the same side that remained exposed all day. The south facade of the block needs to be separated into two, or more, zones, each being provided with a different supply air temperature throughout the year. If not, the lower floors would be excessively cooled by supply air that is suitable for the exposed rooms.

Peak solar heat gains through unprotected clear vertical glazing occur at various times and dates as shown in Table 2.1.

Interior rooms that have no external glazing experience a constant cooling load that is dependent upon internal heat gains rather than the weather. If each building orientation is grouped into separate zones, the time and date of their peak cooling loads will be different.

Zones can be chosen on the basis of the following.

1. Time of occurrence of the peak loads. This enables the air handling plant to provide a suitable air supply temperature to the whole zone.

2. Orientation. Each facade of the building has a unique programme of shading and exposure to solar radiation. Rooms with large areas of glazing are highly vulnerable to the external climate. A building with an irregularly shaped perimeter might have more than one orientation within a zone. A windowless structure is relatively insensitive to cyclic solar radiation heat gains due to its thermal storage capacity and up to 12 hours of time lag for heat transfer from outside to inside.
3. Interior space. Rooms that have no surfaces which are exposed to the external atmosphere have a daily and year-round constant heat load, usually for cooling.
4. Height above ground. Tall buildings are zoned into groups of floor levels due to their exposure or to reduce the size of the air distribution ductwork. Below ground rooms are zoned together.
5. Containment. Areas of a building that must be independent of each other to restrict possible cross-contamination by airborne micro-organisms or radioactive particles are put into separate zones. Examples are kitchens, hospital operating theatres, X-ray rooms, nuclear, chemical and biological production facilities, research laboratories and clean rooms for pharmaceutical and electronic manufacture.

Containment zones are maintained at a static air pressure that is above or below that of the adjacent rooms or the external atmospheric pressure to ensure an air flow in a specific direction that is into or out of the containment zone. Pressurised air locks may be used to separate zones. Table 2.2 gives examples of some applications where a specific flow direction must be maintained by suitable positive or negative air pressure when compared to the surrounding areas.

Air buoyancy, or stack effect, and wind pressure will influence internal air pressure; when less air is extracted than supplied, the room is slightly pressurised and conditioned air will leak outwards into corridors, staircases and eventually outdoors. These air flows are calculated to find the air velocity through gaps around doors and transfer grilles. Conditioned air leaks out of the building. This ensures that the internal conditions remain under control. When more air is extracted than supplied, the room is under a negative pressure and air from surrounding areas or from outdoors, will leak into the room. This dilutes the conditioned air and adds to the room heating or cooling load on the plant. This will be undesirable when close environmental control is essential. The ability of a building to maintain a specified static pressure is dependent upon the quality of construction. Air leakages can be calculated from data used in the design of fire escape routes.

In conclusion, a zone can be a group of areas, rooms, floors, a whole building, a single room or a single module of a building. A single duct air handling system can provide one supply air condition to a whole building. Terminal temperature or volume regulation ensures that each room can be maintained at its desired state. Figure 2.3 shows how a simple building design can be separated into zones.

Table 2.2 Zones static air pressures

Zone use	*Air flow direction*	*Static pressure, Pa*
Conditioned office	out	+10
Kitchen, toilet	in	−10
Operating theatre	out	+10
Air-lock between areas	out	+10
Nuclear, biological, chemical	in	−50
Clean room	out	+10

Data in this table are indicative only

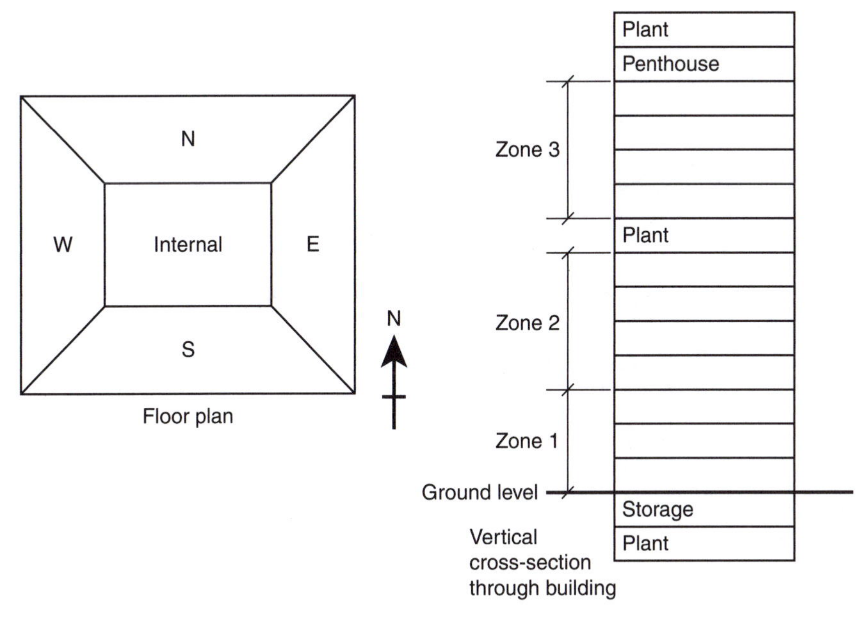

2.3 Air conditioning zones.

Single duct variable air temperature 100% fresh air (SDVATF)

This system is used where contamination produced within the conditioned room is not to be recirculated and a full flushing with outdoor air is to be maintained. A commercial kitchen has a heated supply air duct and hoods over cooking ranges to exhaust the hot steamy and grease-laden air to the outdoor atmosphere as quickly as possible. Two-speed fans can be used so that low ventilation rates will meet comfort needs during food preparation but high extraction can be operated during maximum cooking periods.

An approach to avoiding using recirculated room air can be to pass the incoming outside air through a flat plate heat exchanger that is cooled during summer, and heated in winter, by the outgoing room air. Run around pipe coils or a mechanical refrigeration heat pump system are alternatives. The use of 100% outdoor air and a heat reclaim coil, will substantially increase the summer refrigeration cooling load on the plant and will lose control of the room percentage saturation during winter unless a humidifier is installed.

Single duct variable air temperature with recirculation (SDVATR)

This is a common system adopted for occupied large single volume rooms, offices, atria, theatres, sports halls, swimming pools, factories, clean rooms and mainframe computer facilities. The recirculation allows retention of the conditioned air that has been expensively produced with only the correct quantity of fresh air being admitted. Free cooling is possible by adjustment of the dampers following schedules of air temperatures.

The supply air quantity will be between 4 and around 20 air changes per hour through the room. The amount of fresh air is calculated from the occupancy or process needs. Typical values are 8 litre/s per person for offices where there is no smoking. Where heavy smoking is permitted 25 litre/s might be supplied; however, indoor smoking has disappeared from many locations. Such admittance of fresh air is likely to create one air change per hour. The room air change rate that is required to flush out potentially stagnant pockets of air is more like between 4 and 20 per hour depending upon the design of the air

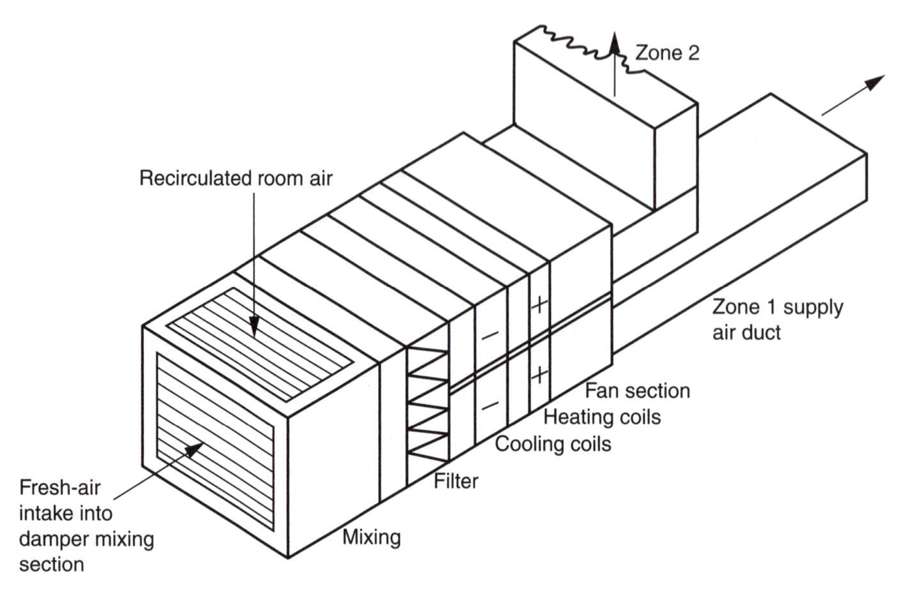

2.4 Two-deck multi-zone air handling unit.

currents produced by the grilles and diffusers. Thus the fresh air proportion of the ducted air supply will be 25% or less; however, this is not a fixed ratio.

The system only satisfies the conditions for one room and is unsuitable for multi-room buildings that have different but simultaneous air temperature requirements. The room can be large but all parts of it are supplied with identical supply air. The volume flow rates to different parts of the room are in proportion to the localised heating or cooling loads but the same average condition exists throughout the space.

Single duct variable air temperature multiple zones (SDVATM)

This is an adaptation of the individual zone single duct system with recirculation but is designed for multi-room applications such as offices. Additional heater coils are installed at the room end of the ducts where they are called terminal reheaters. Each room has an air temperature sensor that modulates the reheat coil water flow control valve to maintain the desired room condition. The air temperature that leaves the air handling unit in the plant room is that for the room requiring the lowest supply air temperature. The other rooms have a few degrees of reheat.

When rooms can be formed into two or four zones that are to be supplied from the same air handling unit, a multi-zone plant can be designed to incorporate two or four outlet ducts. Each duct has a reheat coil and zone air temperature controller. Figure 2.4 shows a multi-zone air handling unit. Zone reheat coils can be in the supply air duct as it enters the zone. Several zones can be connected to a single air handling unit. Each zone has a number of rooms that will be satisfied with one supply air temperature, for example, a row of one person offices along one facade of the building or a similar row of modules in an open plan office. The disadvantages of having distributed hot water reheat coils are that additional flow and return piping, controls and access to ductwork are needed.

Single duct variable air volume (SDVAV)

The single duct variable air volume system has become the standard employed for office buildings due to its economy and controllability when applied to rooms of similar use for comfort air conditioning. It is a cooling system with the air handling unit providing the lowest air temperature required. Reductions in room cooling load due to variation in the weather or the presence of additional heat gains from occupants or electrical equipment are met by changing the supply air quantity with terminal variable air volume controllers until the room air temperature is stabilised. The minimum quantity of supply air permitted corresponds to at least the fresh air need of the occupants, thus the turn-down may be from 100% down to around 20%.

It is sensible to reduce the air flow rate through the building during most of the year when the refrigeration and heating loads are less than their design peak values. When the room air temperature falls, the cool supply air volume is reduced until the desired temperature is reached again. This process is repeated in other VAV controllers at a similar time. The characteristic performance of the centrifugal supply air fan in the main air handling unit is to increase its output static pressure as the volume flow rate is reduced. This is contrary to the room requirements as a higher duct pressure would tend to overcome the control effect of the VAV unit. The main supply air duct static pressure is sensed and used to reduce the speed of the fan by means of a variable frequency inverter. The fan speed is lowered slowly until the set point of the static pressure controller is established. This is also done with the recirculation air fan. Operating the two fans at less than their maximum speeds greatly reduces the electrical power consumed and the noise generated by them.

The air supply flow rate is varied by the following methods.

1. Throttling valve. The conditioned supply air enters the room through a diffuser. This may be a ceiling linear diffuser along the external perimeter. A plenum above the diffuser acts as a header box to distribute the air uniformly along its length. A throttling valve between the plenum and diffuser controls the air flow rate. Figure 2.5 shows a simplified arrangement of three types.
2. Fan. A fan powered VAV terminal box is located in each building module. The fan may be used to deliver a constant volume flow rate into the room and have a variable proportion of cool air from the central plant, modulated by the VAV damper. This ensures the maintenance of the correct air distribution through the room. A terminal fan is an additional capital cost, and there are noise and maintenance considerations. Figure 2.6 shows a simplified arrangement.

Varying the supply air flow has a strong influence upon the distribution of air into the room and the movement of air within it. The supply air is diffused into the room along the length of the external wall and is blown across the ceiling and down the glazing. Airflow along a surface adheres to that surface and forms a thick boundary layer. This is known as the Coanda effect after Henri-Marie Coanda (1885–1972) the Romanian-born aeronautical engineer who made a major contribution to fluidic technology in the 1930s. He observed that a fluid stream creates a lowered static pressure adjacent to the surface thus anchoring the layer to the surface.

Low air flows created during the normal use of a VAV system can lead to discomfort due to a variety of reasons.

Dumping

Supply air that is cooler than the room air temperature is being throttled to, perhaps, 25% of its design flow rate during mild weather. The combination of low jet speed issuing from the diffuser and the greater density of the cool supply air, leads to the air not staying in contact with the ceiling long enough to mix with the warmer entrained room air, and it falls away from the ceiling. The occupants feel cool downdraughts and have cause for complaint. It is necessary to maintain a sufficiently high air velocity jet across the ceiling at the minimum air flow setting to keep the contact. A narrow slot diffuser or a fan powered unit can achieve this.

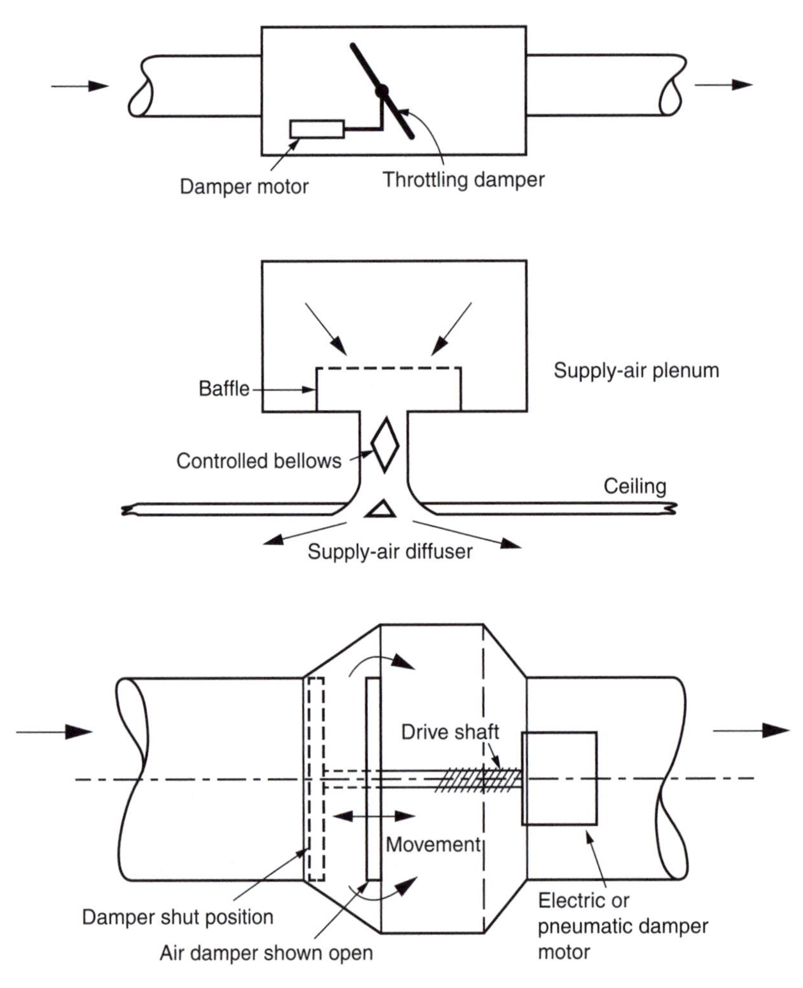

2.5 Variable air volume terminal unit controllers.

Stagnation

Reduced air flows produce fewer changes of the room air per hour and less stimulation of the recirculation eddy currents.

Poor distribution

Parts of the room may be starved of sufficient air circulation and become too warm due to local heat sources such as people, computers, lights and photocopiers.

Dilution

Complaints of stuffiness or smoke-laden air may result if tobacco use is permitted. Odour from the occupants and the materials, solvents and chemicals within the room can become noticeable and can cause a nuisance.

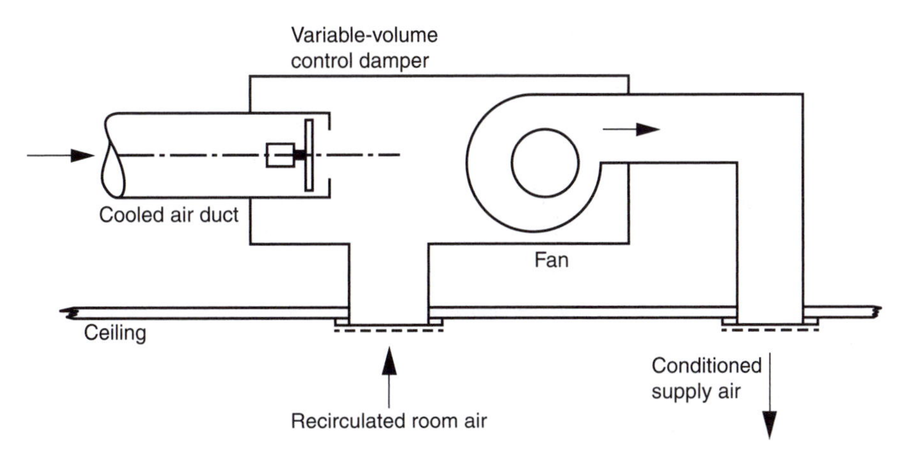

2.6 Fan powered variable air volume terminal unit installed in a false ceiling.

Cases of sick building syndrome (SBS) may be traced to the quantity of air in circulation and its quality. The air volume flow control box of the VAV system is often installed in the false ceiling. It can be situated in a floor void with the supply duct rising to a window-sill-level grille.

Single duct variable air volume and temperature (SDVAVT)

This is the same VAV system but with the addition of terminal reheat coils to adjust the delivered supply air temperature. The air temperature that leaves the air handling plant is determined by the zone requiring the most cooling. The use of reheaters adds to the energy consumption of the system as previously cooled air is raised in temperature. The air pressure loss through the heater batteries and the extra pipes and controls also increases the capital and operating costs.

Single duct variable air volume perimeter heating (SDVAVPH)

The separate perimeter heating system will be radiators, radiant panels, convectors or sill-line convectors on a low pressure hot water distribution with thermostatically controlled modulating valves. This is to allow each room to be maintained at the correct air temperature while being ventilated with a fixed supply air temperature from the air handling plant.

Perimeter heaters can be obtrusive in the occupied space, require careful integration with the furniture, interior design and decoration, need maintenance and result in some loss of floor area. Locating computer workstations near to the heaters has to be avoided due to the possible overheating of the electrical equipment.

Single duct with induction units (SDI)

A single duct supplies fresh air only from an air handling unit in the plant room into each room. This fresh air is termed the primary air supply. An extract air duct removes 90% to 100% of the fresh air supply quantity and discharges it to the atmosphere. Each room or module has an induction unit beneath the window sill, in the floor or in the ceiling. The primary air is introduced into the induction unit through nozzles that increase the inlet air velocity sufficiently to lower the static air pressure within the induction unit casing. Noise

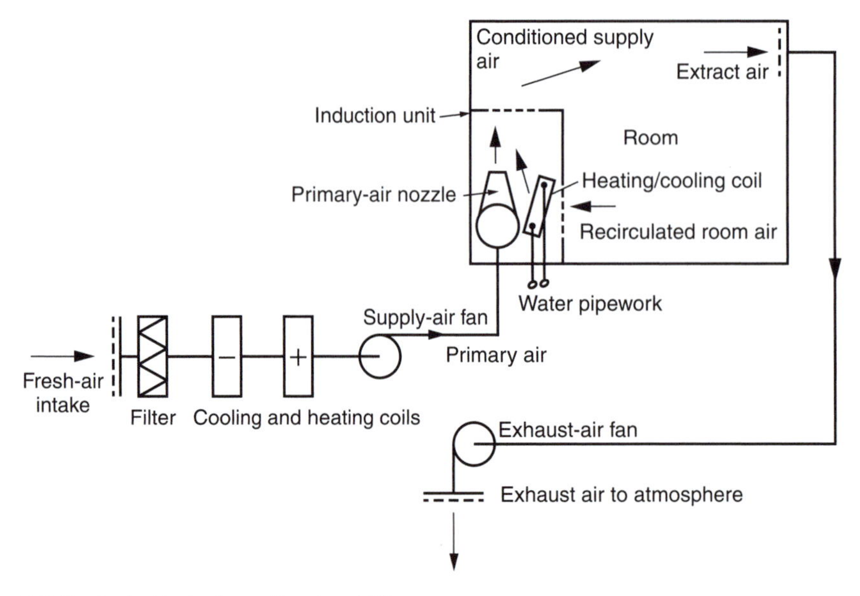

2.7 Single duct induction unit system (SDI).

generation is a problem. Room air is induced to flow through the casing due to the pressure difference. A low pressure hot water heating or chilled water cooling coil conditions the recirculated room air prior to its mixing with the primary air. The mixed air flows into the room through a diffuser. Only the primary air is filtered. The duct systems only passes fresh and exhaust air and so is as small as possible for the building when compared with other designs that require ducted recirculation.

The bulk of the sensible heating and refrigeration room load is met from the induction unit coil. The room moisture content can only be controlled from the moisture content of the fresh air supplied and this restricts the humidity control ability of the system. It would generally not be practical to dehumidify the room air with the cooling coil because condensate drain pipework would be needed. A manually operated damper is fitted to sill-line induction units to enable the occupant to vary the heating or cooling effect by adjusting the recirculated room air flow.

The system is used for office accommodation and a typical installation is shown in Figure 2.7. The water circulation to the heat exchange coil can be of the following types.

Two-pipe changeover

The two-pipe system requires the changeover of the water circulation from the boilers to the refrigeration evaporator by means of a three-way valve in the plant room at a clearly defined season change from winter to spring. Pipe work needs to be zoned so that north and south aspects of the building can have warm and cool water simultaneously.

Two-pipe non-changeover

Only chilled water is circulated to the induction unit room coils. All the heating needed is provided by the heater coil of the plant room air handling unit if a sufficiently high air temperature is possible. 45°C is the maximum available.

Four-pipe

The hot and chilled water circuits are maintained at the correct temperatures separately to provide the maximum availability of heating and cooling at all the induction units throughout the year. Each induction unit has two coils, four water pipes and either two two-way valves or one three- or four-port valve with air temperature control. The system is the most costly option but offers the greatest versatility.

Single duct with fan coil units (SDFC)

A single duct supplies conditioned primary fresh air only from an air handling unit in the plant room into each room. An extract air duct removes 90% to 100% of the fresh air supply quantity and discharges it to the atmosphere. Each room or module has a fan coil unit within the false ceiling, within the floor or below the sill, which recirculates room air and mixes it with the primary fresh air. Heating and cooling coils are fitted into the fan coil unit casing. These are supplied from a four-pipe or a two-pipe changeover, hot and chilled water pipe system. Large heating and cooling loads that occur within the occupied space can be met as there are fewer limitations upon coil duties than with other systems due to the use of a dedicated fan.

The room air can be dehumidified. A drip tray and condensate drain is included in the design. Control of room humidity is poor. Fan coil units are independent of each other. Each fan coil unit creates a zone. The hot and chilled water circulation is supplied at constant temperature from a central plant whereas the primary air ductwork may be zoned.

Drain pipe work connects to the foul drain stack through an air gap and a 75 mm deep water trap. An air filter is fitted within the unit. Each fan coil box has to have a removable ceiling panel for regular maintenance of the filter and for access to the fan and controls. The primary air can be supplied into the room through grilles or diffusers rather than into the fan coil unit. The fan noise produced is related to the acoustic design criteria of the air conditioned room. Figure 2.8 shows a typical configuration.

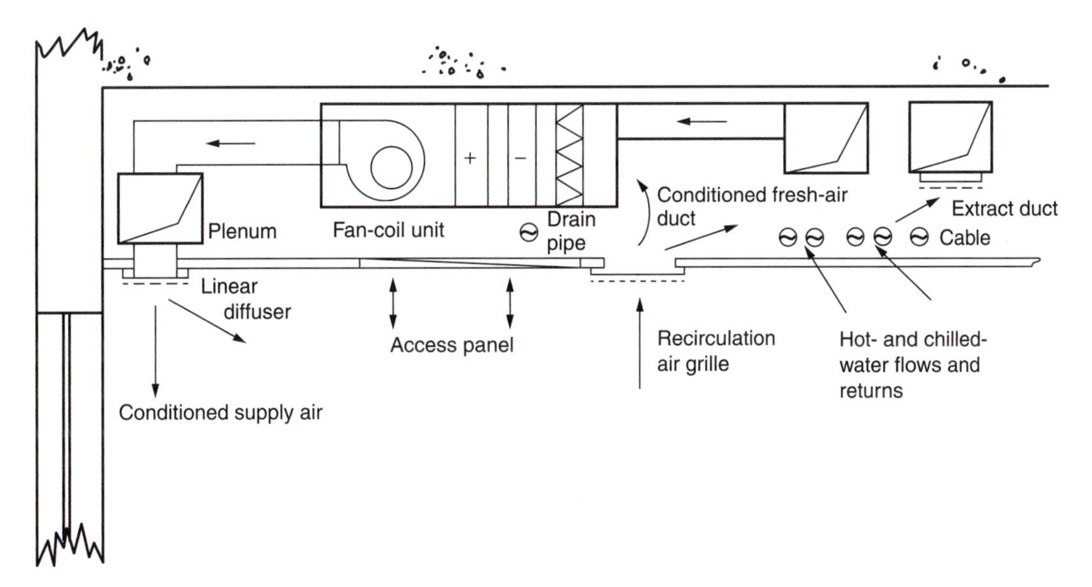

2.8 Fan coil unit air conditioning system (SDFC).

Single duct with reversible heat pump (SDRHP)

A single duct supplies conditioned primary fresh air only from an air handling unit in the plant room into each room. An extract air duct removes 90% to 100% of the fresh air supply quantity and discharges it to the atmosphere. Each room or module has a water source reversible heat pump system within the unit casing that recirculates room air and mixes it with the primary fresh air. The terminal unit can be wall, floor, ceiling or under floor mounted. It is connected to a two-pipe reverse return water circulation system that is maintained at a constant temperature of typically 27°C throughout the year. The water circuit acts as the heat source in winter and the heat sink in summer. The system is ideally suited for buildings that have simultaneous heating and cooling loads, for example, those with large glazing areas that are oriented both south and north. During cool sunny weather, the excess heat gains through the south glazing require the rooms to be cooled and the heat pumps add heat into the water circuit. The north facing rooms are suffering heat losses and the heat pumps extract heat from the water to raise the supply air temperature and heat the rooms.

Surplus heat from the building is ejected into the outside atmosphere through the cooling tower or dry cooler. The water is kept at the desired 27°C. Additional heat is met by the boiler plant to keep the water circulation up to 27°C. A dead band of 1°C or more is maintained between the switching points of the cooling tower and boiler to stop overlap of their operation.

Each room or building module has a self-contained refrigeration system comprising of a reciprocating or rotary compressor, an air heat exchange coil, a water heat exchange coil, a refrigerant reversing valve, controls, filter and recirculation fan. Figure 2.9 shows the arrangement. High noise and maintenance levels are typical. The recirculated room air passing through the coil is heated or cooled by the refrigerant depending upon the position of the reversing valve. During the summer, this coil acts as the refrigerant evaporator. When heating is called for, the room air coil acts as the refrigerant condenser because the reversing valve has exchanged the roles of the two coils.

Humidity control is from the ducted fresh air moisture content as with the other systems described. Close humidity control is unlikely. Condensate drain pipework is needed. The starting and running of refrigeration compressors in each room can be noticeable. They are interlocked electrically to avoid the simultaneous starting of too many and overloading the wiring and current limiters.

Dual duct with variable air temperature (DDVAT)

The dual duct system supplies air to each room, module or zone through a hot duct and a cold duct. A mixing box regulates the relative amount of each source while keeping the supply air volume flow into the room constant. It is a constant volume variable temperature all-air system having no water services outside the air handling plant room. Fresh air is mixed into the recirculated room air in the plant room. Figure 2.10 shows the air handling system and Figure 2.11 the operation of the terminal mixing box.

The dual duct system has two supply air ducts of equal size. Either the hot or cold duct has to pass the full supply air quantity. A recirculation duct passes air from each room back to the plant. The distribution space occupied by three air ducts is much greater than that for water based systems with only fresh air being ducted. Figure 2.12 demonstrates the comparable service shaft space needed for the various systems.

High velocity air distribution ductwork is used when duct spaces are limited but this leads to additional fan and air turbulence noise generation transmitted through the ducts. The room terminal mixing box also acts as a noise attenuator or silencer. Similar treatment may be necessary for the air ducts and plant room. Figure 2.13 indicates where acoustic attenuation may be fitted to an air duct system. Terminal attenuators have the important function of minimising cross-talk between rooms that can be both annoying and reduce security.

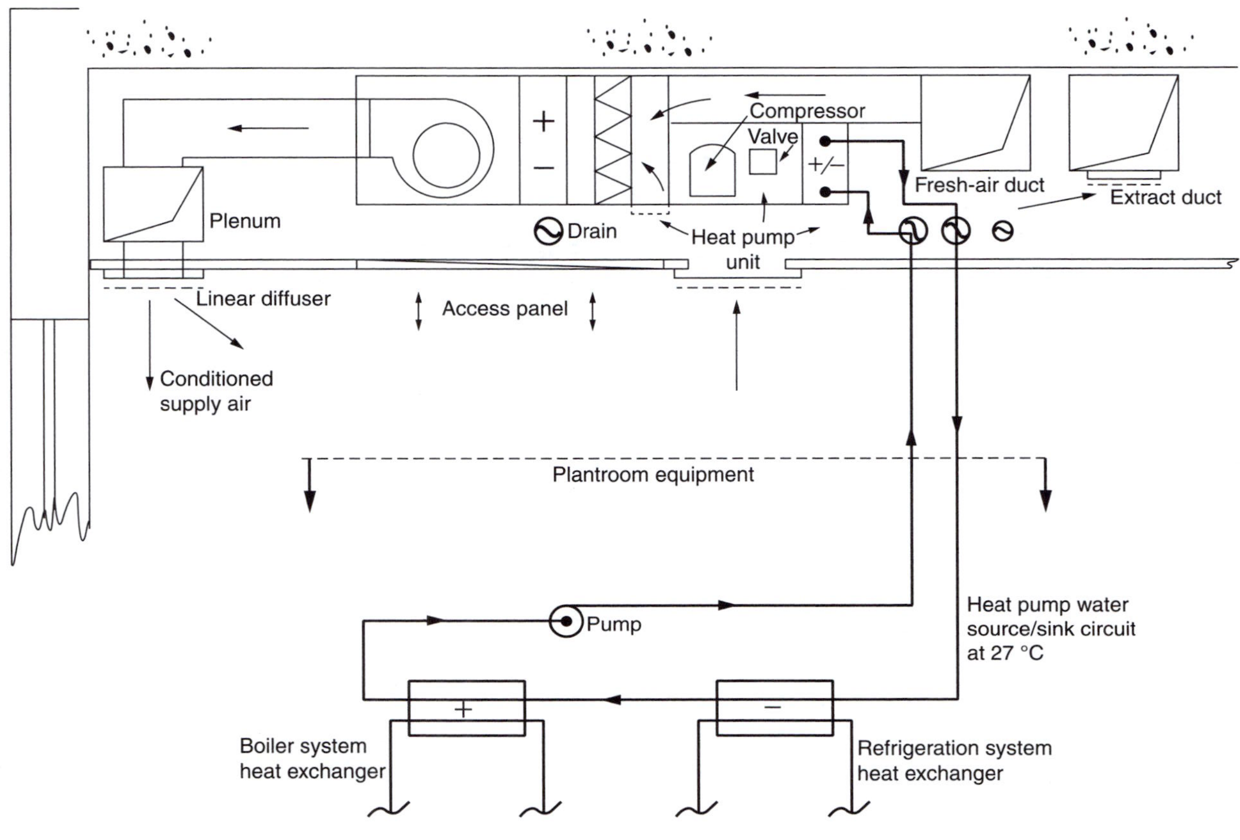

2.9 Single duct system with reversible heat pump (SDRHP).

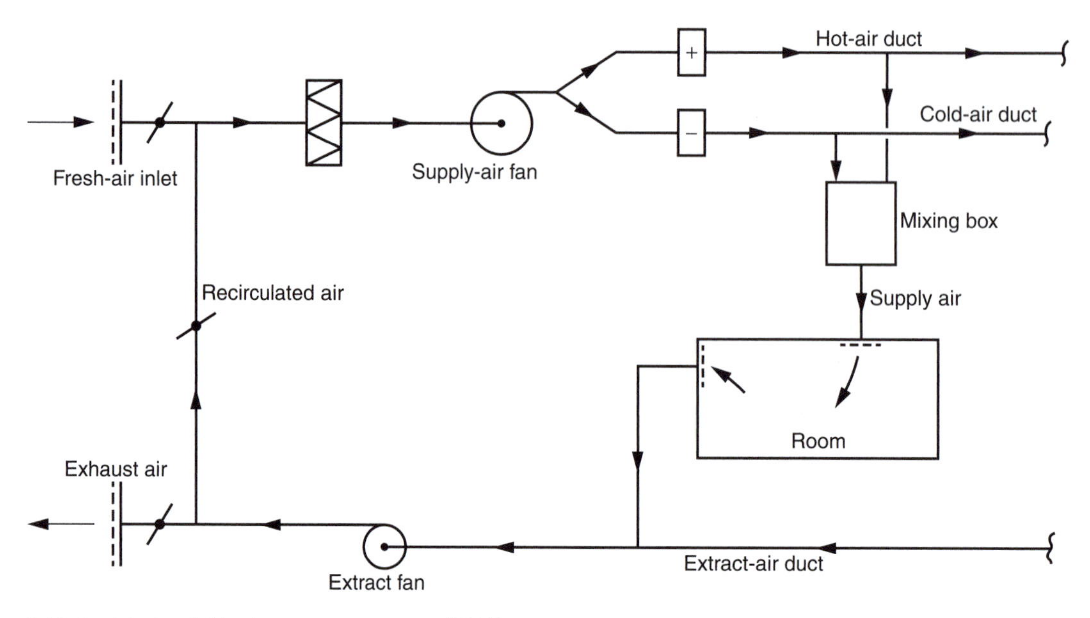

2.10 Dual duct variable air temperature system (DDVT).

The dual duct system provides for simultaneous heating and cooling of adjacent zones and provides the maximum system flexibility. A mixing box can supply a small perimeter room or a large space such as a conference room where the supply air from the box is distributed to several diffusers or grilles. All the ductwork is contained within the ceiling void, as it is impractical anywhere else due to its size. Only the supply and extract grilles and diffusers are visible in the room. An open plan office, airport lounge, banking hall, conference or exhibition area is suitable for the dual duct system, as their occupancy varies during the day and temporary concentrations of the occupants are accommodated easily.

The hot and cold duct air temperatures are scheduled in relation to the external weather conditions for economy and Figure 2.14 shows a simplified illustrative programme. In winter, the hot air duct is maintained at 30°C to be able to offset room heat losses. When the outdoor air reaches 10°C, the reduced heat loss and solar heat gains allow a lower supply air temperature and reduced heating energy consumption. The hot duct temperature is reduced until it is 23°C and the heater coil is fully off. This is the minimum hot duct air temperature that can be achieved by the mixture of recirculated room air and outdoor air at 20°C. Further increases in outdoor air temperature elevate the mixed temperature in the hot duct. During winter, the cold duct air temperature is the product of cold outdoor air mixing with warm recirculated room air and may be 19°C at 0°C outside. Increasing outdoor air temperatures raise the cold duct to 21°C at 10°C externally. At this time of year, solar and internal heat gains produce a cooling load on the plant and the fresh air intake volume can be increased to provide low cost cooling. At around 15°C outside, the refrigeration plant is activated, the fresh air dampers closed to their minimum setting and the cold duct air temperature gradually reduced to its summer value of 13°C.

The supply fan in the air handling plant provides a constant total volume flow rate to the hot and cold ducts. As the mixing boxes vary the quantities taken from each duct, the static pressures at the inlets to the box will vary. This is because reduced air flows in one duct yield a lower pressure drop due to friction while the fan static pressure stays the same, the static air pressure in the duct must rise. The excess static pressure is absorbed by a regulator in each box to maintain system balance and avoid cross-flow between the hot and cold ducts. The regulator is a flexible diaphragm that is exposed to the difference between the static pressures in the room and in the supply duct. It is preset by the manufacturer to allow the design air flow at

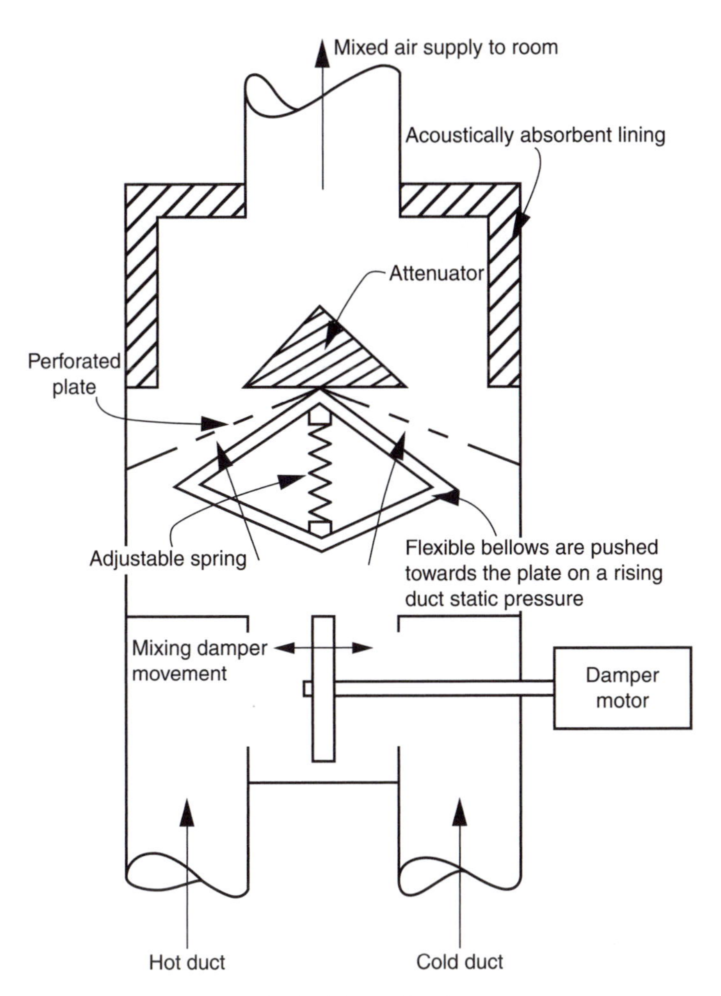

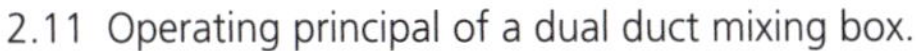
2.11 Operating principal of a dual duct mixing box.

a specific inlet static air pressure of between 200 Pa and 2000 Pa and can be adjusted on site. When in use, an increasing inlet duct air pressure pushes the diaphragm forwards to produce a smaller aperture for the air flow into the room. The increased frictional resistance of the aperture absorbs the excess pressure and maintains a constant supply air flow volume. A reduction in supply duct air pressure allows the diaphragm to move back, open the aperture and provide less frictional resistance.

Dual duct with variable air volume (DDVAV)

The VAV terminal unit consists of a split plenum that receives air from both hot and cold supply ducts. The plenum is fitted within a false ceiling and conditioned air is admitted into the zone through a perimeter slot or other diffuser. The hot air is blown towards the external glazing and may be maintained at a constant volume flow rate throughout the year to counteract the heat gains and losses there. The cold air slot is directed towards the inner part of the room and has its volume flow rate varied according to the cooling requirement. In other respects, the system is the same as previously described for the VAV and DD systems. Figure 2.15 shows the terminal unit.

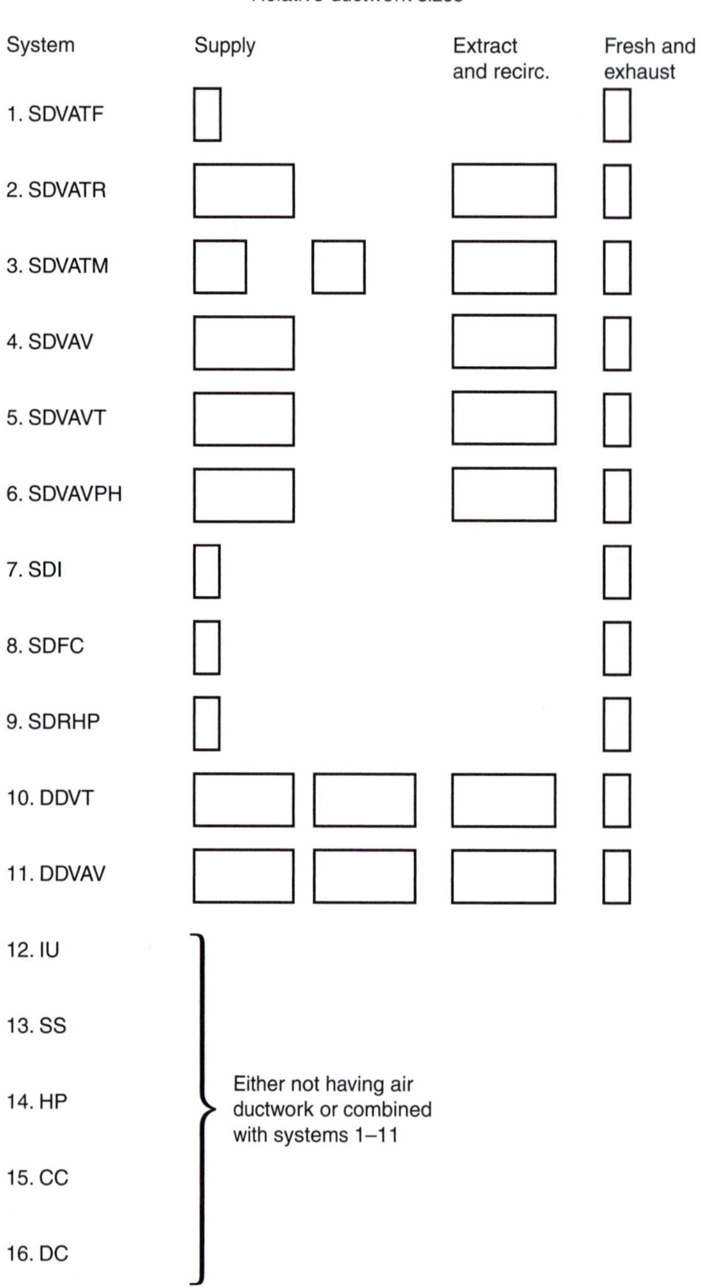

2.12 Comparative air duct spaces.

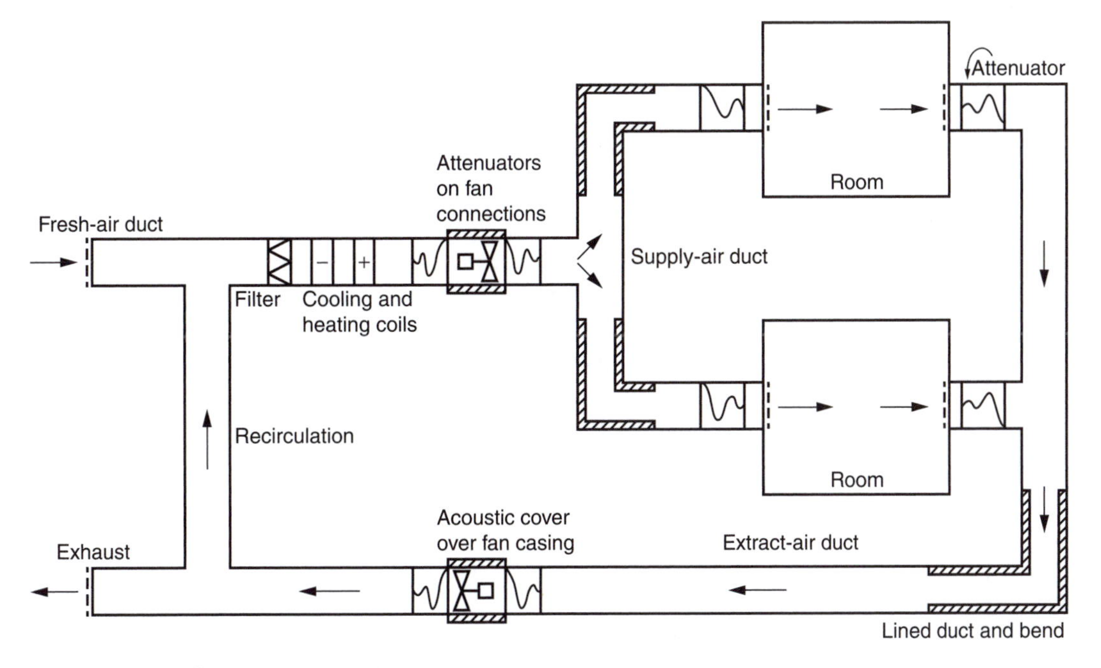

2.13 Likely places for acoustic attenuation.

Independent unit (IU)

These are free-standing air conditioning units containing refrigeration, electrical or water heating, air filtration, controls and a fresh air supply ducted from the local outside air. They are often suitable for installation in existing buildings when a multi-zone ducted system is impractical. They are characterised by their obtrusive position within the conditioned room, compressor and fan noise, low air filtration capacity, prominent appearance on the external building facade and their being influenced by outside traffic noise and wind pressure.

Independent units can be installed through the wall or window to facilitate heat pump operation and provide summer refrigeration and winter heating. They are used in houses, hotels, offices, computer suites, and retail, commercial or industrial buildings. Figure 2.16 shows a through the wall air conditioner during summer operation. The heat extracted from the recirculated room air is raised in temperature by the refrigeration compressor and removed from the building by heating the external air. Figure 2.17 is the same system running with the flow of refrigerant being reversed at the changeover valve and heating the inside air of the building. The refrigerant can only flow one way through the compressor. It can flow in either direction through the expansion device. This is a coil of capillary tube to create a large pressure drop. Flow through the heat exchange coils can be in either direction. The hot refrigerant is now directed to the room air coil to be condensed and this becomes the heating coil. The refrigerant pressure and temperature drop through the expansion device and the outdoor part becomes the cooling coil. Heat is extracted from the outdoor air and raised in temperature to heat the room, making this the heat pump mode of operation. Additional ducted conditioned air ventilation may be used.

The heat rejected from the refrigeration condenser of larger, over 10 kW, independent units can be removed with a piped water system connected to a cooling tower. Figure 2.18 shows the simplified layout of a free-standing unit that may be located within the air conditioned room or concealed and have a ducted air circulation with or without a fresh air inlet.

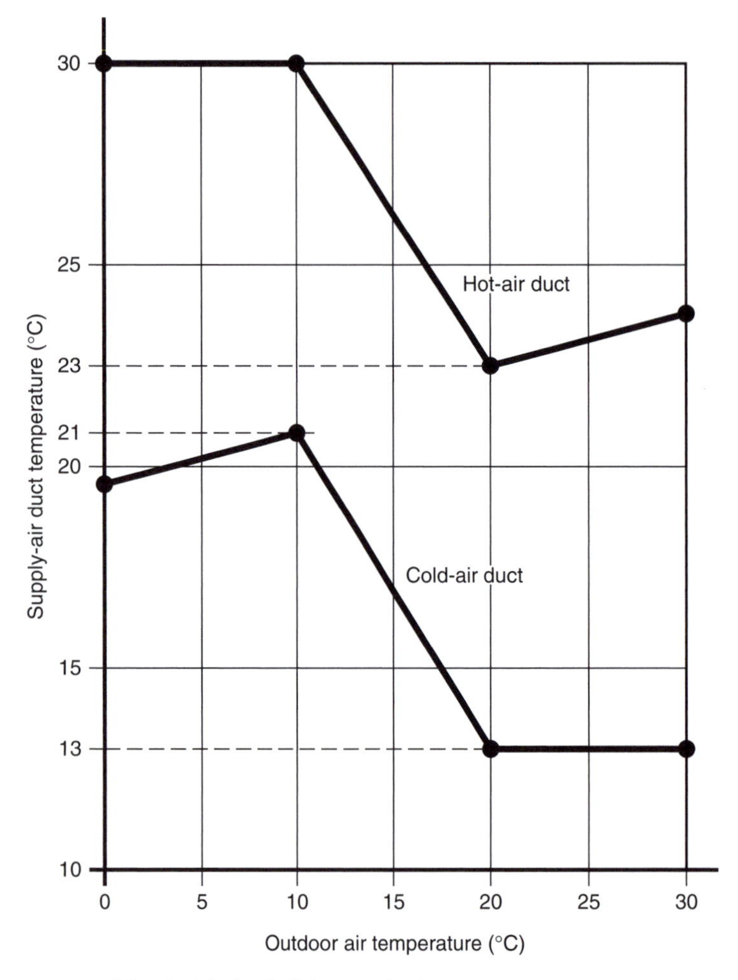

2.14 Possible schedule for dual duct supply air temperatures.

A packaged unit may be fitted with a weather-proof cabinet and installed on a flat roof as shown on Figure 2.19. The refrigerant evaporating temperature in all independent units will normally be lower than the room air dew point and condensation will occur on the room coil, requiring a condensate drain.

Split system (SS)

A refrigeration condensing unit comprising of the compressor, air cooled condenser, condenser fan, controls and weather-proof casing, is located outdoors, usually on the roof of an office, shop, restaurant or computer room. Liquid refrigerant leaves the outdoor unit and is piped indoors to the room cooler that has the expansion valve, refrigerant evaporator cooling coil and controls. Evaporated refrigerant gas is piped back to the condensing unit to be compressed and recycled.

The room cooler has a fan to circulate room air through the heat exchanger and a temperature control that is manually adjustable. A condensate drain is needed from the room cooling and dehumidifying coil. No humidification is possible and room air temperature control is by step control of the fan speed, on/off fan operation, on/off compressor switching, varying the refrigerant volume flow rate (VRV), by compressor

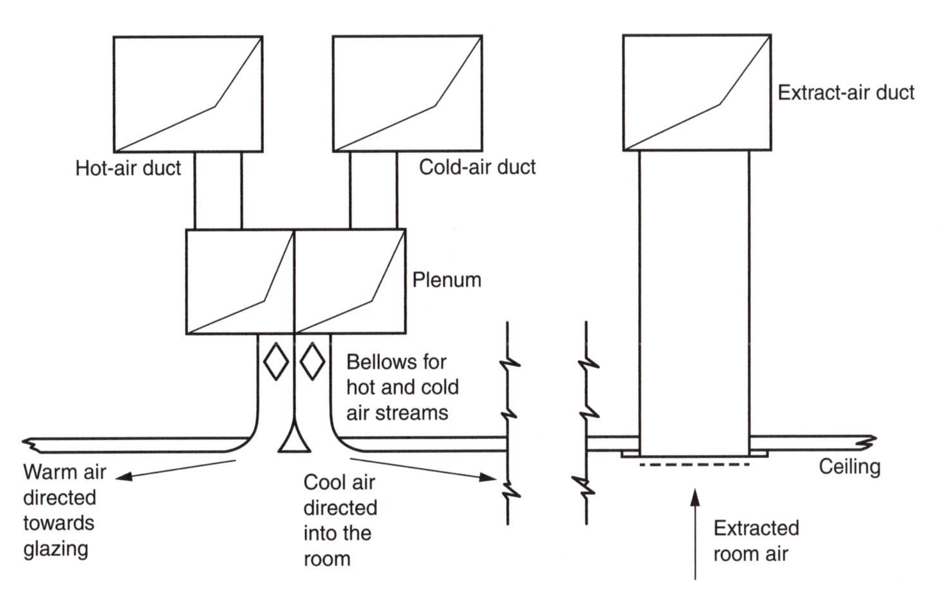

2.15 Terminal unit for DDVAV system.

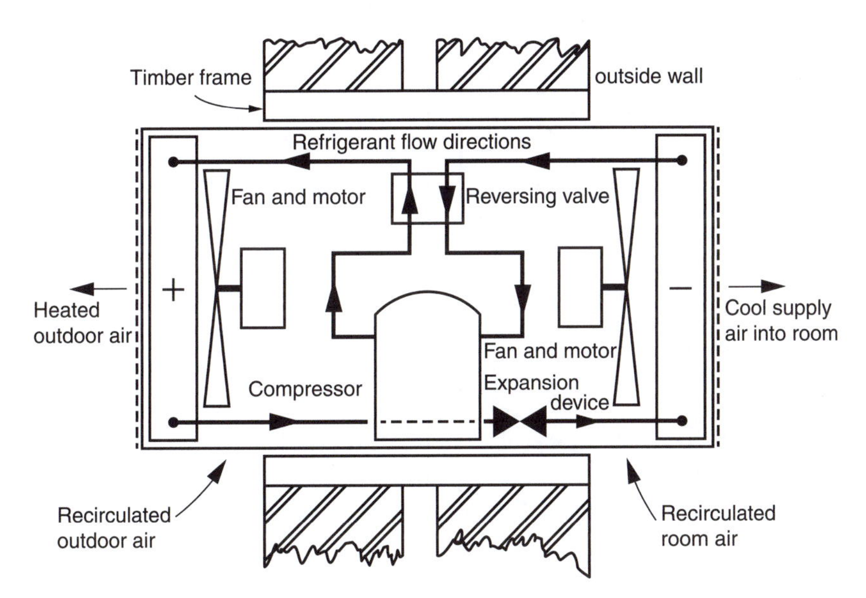

2.16 Through the wall packaged air conditioning unit operating in room cooling mode.

motor speed control with a variable frequency controller (VFC) or multi-stage unloading of the larger compressor systems.

Figure 2.20 shows a typical split system installation. Its advantage is that the noise-producing compressor and condenser fan are located away from the conditioned room. The two units can be separated by several floors of a building. The indoor unit often has high fan and air flow noise. The principle advantage is low cost. They are most suitable for adding into existing buildings.

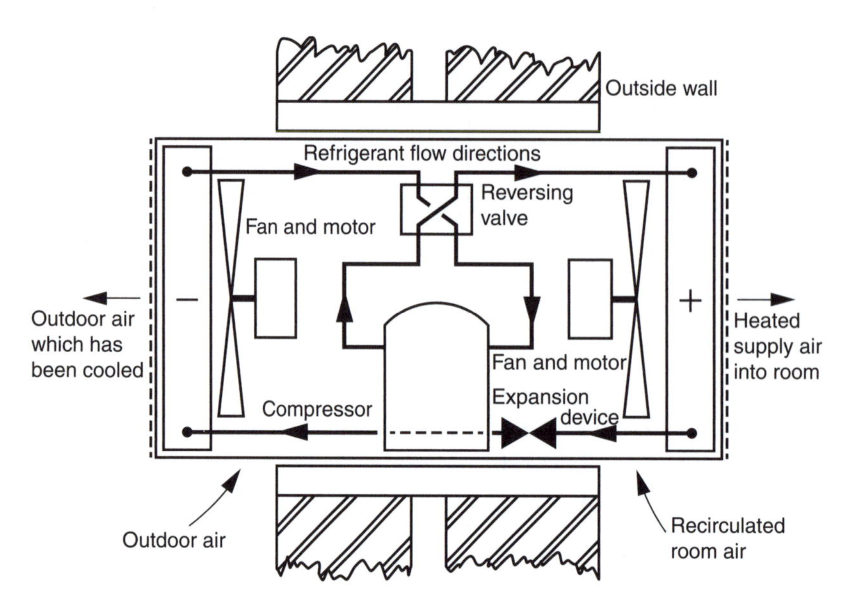

2.17 Through the wall packaged air conditioning unit operating in the room heating mode, this may be described as heat pump mode.

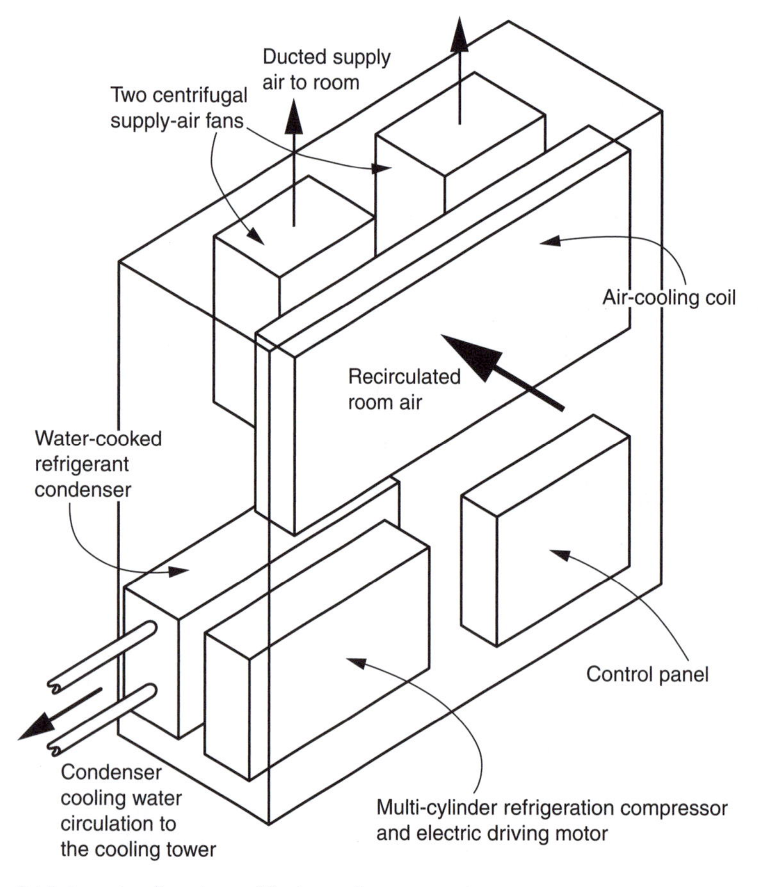

2.18 Free-standing air conditioning unit components.

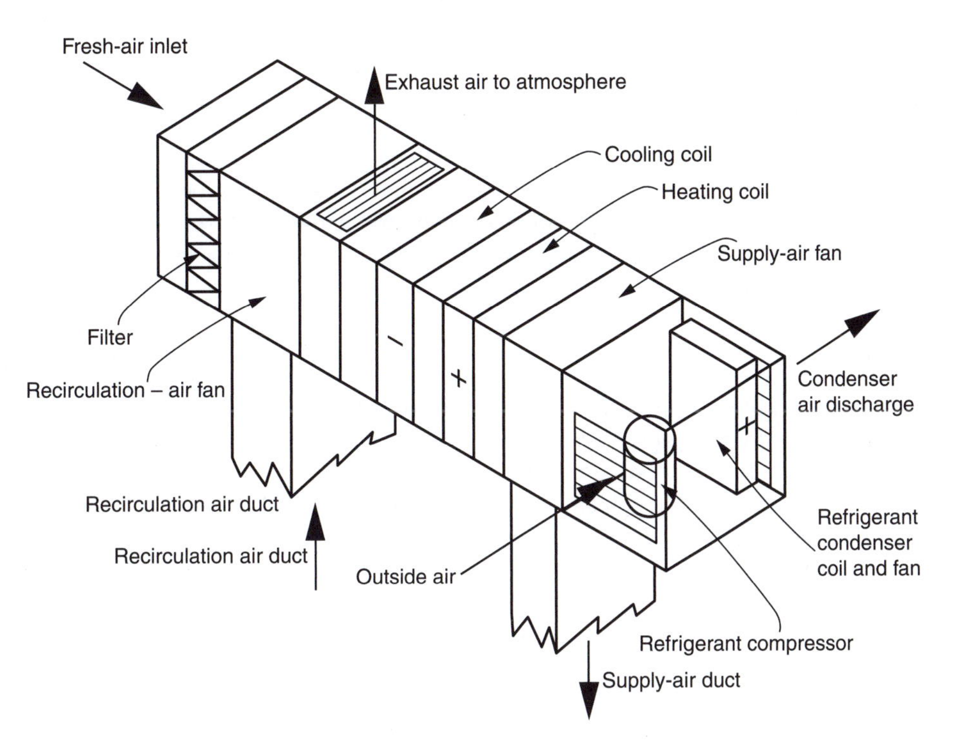

2.19 Roof mounted packaged air conditioning unit.

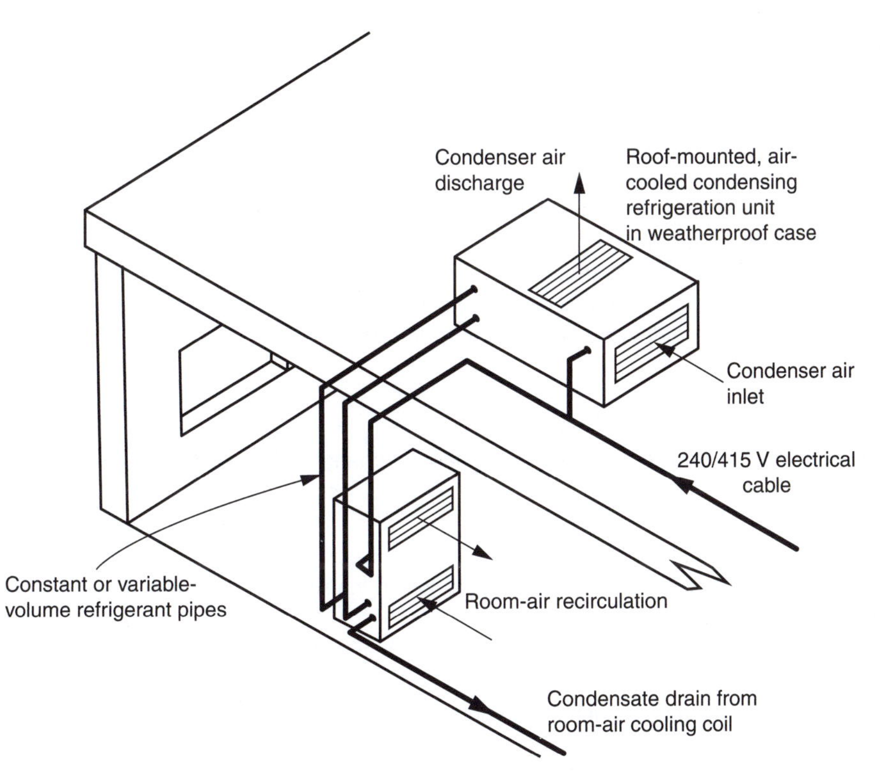

2.20 Split system air conditioning.

Reversible heat pump (HP)

All refrigeration systems are heat pumps. When the useful heat energy is taken from the condenser to warm the interior of the building, such a system is termed a heat pump. These are independent units and split system types used for heating the building from one energy source, electricity. The packaged heat pump can be roof mounted above retail premises to provide year round cooling and heating with a single duct air distribution with recirculation.

Chilled ceiling (CC)

Metal ceiling panels with chilled water pipes attached to them are either suspended below the ceiling or incorporated into the false ceiling design. They provide up to 150 watts/m^2 of convective and radiant cooling in offices and may provide a degree of comfort where solar gains are small. Low cost cooling can be achieved by circulating the water only through an evaporative cooling tower and not using refrigeration. Ventilation, heating and air conditioning systems are dealt with separately. Surface condensation, dust collection, appearance and decoration are attendant problems.

District cooling (DC)

The centralised supply of cooling services is available in a similar manner to district heating. A shopping, residential and office building development can have a central refrigeration plant that circulates chilled water to each rented space. A heat meter integrates the total water flow with the inlet and outlet water temperatures to summate the kWh of cooling energy purchased. Each user is separately billed for the energy consumed. The plant operator provides and maintains the service for the energy charge. The consumer has fan coil air conditioners or ducted air handling as appropriate.

Project building

Figures 2.21 to 2.28 show the design for an office building for Sijoule PLC which requires the whole building to be air conditioned. A set of scale drawings should be created for practical work. The site location is to be chosen by the reader to be locally convenient so that appropriate climate data can be selected. Questions are posed as to the suitability of different systems and, in later chapters, about other aspects of air conditioning. Model solutions are not provided as the designs should be discussed with the tutor and colleagues and use made of reference material, design guides and manufacturers' information.

Airport system

An example was the Euro-Hub airport terminal at Birmingham International Airport (Moss, 1991), which has the following features.

1. Rooftop low temperature hot water boilers.
2. Constant temperature, variable volume heating water circulation to VAV reheat coils, plant air handling units and space heaters.
3. Variable water temperature weather compensated heating circulation to perimeter skirting convectors.
4. Rooftop reciprocating compressor water chilling refrigeration plant.
5. Fan assisted VAV air conditioning terminal boxes in public areas.
6. Ground level offices with fan coil air conditioning.
7. The main concourse area has a single duct constant volume recirculation air conditioning system.
8. Toilet extract ventilation.

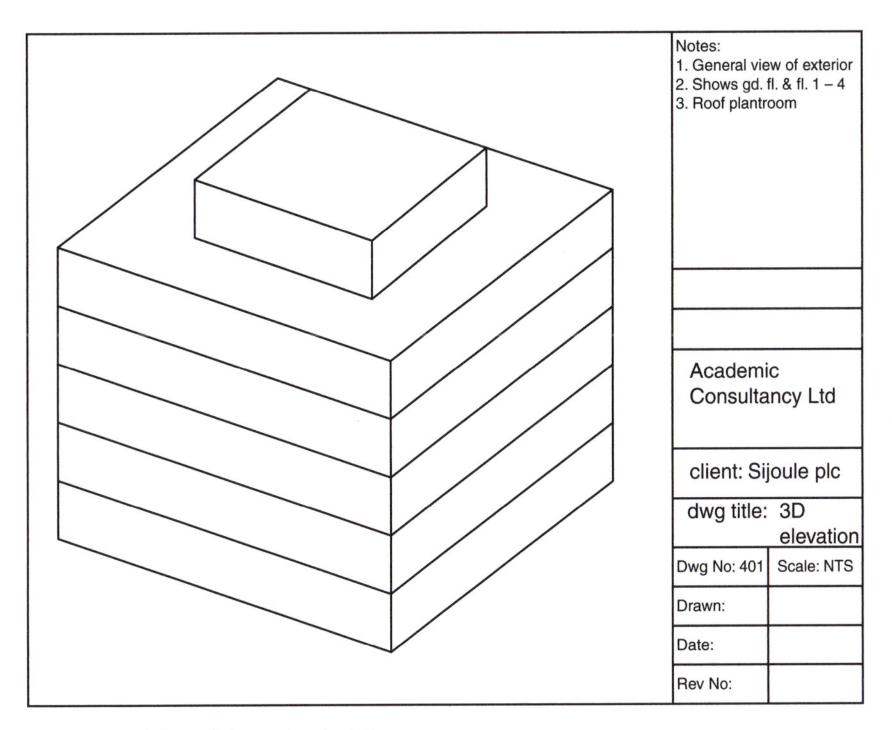

2.21 General view of the project building.

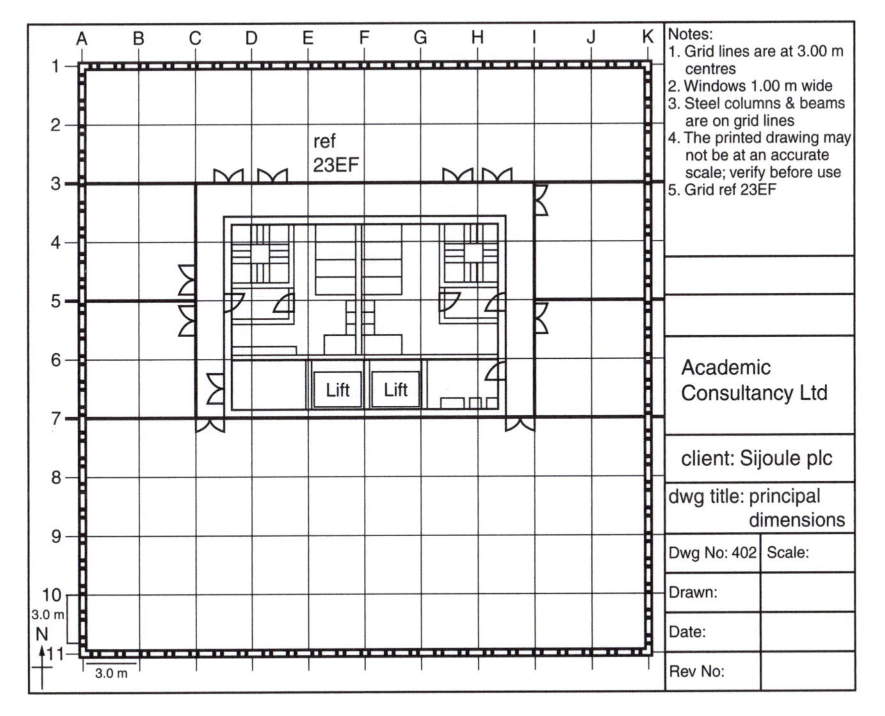

2.22 Grid layout and dimensions.

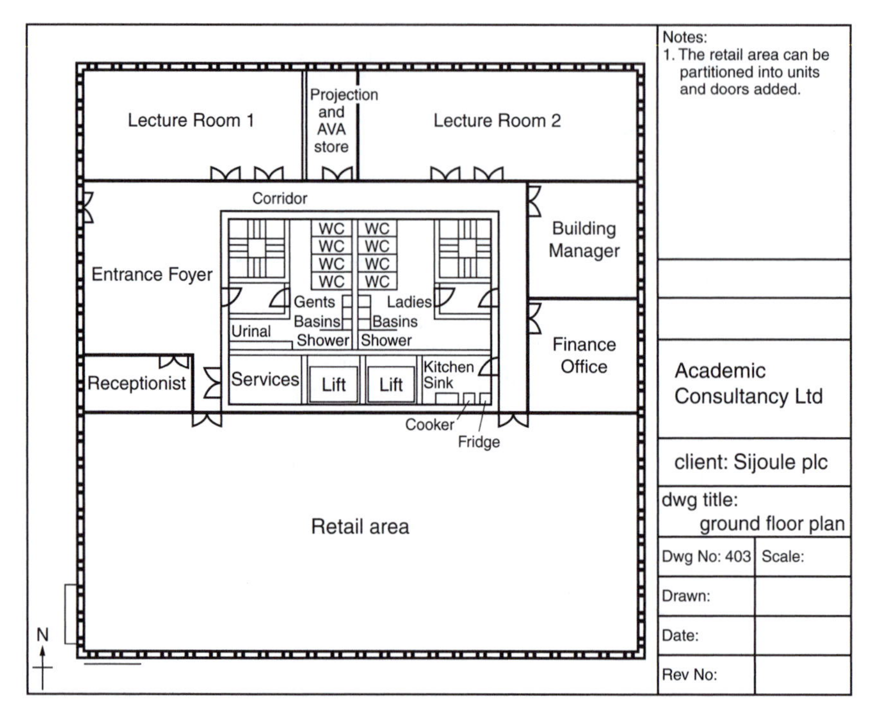

2.23 Ground floor plan.

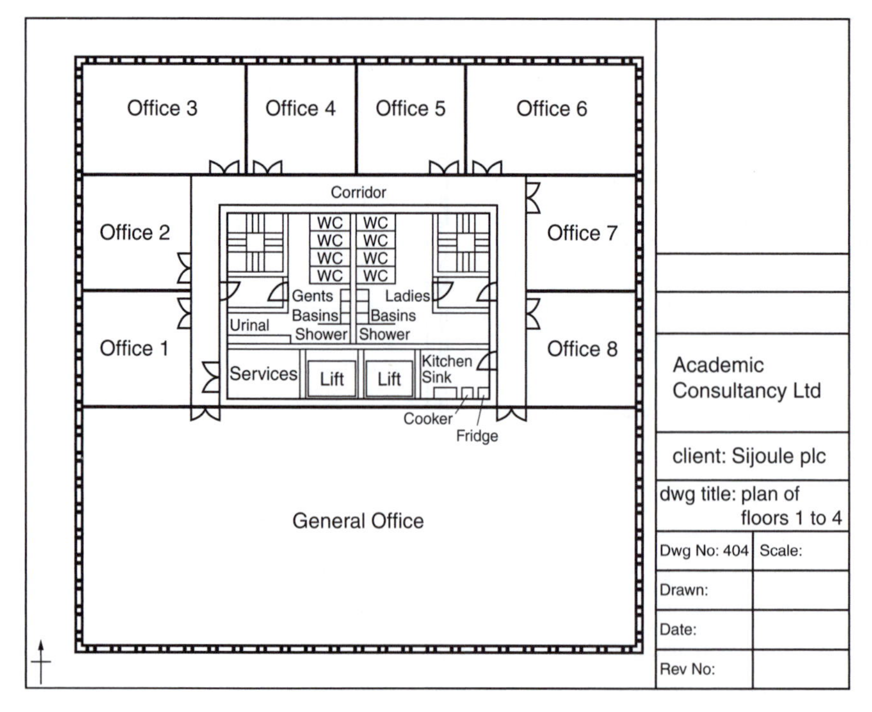

2.24 Intermediate floors.

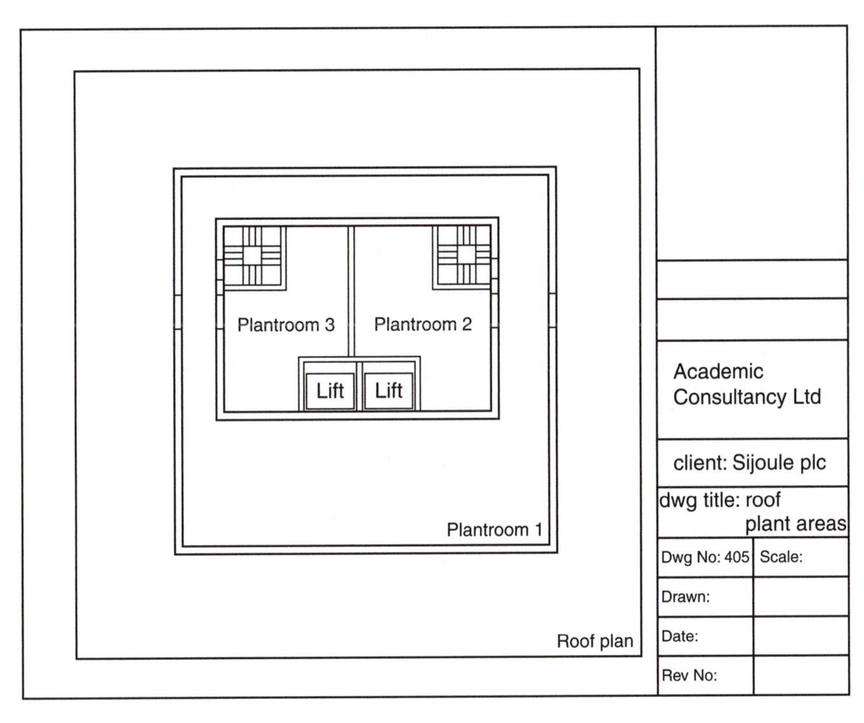

2.25 Roof plant room areas.

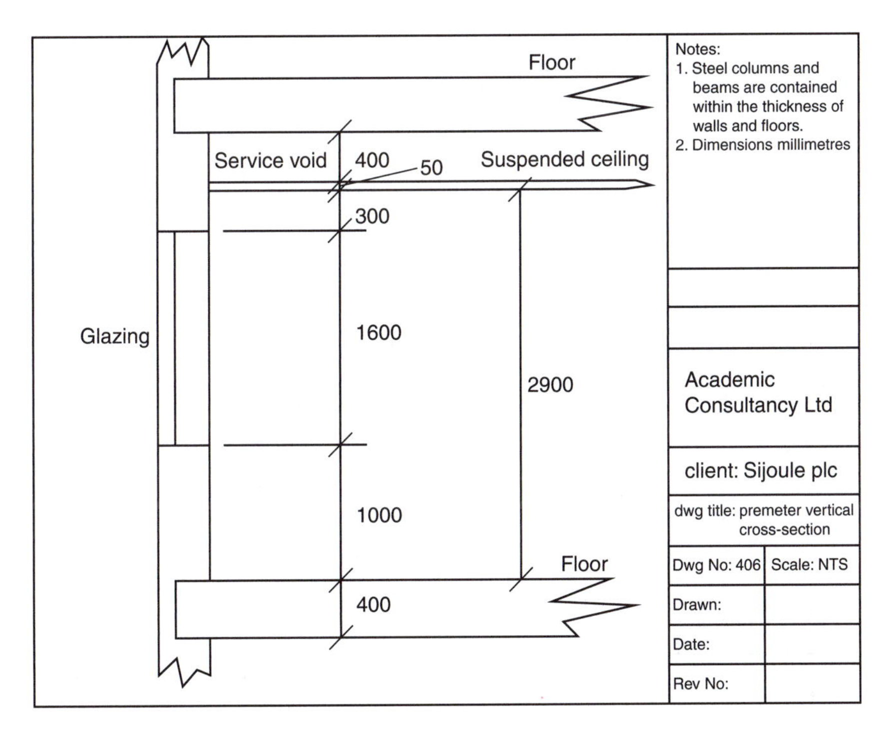

2.26 Perimeter vertical cross-section.

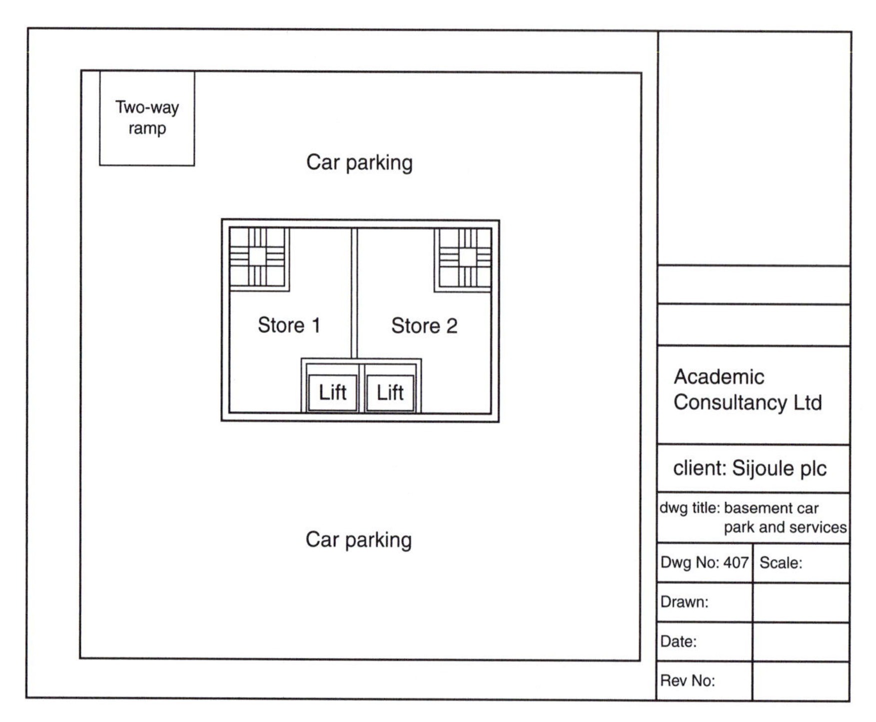

2.27 Basement plan.

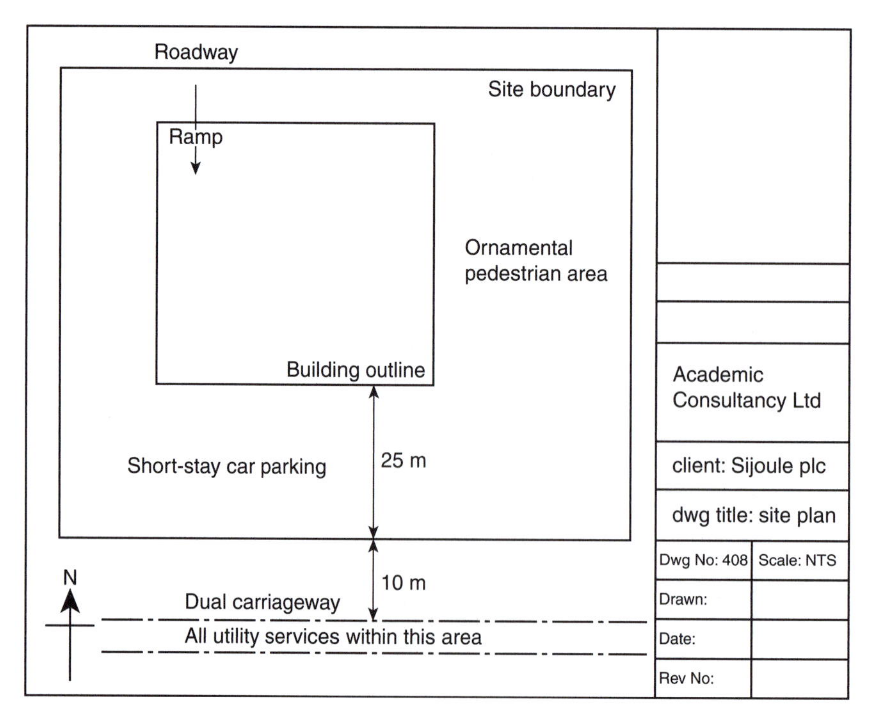

2.28 Site plan.

9. Ventilation plant for kitchens, electric motor rooms and baggage handling areas.
10. The ducted fresh air enters the plant through activated carbon filters to remove the aircraft engine fumes.
11. The terminal building is maintained at positive static pressure above the outside atmosphere to stop the possible ingress of aircraft engine exhaust fumes through boarding gates and external doors.
12. Arrival and departure passenger levels each have four zone air handling units serving the VAV controllers.
13. Each air handling unit has variable speed supply and extract fans that are controlled to maintain constant main air duct static pressures.
14. The fresh air supply is regulated from air quality sensors to adjust for occupancy levels.

Questions

Answers to descriptive questions are within the text of the chapter. The reader is advised to extend the range of study material to the CIBSE Guides, manufacturers' websites and literature and discussions with colleagues.

1. Discuss the statement 'this building is air conditioned' with reference to what this will mean, the systems that must be installed and the possible satisfaction of user thermal comfort.
2. Explain the difference between mechanical ventilation and air conditioning.
3. Define the term low cost air conditioning. State the ways in which it can be achieved.
4. List the reasons for the air conditioning of different categories of buildings, such as residence, office, retail, containment and manufacturing, and write notes to explain each reason.
5. Study the drawings of the Sijoule PLC building and write lists and notes on the heating, ventilating and air conditioning systems that are likely to be needed. List the assumptions that you make on the location and use of the building. Discuss your results with colleagues. This exercise is suitable for groups of students to formulate different proposals and then present them to the class.
6. Summarise in your own words, the considerations involved in making use of outdoor air to cool a building without the use of refrigeration. State the limitations inherent in such designs and the ways in which outdoor air can be cooled without refrigeration.
7. A single duct air conditioning system with recirculation is to have variable quantities of fresh air admitted into an office building when it is between 12°C and 23°C. Outside these air temperatures, the minimum outdoor air of 25% of the supply quantity is to be used. Outdoor air at 14°C can be 100% of the supply air quantity to the rooms. When the fresh air reaches 21°C outside, its proportion must commence reduction towards the minimum at 23°C. Draw a graph of the damper operation to scale and fully explain the sequence of operation of the system during both increasing and reducing outdoor air temperatures.
8. Sketch the air handling arrangement for a single duct variable air temperature recirculation system, state the necessary components and explain how the design is used to satisfy the room cooling and heating loads.
9. List the reasons for creating air conditioning zones and the zones that may be formed for a variety of comfort, industrial process and containment applications.
10. Create suitable air conditioning zones for the Sijoule PLC building, stating the reasons, and draw them onto the plans and elevations.
11. Calculate the peak solar heat gains through the unprotected windows for each side of the Sijoule PLC building using only the glazing heat gain data in the Air Handling Zones section of this chapter. State the dates and times of these occurrences and the influence that they will have on the building peak refrigeration plant capacity.

12. State the reasons for creating different air static pressures in rooms or zones, explain how such differences can be maintained, and state pertinent applications.
13. List the applications of the 16 types of air conditioning system and discuss the suitability of each with colleagues.
14. Form syndicates of four students to propose air conditioning designs for the Sijoule PLC building. Each member of the group is to assume a different role:

    (a) The building owner is to decide the usage of each area and the standard of internal environmental control to be achieved. Set limitations upon the intrusiveness, likely cost, maintenance or appearance of the systems to be proposed.
    (b) The consulting engineer who is engaged for a fee calculated from a fixed percentage of the total cost of the installation is to use the independence of this position to propose designs that will fully satisfy the clients stated requirements.
    (c) The design and installation contractor is engaged in a competitive situation against other, similar companies and is to propose designs that maximise the likelihood of gaining the contract.
    (d) The manufacturer of a wide range of air conditioning products who offer a design service free of charge to the client.

    The three engineers are to formulate proposals for different air conditioning designs and present them to the client with arguments in favour of their company's suitability. Presentations should be made on acetate sheets of the building drawings with coloured sketches of the systems proposed on the plans, sections and elevations for the benefit of the whole student group. It may be possible to integrate such group activity with architectural, building and surveying student groups.
15. 'The single duct air conditioning system provides the basic design for the other configurations that overcome its limitations'. Discuss this statement and state the ways in which it can be adapted for multi-zone applications.
16. The variable air volume system has become very popular for office accommodation. Explain its principles of operation and limitations; including the topics of room air circulation, zone volume control, economy control of the fans, duct air static pressure modulation and the satisfaction of user thermal and aural comfort.
17. State the advantages that can be gained by using a water source heat pump air conditioning system capable of simultaneous cooling and heating of adjacent zones. List suitable applications. Comment upon the maintenance required for such systems.
18. Explain with the aid of sketches and sample graphs how the supply duct air temperatures are controlled in a dual duct air conditioning system to provide the maximum operational economy while satisfying the users' comfort needs.
19. List the applications for independent air conditioning units, split systems, reversible heat pumps, chilled ceilings and district cooling, commenting upon their design characteristics and maintenance requirements. Acquire manufacturers' literature for a variety of equipment and apply them to the Sijoule PLC building.
20. Make sketches of a building you are familiar with and propose suitable designs for air conditioning systems, drawing the components on plans and elevations.
21. Describe with the aid of sketches and manufacturers' literature how the following air conditioning heat recovery devices save energy: recuperator, run-around coil, thermal wheel, heat pipe, regenerator, heat pump. Comment on their suitability for buildings of your choice and their capital and energy cost implications.
22. Explain with the aid of sketches and manufacturers' information how air filters are tested and rated for the following tasks: air flow, face velocity, initial and final pressure drop and dust spot efficiency.

23. Explain what is meant by the air filter terms: dry testing, dust, smoke, mists, impingement, cyclone, washable, diffusion filter and viscous film.
24. State suitable applications for the following types of air filter, giving examples: absolute, dry, viscous, panel, bag, electrostatic, adsorption and mechanical collectors.
25. What are the important factors to be considered and provided for air filter installations?
26. Explain with the aid of sketches or literature how cooling coil face and bypass dampers function, why they may be used and their purpose.
27. List the range of humidifiers available for use in air conditioning systems, sketch their operation and give examples of their use.
28. Discuss the statement, 'evaporative cooling has applications but is not free'. Explain the principles of adiabatic saturation cooling, where it may be applied (world location and type of cooling requirement), and the costs of operating such systems.
29. List the ways in which free cooling may be achieved, sketch and describe suitable installations and where such systems could be utilised.
30. Demonstrate by means of descriptions, sketches and reference material how evaporative pre-cooling can reduce the electrical energy used by a refrigerated air conditioning system; state where such systems might be used, in which climates and for which applications.

3 Heating and cooling loads

Learning objectives

Study of this chapter will enable the reader to:

1. understand the balance between heat gains and losses in an air conditioned building;
2. identify and calculate all the sources of heat gain;
3. understand the cyclic behaviour of heat gains;
4. decide when a cooling system is needed;
5. calculate plant loads in various countries;
6. relate the sun position to a building;
7. manipulate trigonometric ratios for design calculations;
8. calculate heat gains with direct solar and diffuse sky irradiances;
9. understand the terms altitude, azimuth and incidence;
10. calculate solar gains to horizontal, vertical and sloping surfaces;
11. calculate heights and surface areas of buildings using trigonometry;
12. calculate areas of shade;
13. find the design total irradiance;
14. understand the use of sol-air temperature;
15. know the types of glass used and their thermal properties;
16. calculate glass temperature and heat transfer;
17. understand the use of mean heat flow and cyclic swing in heat transfer through opaque structures;
18. use time lag, decrement factor and thermal admittance in heat gain calculations;
19. assess the cooling load of a room and a building;
20. understand the use of sensible and latent heat gains;
21. calculate the peak summertime temperature in a building;
22. find the balance outdoor air temperature for a building;
23. use case studies;
24. understand the difference between plant loads in temperate, hot humid and arid climates.

Key terms and concepts

Absorptivity 79; air temperature 79; altitude 51; azimuth 51; balance temperature 100; decrement factor 85; design total irradiance 76; diffuse irradiance 60; direct irradiance 62; electrical equipment 49; emissivity 79; glass temperature 80; heat gain 50; incidence 51; infiltration 49; intermittent heat gain 49; plant cooling load 84; reflectance 81; sensible and latent heat 73; shading 60; sol-air temperature 79; solar gains 100; surface factor 90; thermal admittance 85; thermal balance 85; thermal transmittance 100; total heat gain 51; wall solar azimuth 51.

Introduction

An air conditioning plant is designed to maintain specified internal air temperature and humidity when the expected design heat gains and losses occur. The CIBSE method of calculation of these plant heat loads is explained with the minimum of background theory. If refrigeration is not to be employed, the summer heat gains will produce an elevated internal air temperature. This summertime indoor temperature is calculated in Chapter 1.

The sources of heat gains are identified and the areas of shading on a building are derived from first principles. Trigonometry is used to find the components of the solar irradiance that is at a right angle to the irradiated surface as this is used to calculate the heat gain. The formulation of design total irradiance is analysed for vertical, horizontal and sloping surfaces. Heat transmission through glazing and the effect of different types of glass are explained.

Sol-air temperature is introduced for the calculation of heat gains through opaque structures. The ideas of transient heat flow into and out of the thermal storage capacity of the walls, roof and floor lead to the use of a 24 hour mean heat flow and a swing, or cyclic variation, to the mean.

The plant cooling load can be found from these methods. Repeated analysis produces the hourly cooling requirements for zone loads and the prediction of the balance temperature for the building. This is a measure of energy efficiency.

Solar and internal heat gains

The thermal conditions within a building result from a balance between:

1. heat loss to the cooler external environment;
2. thermal storage within the fabric of the structure;
3. heat gains from solar radiation through glazing;
4. heat gains from solar radiation onto the opaque parts of the structure such as walls and the roof;
5. conduction heat flows through the glazing;
6. conduction heat flows through the opaque walls and roof;
7. the infiltration or direct injection of hot outdoor air into the rooms;
8. conduction or ventilation heat gains from surrounding rooms;
9. heat gains from the internal use of the building; these are from the occupants, lighting, electric motors, electrical appliances, computers, cooking equipment, water heating for washing purposes, animals, furnaces, industrial processes, gas-fired cooking and water heating, electrical and gas-fired incinerators;
10. heat transfers between hot or cold services pipes or air ducts and surrounding rooms or the movement of hot or cold products between rooms.

These heat transfers are intermittent, meaning that they all change in value during the day and will cause the room air temperatures to be continuously varied. Solar heat gains are cyclical with the Earth's rotation and the annual nature of weather patterns. The calculation of heat losses to find the size of central heating radiators and boilers is usually based upon steady state heat flows from the building on the coldest day. This would be unacceptable for air conditioning cooling loads. The thermal comfort of the occupants or the satisfaction of the conditions for other reasons depends upon the net effect of these heat flows. The combination of large heat gains to an enclosure at a time of high outdoor air temperature and intense solar radiation, together with poor ventilation, may lead to over-heating of the internal air, insufficient removal of bodily heat production, excessive sweating and discomfort.

Air temperature t_{ai} is insufficient to describe the thermal effects produced by the combination of convection and radiation heat exchanges within the building. The external surfaces are heated by solar irradiance thus raising the internal room surfaces above the enclosed air temperature. Discomfort can be produced from the low temperature radiation from the walls and ceiling plus the glare from windows.

Environmental temperature t_{ei} combines some of these,

$$t_{ei} = 0.33t_{ai} + 0.67t_m \text{ °C}$$

t_{ai} = dry bulb air temperature °C d.b.

t_m = mean surface temperature °C

Mean surface temperature is not calculated in this context as the heat gain formulae include appropriate correction factors. Operative temperature, previously termed dry resultant temperature t_c is used to specify the desired comfort condition and is,

$$t_c = 0.5t_{ai} + 0.5t_m \text{ °C}$$

The upper limit for the room environmental temperature for normally occupied buildings is 27°C. External design air temperatures for comfort in offices in London may be chosen as 29°C d.b., 20°C w.b.; higher outdoor air temperatures are often achieved. The indoor limit of 27°C will frequently be exceeded in naturally ventilated buildings due to the combination of the infiltration of outdoor air, solar and internal heat gains. In parts of the world where high radiation intensity and continuously higher external temperatures are common, for example Sidney, 35°C d.b., 24°C w.b., the necessity for refrigeration can be recognised.

Solar heat gains are calculated for the estimation of the peak summertime environmental temperature produced within a naturally ventilated building or, when air conditioning is to be used because of insufficient low cost cooling, for calculation of the refrigeration cooling power needed. The peak solar heat gain through the glazing in summer will often be considerably greater than the sum of all the other heat gains. The heat transfers through the thermal storage effect of the opaque parts of the structure may be small at the time of the peak solar gain through the glazing, or even negative, meaning a net loss from the room. A quick heat gain assessment may justifiably ignore the opaque gain/loss. Solar heat gains to a building are found from:

1. angular position of the sun in relation to each face of the building;
2. intensity of the solar radiation incident upon each external surface;
3. surface areas exposed to the sun;
4. date and the sun time;
5. area of the glazing and walls that are shaded from the direct solar radiation either by nearby buildings or by shading devices or projections from the subject building;
6. type of glazing – clear, heat absorbing, reflective or tinted;
7. provision of external shading devices;
8. internal blinds or curtains for shading or solar intensity reduction;

9. absorption of solar radiation by the opaque parts of the structure; this is related to external surface colour;
10. storage and transmission of solar heat gains through the structure, using transmittance, U W/m^2 K and admittance Y W/m^2 K.

Shaded areas receive only blue sky diffuse solar radiation that is non-directional and equal on all the faces of the building simultaneously. Those areas that are unshaded receive both the directly transmitted, line of sight, and diffuse radiation from the sun. There is a time delay of up to 12 hours between the occurrence of solar radiation falling on an opaque surface and the heat gain appearing within the room and influencing the resultant temperature. Remember that resultant temperature is 50% of the dry bulb air temperature plus 50% of the mean surface temperature. Room surfaces exposed to external walls and roofs will be warmed by the solar heat gains that are stored within the bricks, steel and concrete and subsequently heat the room air.

Total heat gain experienced by a room is found from the sum of:

1. instantaneous gains from the solar transmission through the glazing;
2. heat released from the structure due to solar radiation some hours previously;
3. thermal transmittance through the glazing and opaque structure from the higher outdoor air temperature;
4. infiltration of warmer outdoor air directly into the room;
5. sources of heat generated within the room from people, lights and powered equipment such as computers.

Sun position

It is convenient to consider that the sun moves around the stationary building. The position of the sun relative to the irradiated surface is defined by six angles, as follows.

1. Solar altitude, the vertical angle from the Earth's horizontal plane up to the centre of the sun $A°$.
2. Solar azimuth, the compass orientation of a vertical plane through the sun measured clockwise from north on the plane of the Earth. In plan view, this is the angle between north and the sun position $B°$.
3. Wall azimuth, the compass angle from a line drawn at right angles from a vertical wall to north $C°$. A right angle from a surface is called normal to the surface. A sloping wall or glazing has an azimuth but a flat roof does not.
4. Wall solar azimuth is the difference between the solar and wall azimuth angles $D°$.
5. Slope is the vertical angle from a normal line from the surface to the horizontal $E°$. For example, a vertical wall or window has a slope of 0° and a flat roof has a slope of 90°.
6. Incidence is the combined effect of the other angles and represents the path of the solar radiation striking the surface $F°$.

It is necessary to use trigonometry in the analysis of the solar position in relation to the building and for the calculation of shaded areas. This is a good time to revise the concepts of trigonometric ratios in mathematics. Only three ratios are needed and they are sine, cosine and tangent. Figure 3.1 is used to relate these to a right angled triangle. If design problems produce non-right angled triangles, it may help to divide the shape into right angled triangles before attempting any calculations. Figure 3.1(a) shows a right angled triangle where the points X, Y and Z have angles x, y and 90° and the names of the sides in respect to the angle $x°$ are opposite vertically, adjacent horizontally and hypotenuse of the slope. The opposite and

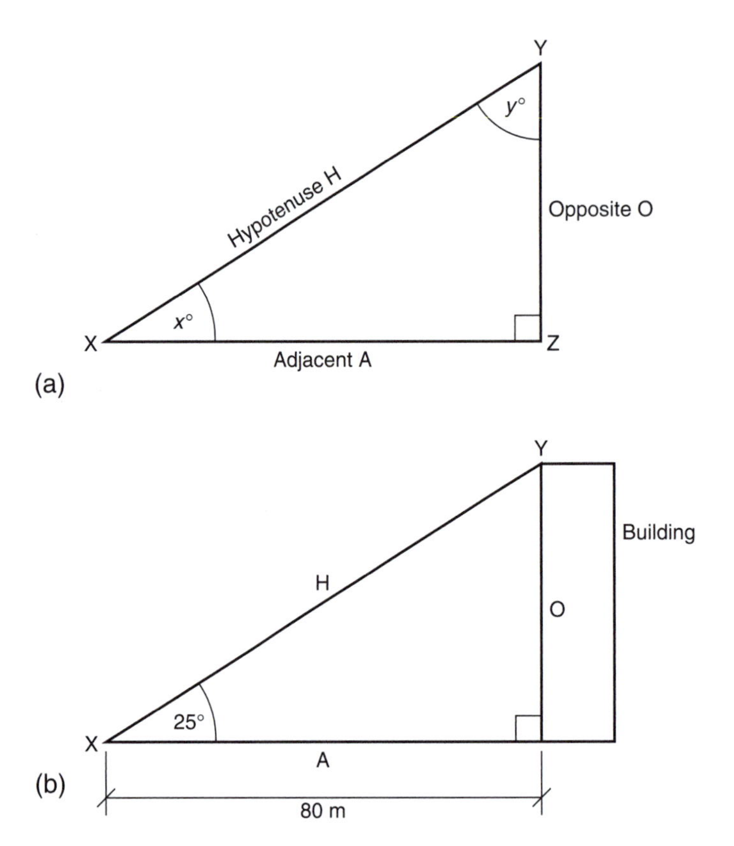

3.1 Trigonometry of a right angled triangle.

adjacent labels would reverse if angle $y°$ were being considered. The trigonometric ratios needed are:

$$\sin x^{O} = \frac{\text{opposite } O}{\text{hypotenuse } H}$$

$$\cos x^{O} = \frac{\text{adjacent } A}{\text{hypotenuse } H}$$

$$\tan x^{O} = \frac{\text{opposite } O}{\text{adjacent } A}$$

EXAMPLE 3.1

Figure 3.1(b) shows a 12-storey building that has been surveyed to find its height. The surveyor measured an angle of 25° from the ground to the top of the building at a distance of 80 m from it. Calculate the building height.

The tangent ratio can be used, $\tan x = \frac{O}{A}$

O = building height H m
A = distance from the building 80 m

$$\tan 25° = \frac{H \text{ m}}{80 \text{ m}}$$

Rearrange the formula to find H m.

$$H = 80 \text{ m} \times \tan 25°$$

Enter 25 into your calculator, making sure that it is set to degree format, and press the tangent key, tan or tan x, $\tan 25° = 0.4663$; multiply this by 80 m and see that the height of the building is 37.305 m. Divide this by 12 storeys and find that each storey is 3.109 m high.

EXAMPLE 3.2

A recent trend has been to design buildings with sloping sides as depicted in Figure 3.2. The maintenance contractor needs to find the surface area of the sloping face of the building for its cleaning cost. The face slopes at 75° to the horizontal and the roof line is at 50° from the ground at a distance of 60 m. Calculate the vertical height of the building, the length of the sloping face and the sloping face area.

Figure 3.2(b) is drawn at a normal to the foot of the building. Divide the triangle XYZ into two right angled triangles by inserting a vertical line YP. The length of YP is the same for both the triangles XYP and YZP. Divide the base of the triangles into the unknown length L and $(60 - L)$. The common vertical side YP is the height of the building H m. Two tangent formulae can be written and solved as simultaneous equations:

$$\tan 50 = \frac{H}{60 - L}$$

$$H = (60 - L) \times \tan 50$$

$$\tan 75 = \frac{H}{L}$$

$$H = L \times \tan 75$$

These can be expressed as:

$$H = (60 - L) \times \tan 50$$

and, $H = L \times \tan 75$

So,

$$(60 - L) \times \tan 50 = L \times \tan 75$$

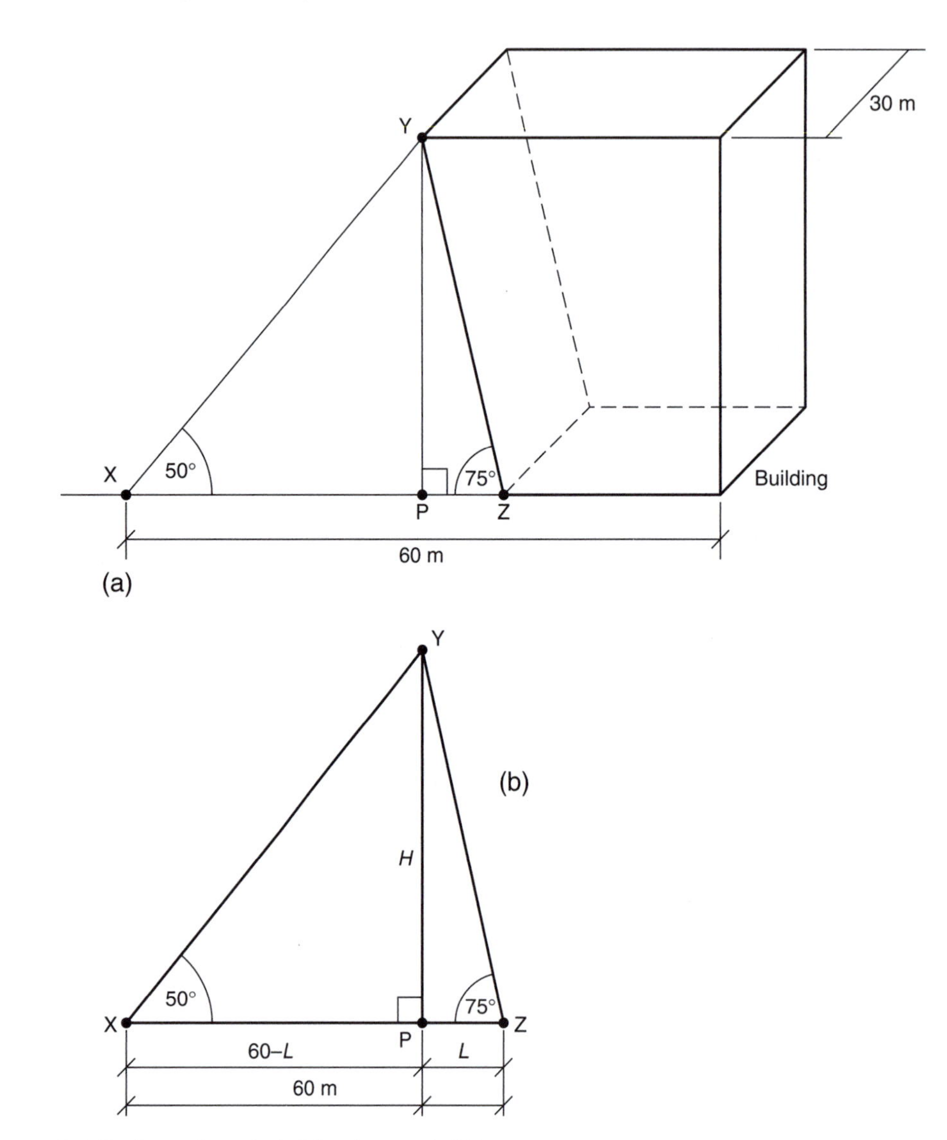

3.2 Sloping surface of a building in example 3.2.

Evaluate the tangents:

$$(60 - L) \times 1.192 = L \times 3.732$$

$$60 \times 1.192 - L \times 1.192 = L \times 3.732$$

$$71.52 = L \times 3.732 + L \times 1.192$$

$$71.52 = L \times (3.732 + 1.192)$$

$$71.52 = L \times 4.924$$

$$L = \frac{71.52}{4.924} \text{m}$$

$$L = 14.525 \text{ m}$$

Now the height can be found:

$$H = L \times \tan 75$$

$$H = 14.525 \text{ m} \times \tan 75$$

$$H = 54.208 \text{ m}$$

And, the length of the sloping face:

$$\cos 75 = \frac{14.525}{YZ}$$

$$YZ = 56.12 \text{ m}$$

$$\text{Area of the sloping face} = 56.12 \text{ m} \times 30 \text{ m}$$

$$\text{Area of the sloping face} = 1683.6 \text{ m}^2$$

The five angles that define the orientation of a surface are shown in Figure 3.3. Tabulated data on the solar positions are available. Latitude gives the vertical angle above or below the Equator of a location on Earth, for example London is 51° 29′ N, meaning 51° and 29 minutes north of the Equator, or 51.5° N as there are 60 minutes in a degree, Sidney is 33° 52′ S, Hong Kong is 22° 18′ N and Singapore is almost on the Equator at 1° 18′ N.

The angle of incidence of radiation on a vertical surface is found from:

$$\cos F = \cos A \times \cos D$$

Where,

F = angle of incidence
A = angle of solar altitude
D = wall solar azimuth angle.

EXAMPLE 3.3

A vertical wall in London faces south east. At 09.00 hours sun time on 22 May the solar altitude is 44° and solar azimuth 113°. Calculate the wall solar azimuth and incidence angles.

A normal from the face of the wall points south east and this has an azimuth angle of 135° from north. The solar azimuth is 113°.

$$\text{Wall solar azimuth} = 135° - 113°$$

$$\text{Wall solar azimuth} = 12°$$

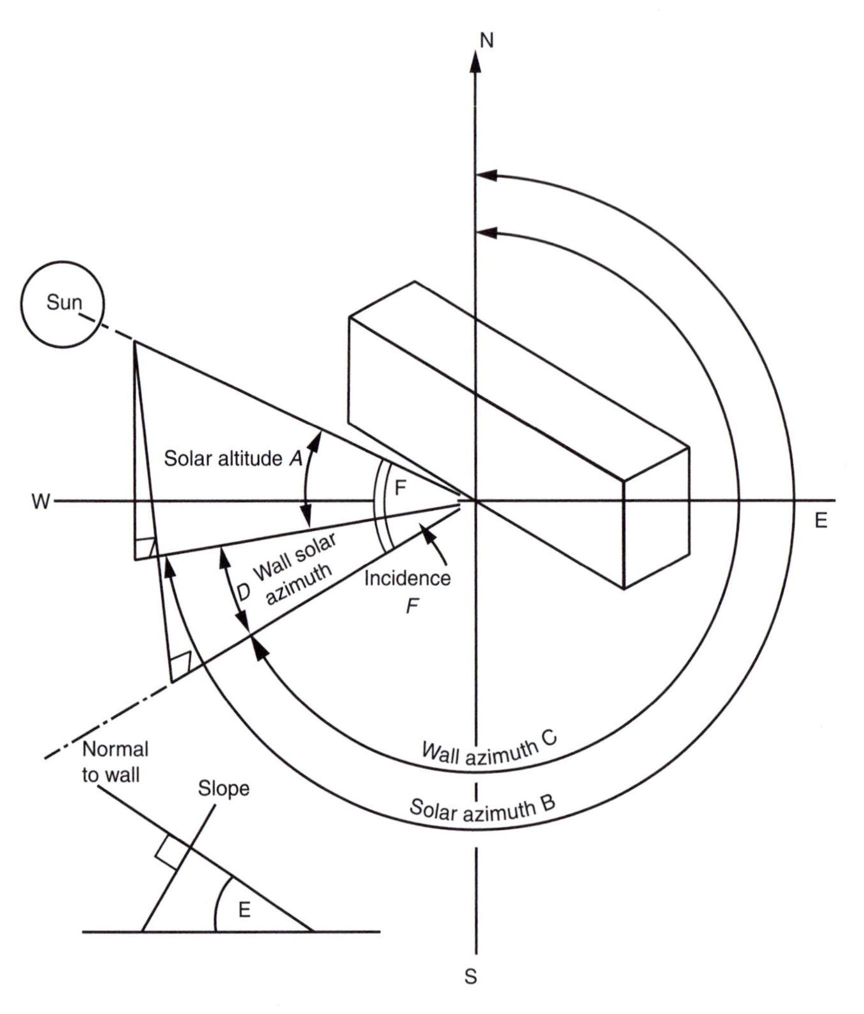

3.3 Solar and wall orientation.

The wall is almost facing the sun in the morning. To find the incidence use:

$$\cos F = \cos A \times \cos D$$

$$\cos F = \cos 44 \times \cos 12$$

$$\cos F = 0.7036$$

Leaving this number in the calculator display, the angle whose cosine is 0.7036 is to be found; this is the $\cos^{-1}$ function.

$$\cos^{-1} 0.7036 = X^{o}$$

Press the $\cos^{-1}$ key and 45.3° is found. Check it is correct by pressing the cosine key to find cos 45.3° = 0.7036. Notice that the incidence is a combination of the solar altitude being altered by the wall solar azimuth.

EXAMPLE 3.4

A wall of an office in Plymouth, latitude 50° 21′N, faces south west and slopes at 10° forwards from the vertical as an architectural feature. The sun time is noon on 21 June, the solar altitude is 64° and the solar azimuth is 180°. The intensity of the direct solar radiation normal to the sun is 900 W/m^2. Prove the formula for solar incidence and calculate the incidence angle upon the wall.

Figure 3.4 shows the plan positions of the sun and wall.

$$\text{Wall solar azimuth angle } D = 225^\circ - 180^\circ$$

$$\text{Wall solar azimuth angle } D = 45^\circ$$

The solar radiation needs to be resolved to a value that is normal to the wall surface. The angle between the lines normal to the sun and normal to the wall is the incidence angle, F.

View 1 would show a vertical triangle along a north south line and the sun's altitude of 64° as depicted in Figure 3.5.

$$\cos 64 = \frac{Z}{I}$$

$$Z = I \cos 64$$

From Figure 3.4,

$$\cos 45 = \frac{Y}{Z}$$

$$Y = Z \cos 45$$

Z is known

$$\text{side } Y = I \cos 64 \times \cos 45$$

The face of the building is tilted downwards 10° so the component of the solar radiation that is normal to it is line X. Line Z has an equal length along the ground as shown and this becomes the hypotenuse of the lowest triangle in Figure 3.6.

$$\sin 80 = \frac{X}{Y}$$

And, as YY is known,

$$X = I \cos 64 \times \cos 45 \times \sin 80$$

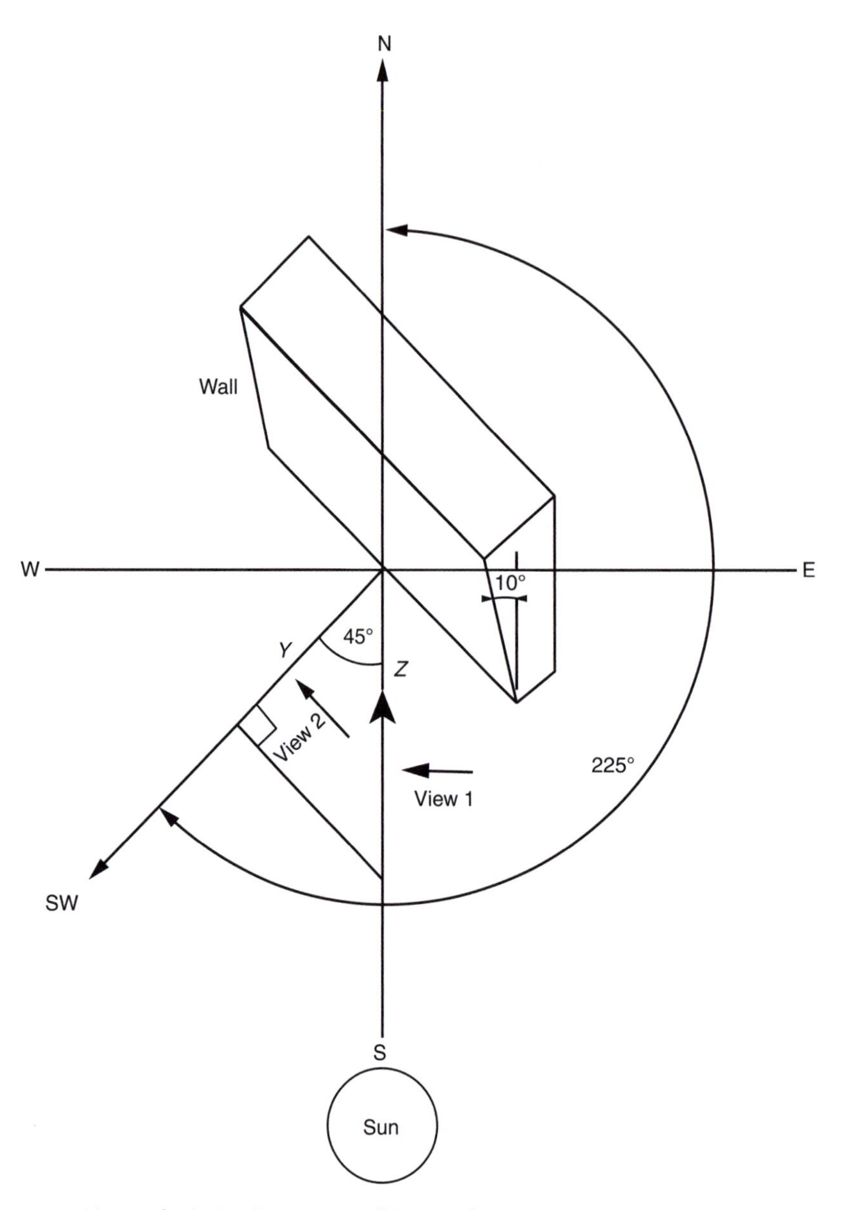

3.4 Incidence of solar irradiance on a wall in example 3.4.

Where side X represents that component of the solar radiation that is normal to the face of the wall I_X.

$$I_X = 900\frac{\text{W}}{\text{m}^2} \times \cos 64 \times \cos 45 \times \sin 80$$

$$I_X = 900\frac{\text{W}}{\text{m}^2} \times 0.438 \times 0.707 \times 0.985$$

$$I_X = 274.5\frac{\text{W}}{\text{m}^2}$$

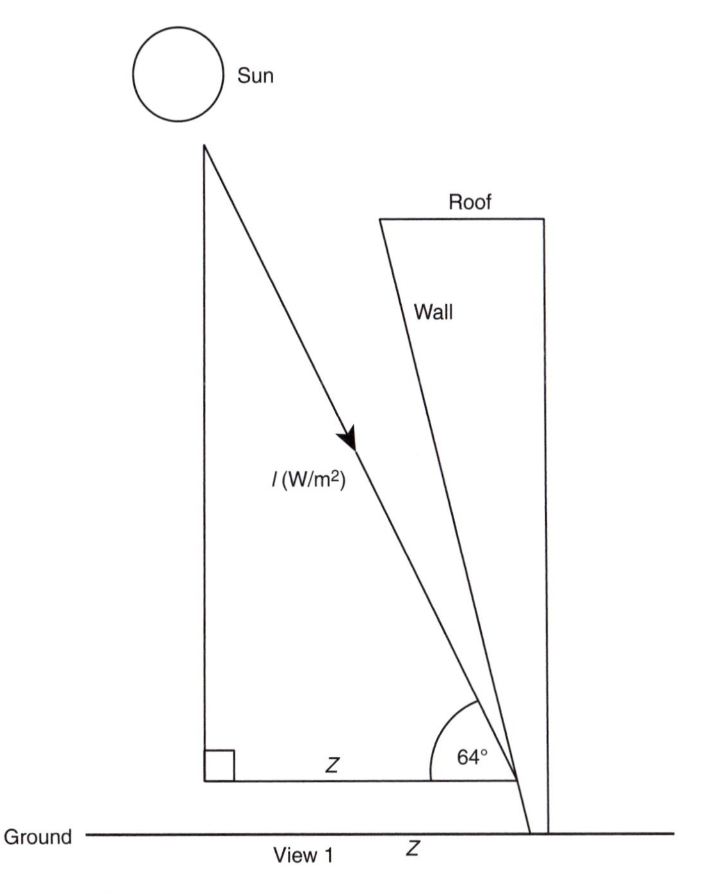

3.5 Solar altitude for example 3.4.

Referring to Figure 3.3, the incidence angle F is that between the normal from the surface I_X and the normal from the sun I. This triangle slopes outward from the surface of the paper and,

$$\cos F = \frac{I_X}{I}$$

$$\cos F = \frac{I \cos 64 \times \cos 45 \times \sin 80}{I}$$

$$\cos F = \cos 64 \times \cos 45 \times \sin 80$$

$$\cos F = 0.438 \times 0.707 \times 0.985$$

$$\cos F = 0.305$$

And to find the incidence angle,

$$\cos^{-1} 0.305 = 72.2°$$

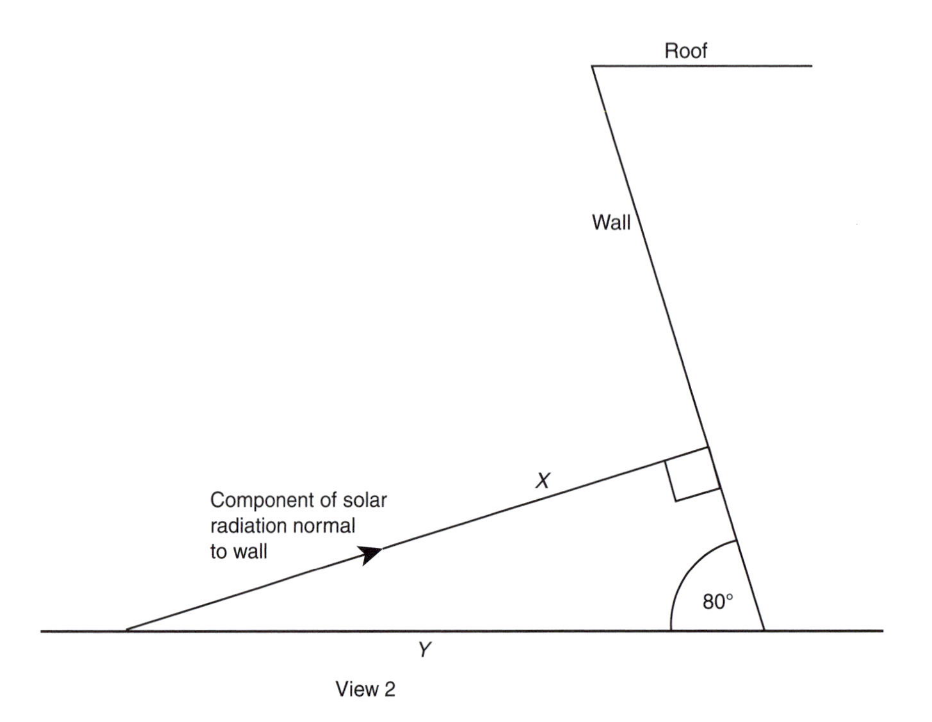

3.6 Solar incidence on the sloping wall in example 3.4.

Bearing in mind that,

$64° =$ solar altitude angle A

$45° =$ wall solar azimuth angle D

$80° =$ slope angle of the surface E

Then,

$$\cos F = \cos A \times \cos D \times \sin E$$

And this can be generally used for other problems.

Shading effects

Shade upon windows and walls greatly reduces the incident solar irradiance by eliminating the direct line of sight heat flow and only leaving the blue sky diffuse irradiance. Deliberate shading is produced by external louvres, balconies, verandas and protruding columns while fortuitous shading may be temporarily created by nearby buildings, vegetation or landscape. A window that is recessed from the face of the building will be partly shaded from the side due to the wall solar azimuth and from the top due to the combination of solar altitude and wall solar azimuth. Figure 3.7 shows that the width of shade from the side wall W can be found from the wall solar azimuth angle D and the depth of the recess R such that:

$$\tan D = \frac{W}{R}$$

$$W = R \tan D$$

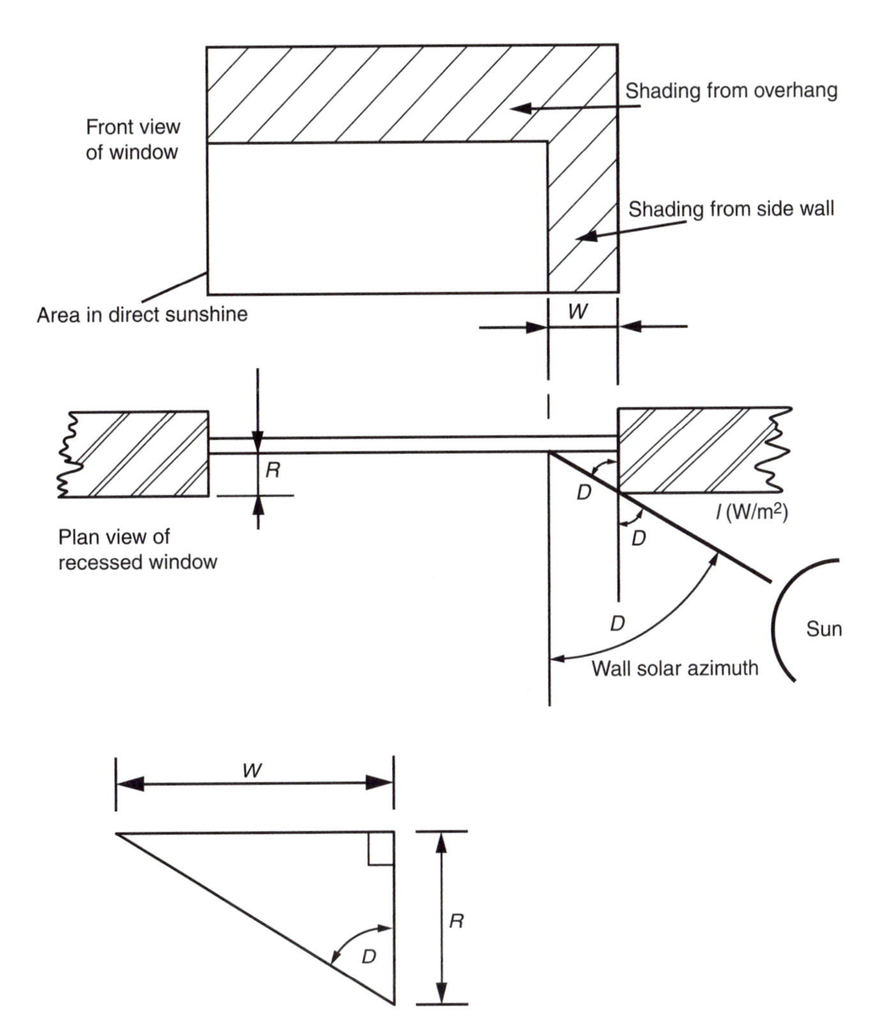

3.7 Angles for calculation of vertical shading.

EXAMPLE 3.5

A window is 2.5 m long and 1.5 m high and has a vertical column alongside projecting 250 mm from the face of the building. The wall azimuth is 220° and the solar azimuth is 180°. Calculate the areas of glass exposed to direct and diffuse irradiances from the shade cast across the side.

Figure 3.7 shows the shading angles.

Wall solar azimuth $D = 220° - 180°$

Wall solar azimuth $D = 40°$

Width of side shade $W = 250 \text{ mm} \times \tan 40$

$$\text{Width of side shade } W = 209.8 \text{ mm}$$

$$\text{Width of side shade } W = 0.21 \text{ m}$$

$$\text{Side shaded area} = 0.21 \text{ m} \times 1.5 \text{ m}$$

$$\text{Side shaded area} = 0.315 \text{ m}^2$$

$$\text{Total area of glazing} = 2.5 \text{ m} \times 1.5 \text{ m}$$

$$\text{Total area of glazing} = 3.75 \text{ m}^2$$

$$\text{Unshaded area of glazing} = (3.75 - 0.315)\text{m}^2$$

$$\text{Unshaded area of glazing} = 3.435 \text{ m}^2$$

Thus the glass exposed to direct irradiance is 3.435 m^2 while the glass exposed to blue sky diffuse irradiance is the whole window area of 3.75 m^2.

Vertical shading caused by the overhanging projection of a balcony, external blind, slats or louvres and recessing of the glazing is calculated from a combination of the wall solar azimuth and the solar altitude. The greater the wall solar azimuth, the deeper the shadow for a constant value of solar azimuth, and this can be tested with a card folded into an L and sunshine or a lamp by turning the vertical surface at increasingly acute angles from the light source. Figure 3.8 shows the basis for calculating the height of the window's horizontal shaded portion.

From the plan view,

$$\frac{R}{Y} = \cos D$$

$$Y = \frac{R}{\cos D}$$

Where Y is the length of the horizontal component of the irradiation; view AA reveals a vertical triangle where,

$$\frac{h}{Y} = \tan A$$

$$h = Y \tan A$$

Substitute for Y,

$$h = R \frac{\tan A}{\cos D}$$

EXAMPLE 3.6

A window 2.5 m long and 1.5 m deep is recessed 250 mm from the face of a building. Calculate the unshaded glass area at 10.00 hours GMT when the solar altitude is 52° and the solar azimuth is 131° on 21 May. Wall azimuth is 195°. Calculate the unshaded area of the glass.

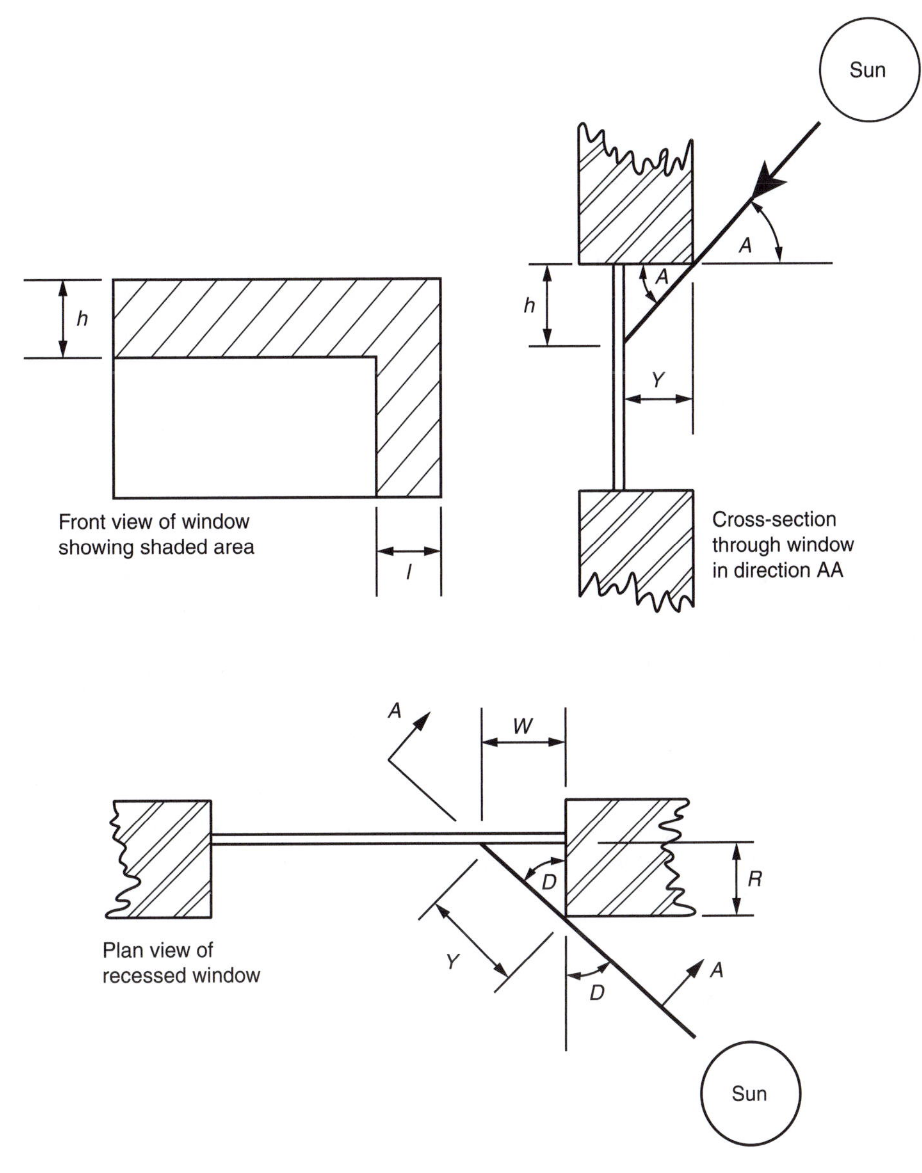

3.8 Derivation of horizontal shading depth.

Side shade length $W = R \tan D$

Wall solar azimuth $D = 195^\circ - 131^\circ$

Wall solar azimuth $D = 64^\circ$

$$W = 250 \tan 64$$

$$W = 512.6 \text{ mm}$$

$$W = 0.513 \text{ m}$$

$$h = R\frac{\tan A}{\cos D}$$

$$h = 250\frac{\tan 52}{\cos 64}$$

$$h = 729.9 \text{ mm}$$

$$h = 0.73 \text{ m}$$

$$\text{Length of unshaded area} = (2.5 - 0.513)\text{m}$$

$$\text{Length of unshaded area} = 1.987 \text{ m}$$

$$\text{Height of unshaded area} = (1.5 - 0.73)\text{m}$$

$$\text{Height of unshaded area} = 0.77 \text{ m}$$

$$\text{Unshaded area} = 1.987 \times 0.77 \text{ m}^2$$

$$\text{Unshaded area} = 1.53 \text{ m}^2$$

Areas of shade on a window or a complete building can be found by calculation or by illuminating a scale model of the subject building and its surroundings with a lamp positioned at the correct altitude and azimuth angles corresponding to the time of day. The refrigeration cooling load and the size of each room terminal cooling unit are calculated from the peak heat gain that normally occurs at the time of the greatest solar irradiation upon the unshaded glazing. The deliberate use of shading and the calculation of adventitious shading from nearby buildings can significantly reduce the refrigeration plant load. Compare the exposure of a building to solar irradiance at two different locations using the data in Table 3.1 (CIBSE, 2006).

Notice the reversal of the dates of the seasons between the northern and southern latitudes and the greater solar altitude when closer to the Equator where it is overhead at noon on 21 March and

Table 3.1 Solar positions

GMT, h	*Near Brisbane 30° S*		*Near London 50° N*	
	21 December		*21 June*	
	Altitude	*Azimuth*	*Altitude*	*Azimuth*
06	12	69	18	74
07	24	76	27	85
08	37	82	37	97
09	50	88	46	110
10	62	96	55	128
11	75	112	61	151
12	84	180	64	180
13	75	248	61	209
14	62	264	55	232
15	50	272	46	250
16	37	278	37	263
17	24	284	27	275
18	12	291	18	286

Data in this table are indicative only.

21 September. Overhanging roofing traditionally provides completely shaded windows near Brisbane and this also serves to collect rainwater for storage. Variations in shade across the facades of a building occur throughout the day and this has an effect upon the cooling load that has to be matched by the automatic control system.

EXAMPLE 3.7

On 20 February at noon sun time, the solar altitude is 29°. Figure 3.9 shows the relative positions of two nearby buildings. Calculate the area of shade on building *A* caused by building *B*.

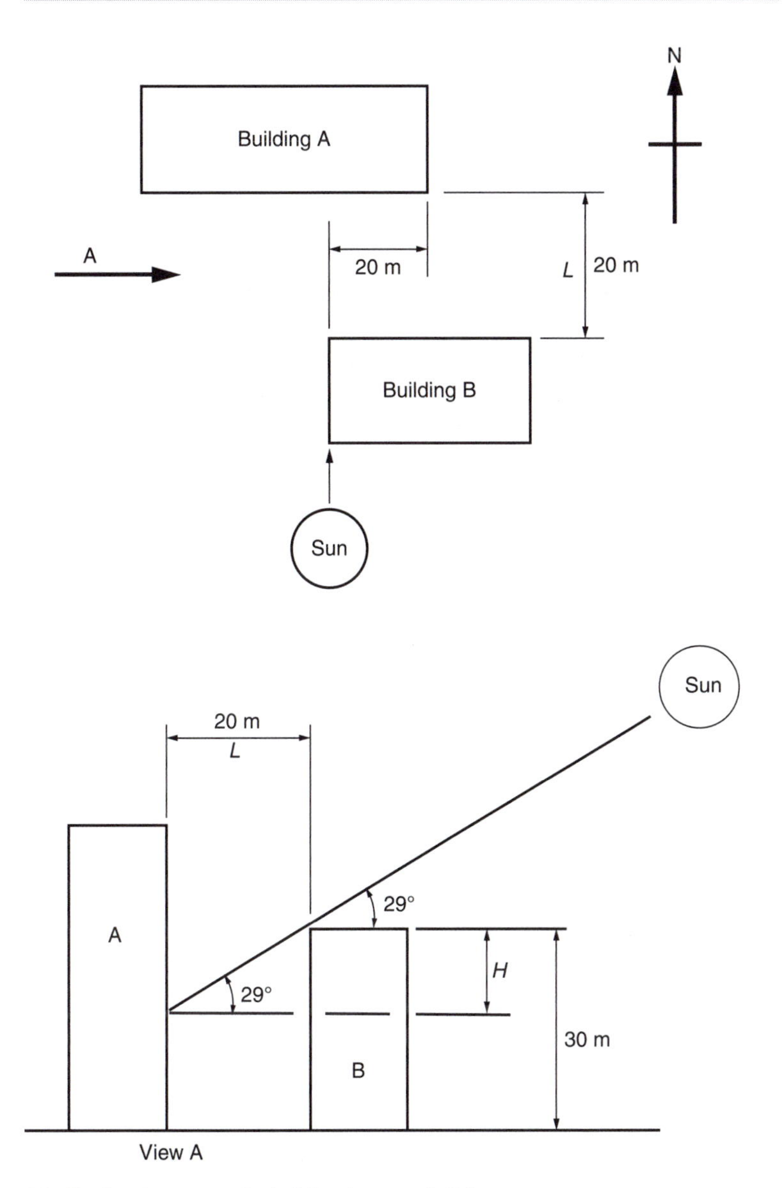

3.9 Shading from a nearby building in example 3.7.

At noon the sun is due south so shade width is 20 m.

Height of shade on building $A = (30 - H)\,\text{m}$

$$\tan 29 = \frac{H}{L}$$

$$H = L \tan 29$$

$$H = 20 \tan 29$$

$$H = 11.09 \text{ m}$$

Shaded area $= 20 \text{ m} \times (30 - 11.09) \text{ m}$

Shaded area $= 378.2 \text{ m}^2$

EXAMPLE 3.8

Figure 3.10 shows the site plan and elevation of building *A* that is to be air conditioned, and building *B* that is across the street and will cast shadows onto *A*. The site is at a latitude of 50°N. Calculate the solar irradiance normal to each surface of building *A* at 15.00 hours GMT on the 21 March when the solar altitude is 27°C and solar azimuth is 232°C, and the areas of roof and walls that are exposed to direct and diffuse irradiance. The basic direct solar irradiance is 700 W/m^2 and the basic diffuse irradiance is 72 W/m^2.

Figure 3.11 shows the geometry needed to find the areas of shade on the roof and front face of building *A*. The leading edges of building *B* cast the shadows.

In plan,

$$\tan 58 = \frac{Y}{20}$$

$$Y = 20 \tan 58$$

$$Y = 32 \text{ m}$$

From view *Z* the true length of the horizontal component *T* of the irradiance upon the roof can be found.

$$\tan 27 = \frac{15}{T}$$

$$T = \frac{15}{\tan 27}$$

$$T = 29.439 \text{ m}$$

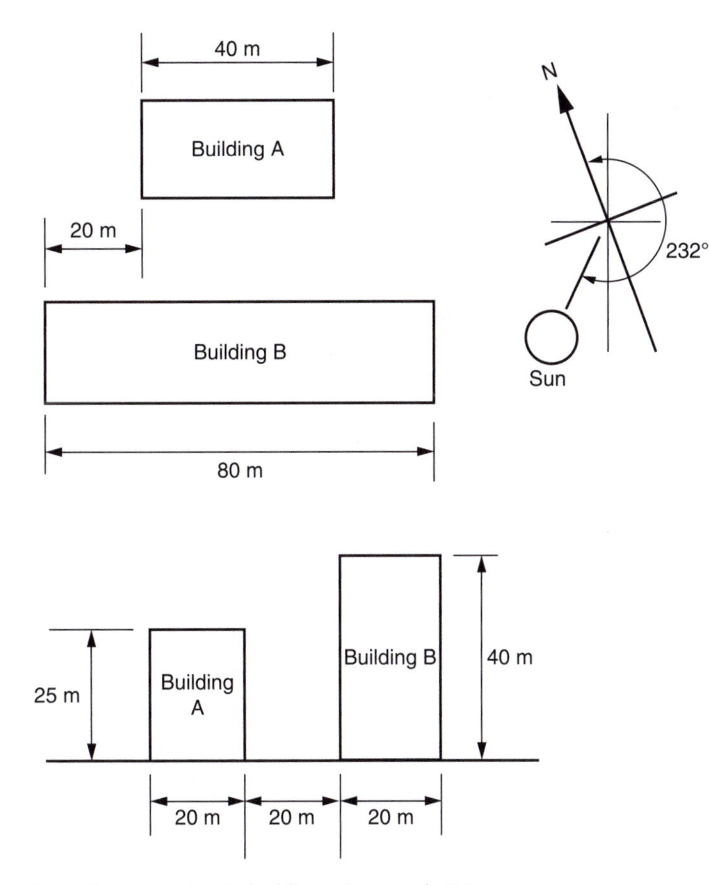

3.10 Shading cast onto building A in example 3.8.

The leading edge of the roof of building *B* casts a shadow onto the roof of building *A*. The depth of this shadow is found in plan as:

$$\sin 58 = \frac{S}{29.439}$$

$$S = 29.439 \sin 58$$

$$S = 24.966 \text{ m}$$

The shadow projects,

$$(24.966 - 20) = 4.966 \text{ m}$$

$24.966 - 20$ m onto the roof, 4.966 m.

The front face of *A* is completely shaded.
The north west face and roof of *A* is partly in direct irradiance.
The south east and north east faces of *A* are completely shaded from direct irradiance.

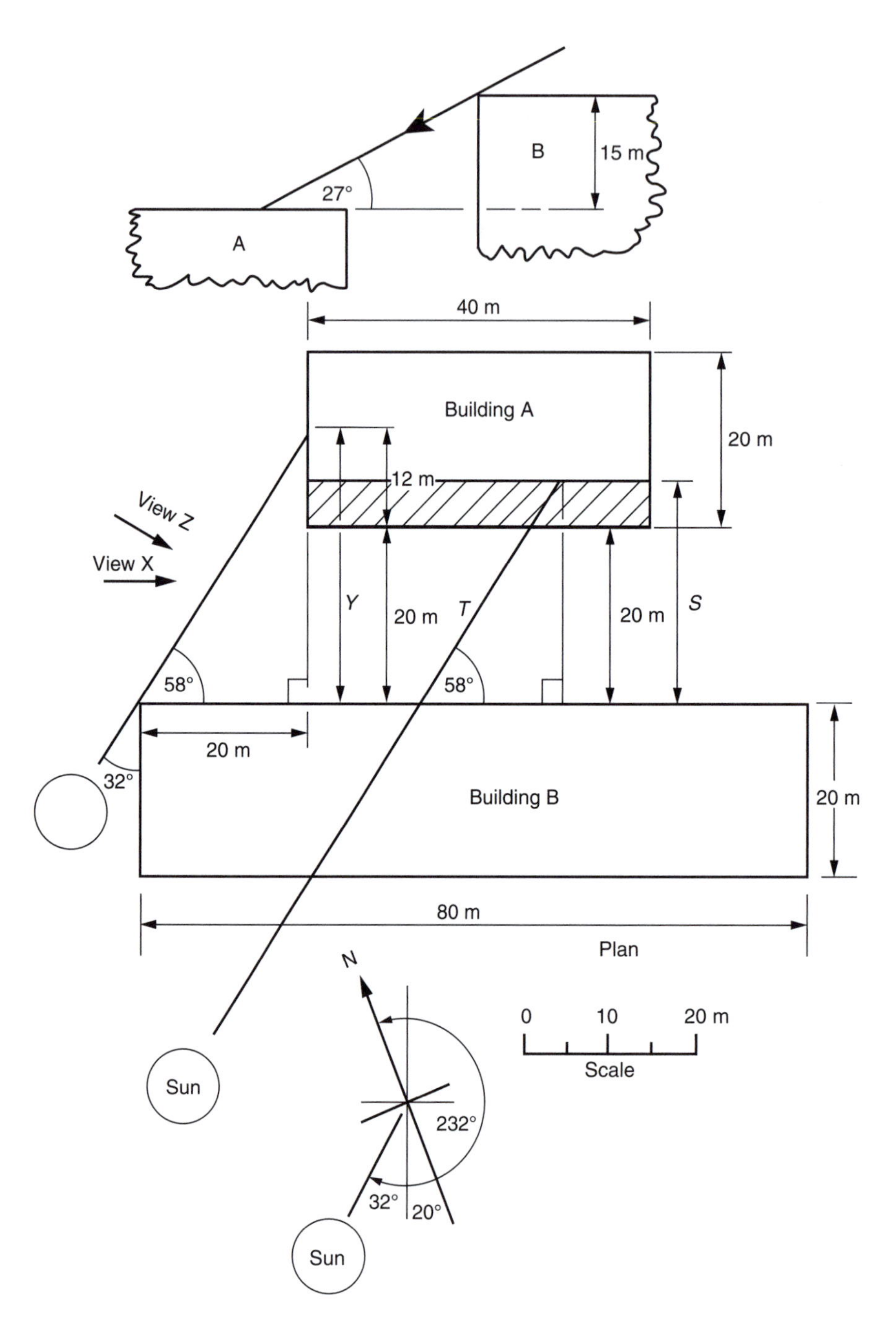

3.11 Solution of shading example 3.8.

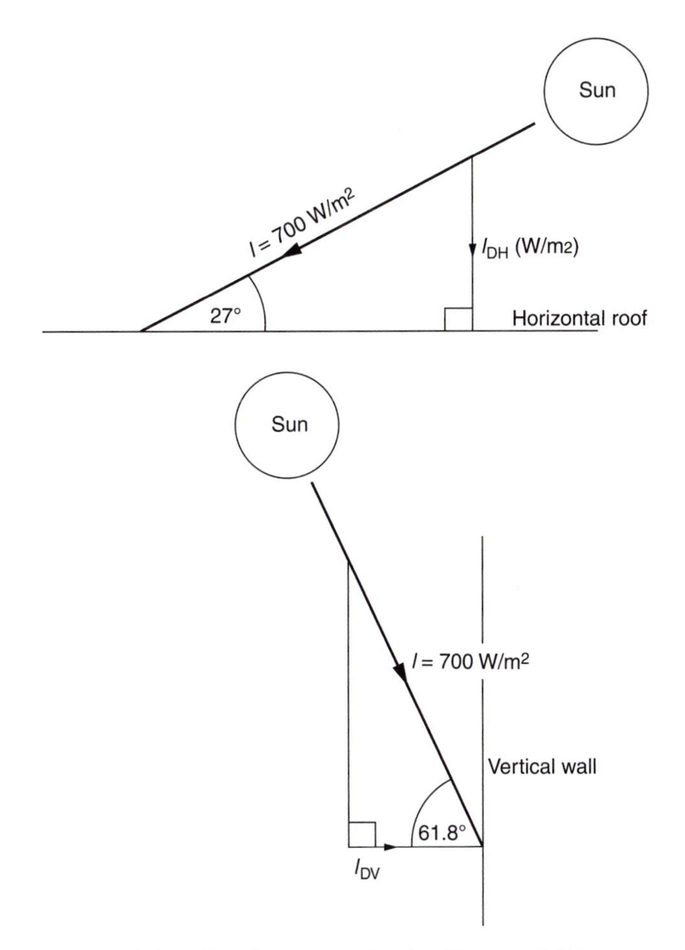

3.12 Resolution of irradiance on roof and wall in example 3.8.

The roof receives direct irradiance on the horizontal surface that is the vertical component of the basic direct irradiance of 700 W/m^2 as indicated in Figure 3.12.

$$I_{DH} = 700\frac{\text{W}}{\text{m}^2} \times \sin 27$$

$$I_{DH} = 318\frac{\text{W}}{\text{m}^2}$$

A normal from the left wall of building *A* faces 20°N of west and has a wall azimuth of 290°C and the direct solar irradiance upon the exposed part of the vertical surface is I_{DH}.

Wall solar azimuth angle $D = 290° - 232°$

Wall solar azimuth angle $D = 58°$

From earlier work, $\cos F = \cos A \cos D$

$$\cos F = \cos 27 \cos 58$$

$$\cos F = 0.472$$

$$\cos^{-1} 0.472 = 61.8°$$

The angle of incidence between the vertical wall and the irradiance is 61.8°C. The horizontal component that is normal to the wall can be found from,

$$I_{DV} = I \cos F$$

$$I_{DV} = 700 \frac{W}{m^2} \times 0.472$$

$$I_{DV} = 330 \frac{W}{m^2}$$

The shaded surfaces receive diffuse sky irradiance of 72 W/m^2.

Around the world

Air conditioning systems are used widely. While the focus of this book may be seen as UK and CIBSE based, the basic principles apply wherever air conditioning is used. Consider the application locations:

1. Spacecraft, satellite space stations and eventually, sealed environments on planets other than Earth; external climate at absolute zero temperature and may have no air;
2. Vehicle transport, ships, passenger cruise liners, naval ships and submarines; these mobile air conditioned vehicles may pass through every possible environment including undersea;
3. Antarctica (*CIBSE Journal* April 2013, Halley VI); −50°C d.b. deep snow all year and long periods of no sunshine;
4. Europe, ranging from the Arctic Circle −50°C d.b. to the Mediterranean 40°C d.b;
5. Continent of America from the Arctic Circle through the Equator and down to near the Antarctic;
6. Africa and Asia, mainly tropical, hot and wet all year; 32°C and 70% humidity all year;
7. Many islands with few power sources;
8. Australia ranging from Alpine Snowy Mountains at −10°C d.b., through tropical, to desert 50°C d.b.;
9. Large islands such as New Zealand and the UK that have a temperate maritime climate;
10. A very wide range of developing nations from impoverished subsistence farming societies all the way up to luxurious.

Can you think of any more categories? A useful way to evaluate the differences between climate locations is to observe how much cooling, heating, humidification and dehumidification is needed to change the external air into normally comfortable indoor conditions of 24°C d.b., 50% humidity. A large proportion of the world's population live within the tropics of ±23.5°C latitude from the Equator where the climate varies from hot dry Saharan desert, through dense wet equatorial Amazonian jungle, to the vastness of Asia. These also vary between the poorest areas having little access to electricity, relying on passive comfort control measures, with no access to food refrigeration, to the richest parts of the world. Have a look around on Google Earth and stand in a few locations new to you; such travel at no cost is highly educational.

Counting the number of days in climate regions where 26.7°C d.b. and 37.8°C d.b. are exceeded (The Australian Institute of Refrigeration, 1997) is an aid to selecting cooling systems but is not conclusive. Australia's green edge land has less than 10 days a year above 37.8°C d.b., while parts of the hot centre have 60–100 days. Modern commercial buildings and house design requires active ventilation and cooling systems. Society has grown to accept this as normal practice. The occupants of many buildings before the

1960s coped with many days of discomfort with passive design and more tolerance than today's modern society. Just remember, there are many millions of people who do not have air conditioned homes, transport, workplaces or even refrigerators.

Cooling loads at different locations can be compared from their outdoor air conditions. Use a psychrometric App or data source for the following example.

EXAMPLE 3.9

Find the moisture content and specific enthalpy for indoor comfort conditions of 24°C d.b., 50% humidity and compare them with the design cooling loads for London and Hong Kong.

Refer to data tables or an App for air conditions.
Room conditions 24°C d.b., 50% RH, 17°C w.b., moisture content 0.00920 kg/kg, specific enthalpy 47.6 kJ/kg.
Take the following design conditions;
London 28°C d.b. and 20°C w.b., 48% RH, moisture content 0.01135 kg/kg, specific enthalpy 57.13 kJ/kg and latitude 51.5°N.

$$\text{Dehumidification} = (0.01135 - 0.00920)\,\frac{\text{kg}}{\text{kg}}$$

$$\text{Dehumidification} = 0.00215\,\frac{\text{kg}}{\text{kg}}$$

$$\text{Dehumidification} = \frac{0.00215}{0.00920} \times 100\%$$

$$\text{Dehumidification} = 23.4\%$$

$$\text{Total cooling load} = (57.13 - 47.6)\,\frac{\text{kJ}}{\text{kg}}$$

$$\text{Total cooling load} = 9.53\,\frac{\text{kJ}}{\text{kg}}$$

$$\text{Total cooling load} = \frac{9.53}{47.6} \times 100\%$$

$$\text{Total cooling load} = 20\%$$

Hong Kong 33.2°C d.b. and 27.8°C w.b., 66.5% RH, moisture content 0.02150 kg/kg, specific enthalpy 88.49 kJ/kg and latitude 22.1°N.

$$\text{Dehumidification} = (0.02150 - 0.00920)\,\frac{\text{kg}}{\text{kg}}$$

$$\text{Dehumidification} = 0.0123\,\frac{\text{kg}}{\text{kg}}$$

$$\text{Dehumidification} = \frac{0.0123}{0.00920} \times 100\%$$

$$\text{Dehumidification} = 133.7\%$$

$$\text{Total cooling load} = (88.49 - 47.6)\frac{\text{kJ}}{\text{kg}}$$

$$\text{Total cooling load} = 40.89\frac{\text{kJ}}{\text{kg}}$$

$$\text{Total cooling load} = \frac{40.89}{47.6} \times 100\%$$

$$\text{Total cooling load} = 85.9\%$$

To compare the two locations,

$$\text{Hong Kong dehumidification} = \frac{133.7\%}{23.4\%} \text{ or } \frac{0.0123}{0.00215}$$

$$\text{Hong Kong dehumidification} = 5.7 \text{ times that in London}$$

$$\text{Hong Kong cooling coil load} = \frac{85.9\%}{20\%} \text{ or } \frac{40.89}{9.53}$$

$$\text{Hong Kong cooling coil load} = 4.3 \text{ times that in London}$$

Use the downloadable file *around the world.xls* to compare a wide variety of world locations.

As the world's population develop their economies, the need for air conditioning will increase, along with its attendant power demand. Those who think of fossil fuel atmospheric emissions, have to come to terms with economic development around the world by those who do not now have air conditioning, food refrigeration, TV and smart phones. Make up your own mind about how the world can develop and be sustainable into future generations.

The extent of the cooling season determines energy use and running costs for air conditioning. Tropical regions have a cooling season that lasts all through the year, with no winter as compared with Europe. Many tropical regions have few periods of clear blue sky and bright sunlit conditions. Peak summer has the sun directly overhead and this coincides with the wet or monsoon season. Variations in monthly design conditions can be minor with year-round cooling and dehumidification plant loads producing consistently high energy usage and power demand. Darwin, Australia, for example, has two separate peak design dry bulb periods of clear sky days during April and October (The Australian Institute of Refrigeration, Air-Conditioning and Heating, 1997).

Although tropical clear skies may be absent, diffuse radiation can be five times the intensity of clear blue sky intensity, maintaining solar radiation heat gains. Solar shading can still be necessary. Significant overhanging roof eaves and verandas provide radiation and rain protection and highly ventilated work and recreation spaces. Ventilation air movement at higher velocity aids personal moisture removal and improves comfort in humid environments. An indoor humidity of 60% is generally recommended as it is relevant to the outdoor high levels and achievable with air conditioning plant. Cooling coil dew point temperature controls dehumidification and very low supply air moisture content levels may be impractical in wet environments.

An indoor operative temperature range of 23°C to 26°C is preferred for those wearing lightweight clothes, clo 0.5. When wearing a business suit, clo 1.0, 24.5°C is needed. Operative temperature approximates to the centre of a 40 mm painted globe freely exposed to room air temperature, radiations, heat exchanges and air velocity, as are the room occupants.

Indoor air velocity around sedentary people in a humid climate may be preferred to be around 1.0 m/s as just acceptable; at this speed, paper on a desk can be lifted but probably not blown away. While outdoors in 30°C to 35°C and 70% humidity, up to 2.5 m/s decreases operative temperature and improves comfort, at least reducing discomfort. We tend to adapt when exposed to a new climate for an extended stay or a period of years, coming to terms with an acceptance of local conditions. When going from a cooled indoor 24°C and 50% humidity to outdoors in blazing sunshine at above 35°C, it feels like walking into an oven and shade is sought to minimise solar radiation heating and sunburn. In locations such as Hong Kong, any outdoor activity is done slowly due to the hot humid air at all times; Tai Chi groups meet in parks on a Sunday morning, and that is enough outdoor activity for most people, in addition to the problem of the minimum availability of outdoor space for any activity. Outdoor high humidity is probably never thought of as comfortable. Walkers, runners and cyclists travelling to work in for example Brisbane, in January, make sure they arrive soon after sunrise; also darkness falls early in the evening in the tropics. It is the contrasts in air conditions that we notice most.

Hot dry outdoor air at 35°C d.b., 21°C w.b., 25% humidity in such locations as Adelaide, and 41°C d.b., 26°C w.b., 28% humidity in Dubai, is often passed through an evaporative cooler where direct injection of mains water provides latent heat transfer from the air during water evaporation. Latent heat removal causes lowering of sensible heat in the air and produces a reduction of air dry bulb temperature down towards the original wet bulb temperature of the incoming air. A limiting factor is the available contact between the wetted surfaces of the filter material in the evaporative cooling unit and the incoming hot dry air. Not all the air can be saturated with water, so 100% humidity cannot be achieved. Something like 50–60% saturation is likely to be achieved, meaning that air supplied into the occupied room may be around 26°C d.b., 60% humidity in Adelaide, while somewhere like Dubai might achieve 30°C d.b., 70% humidity. These are not comfortable air conditions even with high airflow and air velocity around the occupants, but it is better than no cooling at all and is very low cost in terms of a small fan and a small amount of water consumed. Water spray evaporative coolers are not used in hot humid tropical climates as adding moisture into the air would create little cooling effect and increase room humidity. Applications are usually limited to single- or two-storey residences, retail fast food shops, offices and some industrial buildings. Supply air ducts between the evaporative cooling unit and the room outlet supply grille must be short to minimise air reheating as the ducts pass through the roof or other unconditioned spaces and condensation forms within the duct. At least annual cleaning of the evaporative cooling unit and supply air ducts is recommended as filtration of dust and debris is not very fine in the unit. Air is often exhausted from rooms through open windows and doors but can be with ducted exhaust air and fan systems. Complex and multi-storey buildings often require a sealed building with refrigerated air conditioning and piped chilled water from a central plant to dispersed air handling units.

Flow of moisture is always from warm humid air at high vapour pressure towards air at lower vapour pressure, cooler and lower moisture content. Moisture flows through porous building materials, walls, timber, plasterboard and thermal insulation. Centrally heated buildings in cold climates pass moisture from indoor sources, evaporation and exhalation from occupants, washing and water surfaces, through the structure to outdoors. Condensation on windows and within walls and roofs can be problematic, leading to mould, rot, ill health and long term damage to the building, bathrooms, kitchens, decorations and cupboards on external walls.

Hot humid tropical climates have a higher outdoor air vapour pressure than within a cooled air conditioned building; the same effects take place but in the reverse direction. Vapour barriers, low permeance membranes, within walls and roofs and around air ducts, reduce condensation risks in both types of climate.

Thermal bridges formed by concrete and steel structural columns, beams and aluminium window frames accelerate heat loss from warm buildings in cool climates. These are condensation sites. Moisture flow is outwards to the colder external environment, eventually harmlessly evaporating in the outdoor air. In hot humid climates, cooled buildings suffer moisture ingress, driven inwards, especially through gaps in joints and corners of vapour barriers. This happens inwards into air conditioned spaces, air ducts and air handling units leading to internal condensation, mould, rust and damage. Cold tracking thermal bridges can produce condensation on the outer warm humid side of air ducts, water chillers, chilled water pipes, pumps, valves and air handling units. Adequate thermal insulation, vapour sealing and avoidance of cold bridges are needed to avoid long term plant degradation and water pooling. Tropic-proofing of air conditioning plant and associated electrical systems requires a great deal of care and specialised local knowledge.

Ensuring continued satisfactory performance of complex air conditioning systems, BMS computer systems, plant and equipment continues long after the installation has been handed over and the design team and contractors have left the locality. This applies equally to small packaged air conditioning units which are also at risk of being installed and maintained by relatively untrained personnel.

Plant rooms in extremely cold or tropical climates are workspaces requiring manual attendance on a regular basis. They have to be ventilated and maintained at comfortable working conditions as well as for the protection of mechanical plant, electric motors, switchboards and complex computer control panels, for example, in lift motor rooms. Excesses of air temperature, humidity or rain ingress, can rapidly damage expensive equipment and cause system failures. A continuously operating air handling unit (AHU) is often provided to condition a plant room even if the main plant runs intermittently. A large AHU might have a spill supply air duct outlet into the plant room for conditioning. Extreme climates, as discussed, have continuous plant operation, as is the case with, for example, planes, ocean passenger ships and military vessels where switching off to save energy are not options. Electrical and electronic equipment is susceptible to damage outside the range of 10°C to 40°C or subject to condensation, overheating or moisture damage. Plant rooms can become excessively hot from the climate, solar radiation plus heat generated by motors, refrigeration and air compressors, heater pipes, pumps and lighting.

Mechanical and electrical services switchboards, control system cubicles, plant isolating switches, computer servers, pump and lift electric motors and valve motors are tropic-proofed against ingress of moisture. Switchboards and motors may have electric heaters to drive out moisture. Corrosion protection of exposed metalwork is always needed.

Cyclones affect many tropical areas. Exterior location of plant is avoided, or, it is built to withstand potential wind velocities of up to 250 km/h. Air ducts exposed to high wind environments have internal bracing. External plant is protected with screening against wind pressure and flying objects. Air intake and exhaust grilles are designed to withstand extreme wind and monsoon rain. Flash flooding is a risk. Plant is fixed to plinths, and sump pumping and adequate storm water drainage is used. Basement plant room are avoided.

Air conditioning systems in hot humid climates run continuously with daily and seasonal temperatures that vary minimally, day and night. Matching refrigeration plant capacity to the demand for cooling from air handling unit coils is critical. Over-sized chillers cycle cylinders and compressors on/off to maintain the set point chilled water temperature within a controllable band of maybe ±1°C. Bringing compressors back on line takes time to reduce chilled water temperature back to its design value of, say, 6°C. A small drop in chilled water temperature could cause a significant reduction in dehumidification capacity and make room conditions fluctuate noticeably. This may not be problematic in a hot dry or temperate climate, but loss of humidity control in a humid climate could easily lead to discomfort; a chilled water buffer tank smoothes the difference between input cooling power and demand.

EXAMPLE 3.10

The Shard case study later in this chapter assesses its air conditioning plant load at London Bridge. Relocate The Shard to Dubai and assess the air conditioning design load and annual energy use. Find the climate data, comment on what you observe and explain the results.

Table 3.2 Dubai climate

Month	*t_{ao} °C d.b. high*	*t_{ao} °C d.b. low*	*RH%*
January	24.0	14.3	65
February	25.4	15.4	65
March	28.2	17.6	63
April	32.9	20.8	55
May	37.6	24.6	53
June	39.5	27.2	58
July	40.8	29.9	56
August	41.3	30.2	57
September	38.9	27.5	60
October	35.4	23.9	60
November	30.5	19.9	61
December	26.2	16.3	64

Dubai city in the UAE is on the gulf coast of a flat sandy desert region. Climate is hot desert, windy and always humid. Most days are sunny as rain falls for around 5 days a year. Summers reach 49°C d.b., 29°C at night, sea temperature rises to 37°C and humidity rises to 90% from seawater evaporation. Winters have a daytime of 23°C d.b. and 14°C at night. Approximate monthly design conditions are indicated in Table 3.2 (Wikipedia).

Dubai has no winter heating period. There are no days in the year when free cooling can be used, that is, turning off the water chillers, cooling towers and pumps, while admitting uncooled outside air into the building, as it is always too warm and too humid, for example causing discomfort overnight in January, the coolest month. Residents might not have air conditioned homes but The Shard is a prestige building.

Relocating The Shard's 87 floors and 306 m height to Dubai would put this iconic London skyscraper into a wider perspective. It would be seen as a modest effort, not very impressive or strange. Dubai city has over 900 high rise buildings of recent (post 1970s) construction. Burj Khalifa has 163 floors and 828 m height; Princess Tower has 101 floors and 413 m height; while Burj Al Arab has 60 floors and 321 m height.

Use the file *around the world.xls* to see that the AHU cooling coil load is 339% of the London climate, including 270% greater dehumidification moisture removal.

The Shard case study calculates the design peak heat gain to be 7973 kW for the cooling plant to deliver a supply air flow rate of 727 m^3/s and creating 12 air changes/h through the building. Use the downloadable file *Shard in Dubai.xls* to estimate the energy requirements and annual air conditioning energy cost in comparison with London using the same utility rates.

Take peak outside air design temperature t_{ao4} as 41°C d.b., although you may wish to test what happens on a 49°C day.

Minimum outside air temperature t_{ao3} taken as 14.3°C d.b.

$$\text{Dubai } Q = 727 \frac{\text{m}^3}{\text{s}} \times 339\%$$

$$\text{Dubai } Q = 2465 \frac{\text{m}^3}{\text{s}}$$

Occupancy and other data remain the same. No gas heating is used. AHU fans might be double the power of the London location, 1000 kW, this might be an underestimate. A summary of approximate data is given in Table 3.3.

In conclusion example 3.10 shows a simplified method of using calculation files to assess air conditioning cooling loads in buildings and costs around the world. The user can enter data from any source; and add

Table 3.3 The Shard in Dubai and London

Energy use	*London*	*Dubai*
AHU cooling load	8 MW	27 MW
Air conditioning electrical demand	4 MW	12 MW
Air conditioning annual cost	£3M	£9M

other climates and locations to make comparisons. As this is a book on air conditioning for buildings, it is not necessary to consider extreme cold climates here as these bring heating and ventilation requirements.

Design total irradiance

The design total solar irradiance upon a vertical surface I_{TV} comprises the direct line of sight I_{DV} the diffuse sky irradiance I_{dV} plus that reflected from the ground. The first suffix is for the total *T*, direct *D* or diffuse *d* irradiance. The second suffix denotes a horizontal surface *H* or a vertical wall *V*. There may be correction factors needed for altitude and ground reflectance. Ground reflected irradiance is 0.2 of the total of the direct and diffuse values in temperate and humid tropical localities but 0.5 in arid regions. Tabulated irradiance data used for design may be exceeded on 2.5% of occasions each month. The design total irradiance upon a vertical surface I_{TV} is:

$$I_{TV} = I_{DV} + I_{dV} \frac{W}{m^2}$$

$$I_{TV} = \text{design total irradiance on a vertical surface } \frac{W}{m^2}$$

These are found in CIBSE, (2006), Table A2.30 for the London area and have peak values for a vertical east wall at 07.30 h on 21 June,

$$I_{DV} = 579 \frac{W}{m^2}$$

$$I_{dV} = 189 \frac{W}{m^2}$$

$$I_{TV} = I_{DV} + I_{dV} \frac{W}{m^2}$$

$$I_{TV} = 579 + 189 \frac{W}{m^2}$$

$$I_{TV} = 768 \frac{W}{m^2}$$

Design total irradiance upon a horizontal surface I_{TH} is the sum of the direct and diffuse irradiances:

$$I_{TH} = I_{DH} + I_{dH} \frac{W}{m^2}$$

A roof at 12.30 h on 21 June in the London area has,

$$I_{TH} = I_{DH} + I_{dH} \frac{W}{m^2}$$

$$I_{TH} = 738 + 146 \frac{W}{m^2}$$

$$I_{TH} = 884 \frac{W}{m^2}$$

A sloping surface has a combination of I_{DH} and I_{DV} depending upon the inclination from the horizontal, angle S. When S is between 0°C and 90°C, the design direct irradiance upon the sloping surface I_{DS} is found from:

$$I_{DS} = I_{DH} \times \cos S + I_{DV} \times \sin S \ \frac{W}{m^2}$$

Only the direct irradiance has angular effects. The diffuse I_{dH} exists equally in all directions and is added to the direct components to form the total irradiance,

$$I_{TS} = I_{DS} + I_{dH} \ \frac{W}{m^2}$$

EXAMPLE 3.11

Prove the formula for the irradiance upon a sloping surface. Calculate the I_{TS} for a roof pitch of 20° facing south west at 14.00 h on 23 July in south east England when I_{TH} is 715 W/m^2 and I_{TV} is 580 W/m^2 and I_{dH} is 230 W/m^2.

First, find the direct solar irradiances on the horizontal and vertical surfaces from the tabulated totals:

$$I_{TH} = I_{DH} + I_{dH} \frac{W}{m^2}$$

$$I_{DH} = I_{TH} - I_{dH}$$

$$I_{DH} = 715 - 230$$

$$I_{DH} = 485 \frac{W}{m^2}$$

Second, find the direct solar irradiance on a vertical surface from,

$$I_{TV} = I_{DV} + I_{dH} \frac{W}{m^2}$$

$$I_{DV} = I_{TV} - I_{dH}$$

$$I_{DV} = 580 - 230$$

$$I_{DV} = 350 \frac{W}{m^2}$$

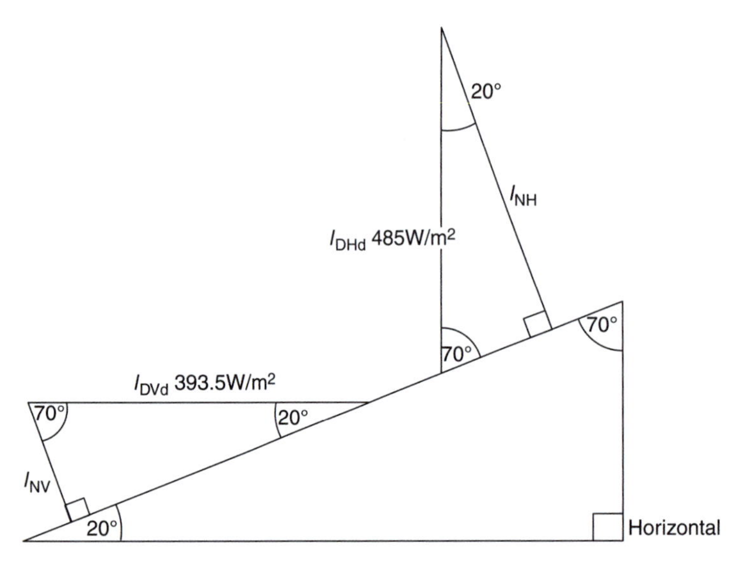

3.13 Finding the direct solar irradiance upon a sloping surface in example 3.11.

The direct irradiance normal to the sloping surface will be a combination of the horizontal and vertical components as shown in Figure 3.13. Let I_{NV} and I_{NH} be the components that are normal to the surface from the vertical and horizontal irradiances. The three triangles have equal angles.

$$\sin 20 = \frac{I_{NV}}{I_{DV}}$$

$$I_{NV} = I_{DV} \sin 20$$

$$\cos 20 = \frac{I_{NH}}{I_{DH}}$$

$$I_{NH} = I_{DH} \cos 20$$

The sum of I_{NV} and I_{NH} is the direct irradiance normal to the sloping surface, I_{DS}

$$I_{DS} = I_{NV} + I_{NH}$$

$$I_{DS} = I_{DV} \sin 20 + I_{DH} \cos 20$$

This proves the previously stated formula,

$$I_{DS} = I_{DV} \sin S + I_{DH} \cos S$$

$$I_{DS} = I_{DV} \sin 20 + I_{DH} \cos 20$$

$$I_{DS} = 350 \sin 20 + 485 \cos 20 \frac{\text{W}}{\text{m}^2}$$

$$I_{DS} = 575 \frac{\text{W}}{\text{m}^2}$$

The total irradiance on the sloping surface is the sum of the direct and diffuse values:

$$I_{TS} = I_{DS} + I_{dH} \ \frac{W}{m^2}$$

$$I_{TS} = 575 + 230 \ \frac{W}{m^2}$$

$$I_{TS} = 805 \ \frac{W}{m^2}$$

Sol-air temperature

Sol-air temperature t_{eo} is that outdoor temperature that, in the absence of solar irradiance, would give the same temperature distribution and rate of heat transfer through the wall or roof as exists with the actual outside air temperature and solar irradiance (CIBSE, 2006). Thus it is an artificial, calculated temperature and it is used in the calculation of steady and transient heat transfers into the building from the continually changing hot external climate. Solar heat gains to opaque, solid, building components, due to their radiation absorptivity, raise the temperature of the solid and this can easily be verified by touch. The heated surface transmits low temperature long wave radiation to nearby cooler surfaces and to the sky in proportion to its emissivity, $(-0.9I_l)$. The absorptivity and emissivity of dark coloured matt finishes on brick, concrete and asphalt are both 0.9 and the long wave radiation rate is 93 W/m^2 when the sky is cloudless at times of maximum solar heat gains to the building.

$$t_{eo} = t_{ao} + R_{so}\left(0.9I_{TH} - 0.9I_l\right) \ ^\circ\mathrm{C}$$

$$R_{so} = \text{external surface film thermal resistance}, 0.07 \frac{m^2\ K}{W}$$

EXAMPLE 3.12

The peak I_{TH} for a roof in south east England of 850 W/m^2 is expected to occur at noon on 21 June when the outdoor dry bulb air temperature t_{ao} is 19°C and the I_l is 93 W/m^2. Calculate the sol-air temperature for the roof.

$$t_{eo} = t_{ao} + R_{so}\left(0.9I_{TH} - 0.9I_l\right) \ ^\circ\mathrm{C}$$

$$t_{eo} = 19 + 0.07 \times (0.9 \times 850 - 0.9 \times 93) \ ^\circ\mathrm{C}$$

$$t_{eo} = 66.7^\circ\mathrm{C}$$

EXAMPLE 3.13

The I_{TV} for a south facing vertical wall in south east England of 375 W/m^2 occurs at 15.00 hours on 22 August when the outdoor dry bulb air temperature t_{ao} is 21.5°C and the I_l is 17 W/m^2. Calculate the sol-air temperature for the dark coloured wall.

$$t_{eo} = t_{ao} + R_{so}\left(0.9I_{TH} - 0.9I_l\right)°C$$

$$t_{eo} = 21.5 + 0.07 \times (0.9 \times 375 - 0.9 \times 17)°C$$

$$t_{eo} = 44°C$$

Heat transmission through glazing

Glass is formed from a molten mixture of 70% silica (SiO_2), 15% soda (Na_2O), 10% lime (CaO), 2.5% magnesia (MgO), 2.5% alumina (Al_2O_2) or other metallic elements to give it particular properties, such as gold (Au) and selenium (Se) to make it photosensitive. An oil-fired furnace liquefies the components at up to 1590°C that are then floated onto a bath of molten tin and progressively cooled. This is the float glass method in current use. Glass fibre and glass wool are made by passing the hot liquid through fine orifices. Glass is manufactured in 4, 6, 10 and 12 mm thicknesses as clear float, modified float, toughened and laminated depending upon type, safety requirements and wind loading. Clear float glass is the most common but body tinted grey, bronze, blue or green can be chosen for aesthetic, day-lighting or thermal reasons. Colour pigments are added during the molten stage and the full thickness of the glass is coloured. Metallic particles added to the surface during the fluid stage of production form a reflective surface on one side of the glass. Reflected light may give a silver colour rendering, but transmitted light may appear as bronze. Low emissivity surfaces improve the thermal insulation properties of the glass.

Glass is transparent to high temperature, high frequency solar irradiance but is poor at passing long wave radiation from low temperature surfaces within the building, thus creating the greenhouse effect. The glass surface that has the reflective coating is often located on the inner face of double glazed units for protection. Body tinted glass is good at absorbing solar irradiance and reducing heat flow into the building; however, increased glass temperature and thermal expansion are caused. Freon gas can be used to fill the space within sealed double glazing units as its reduced convection heat transfer ability lowers the U value.

The area of glazing used may be limited by statutory regulations for energy conservation reasons. Whether large or small areas are advantageous depends upon the annual balance between the adventitious solar heat gains, the internal heat generation from people, lights and equipment and the heat losses. The availability of natural internal illumination through the glazing is an important part of the energy balance, with reductions in the electrical energy used for artificial lighting. Whether large glass areas are advantageous or not can be analysed from the net energy consumption of the whole building on a monthly and annual basis.

Figure 3.14 shows the proportions of the solar energy flows through glass due to its three properties of transmissivity T, absorptivity A and reflectivity R. Glasses have different transmittances for direct and diffuse irradiance. Some irradiance that is absorbed by the glass is radiated into the room and a total transmittance is given. These properties remain constant until the solar incidence angle exceeds about 45° when they rapidly diminish to zero at 90° incidence. Double glazing may consist of tinted or reflective glass in the exterior layer and a clear float inner pane. The outer glass is raised to a higher temperature by being insulated from the cooler interior of the building and sustains greater thermal expansion. The overall direct solar transmittance will be the multiplication of the two glass types such as 0.86 for the clear and 0.23 for the reflective, equalling 0.2. The outward view through tinted glass is not seriously affected and may be hardly noticeable. Reflective glass provides a high degree of internal privacy for the occupants of the building. The exception is at night when interior lighting and external darkness causes the viewing direction to be reversed. The interior is clearly visible from the street. Typical values are given in Table 3.4.

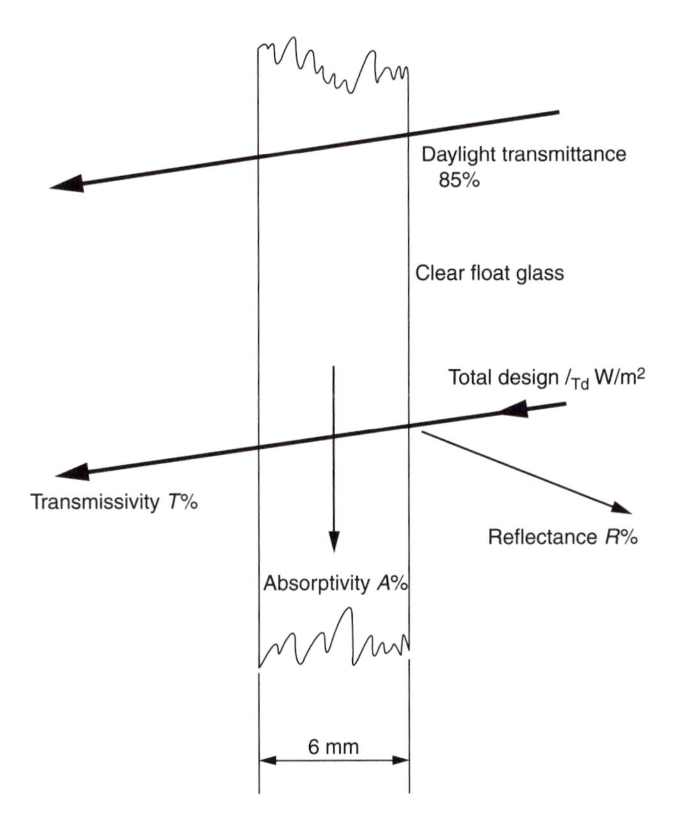

3.14 Properties of glass.

Table 3.4 Performance data for glass

Glass type	*Light*	*Direct solar irradiance*			*Total*
	T	*T*	*A*	*R*	*T*
4 mm clear	0.89	0.82	0.11	0.07	0.86
6 mm tinted	0.5	0.46	0.49	0.05	0.62
6 mm reflective	0.1	0.08	0.6	0.32	0.23

Data in this table are approximate and for use within this book only

EXAMPLE 3.14

A window facing south west is 3 m long and 2.75 m high and has a shaded area of 0.2 m^2. Solar altitude is 43.5°, solar azimuth is 66° west of south, direct solar irradiance normal to the sun is 832 W/m^2 and diffuse irradiance is 43 W/m^2. The transmissions of direct and diffuse irradiances are 0.85 and 0.8. Calculate the solar heat gain transmitted through the glass.

Glass azimuth is 45° west of south

$$\text{Glass solar azimuth} = 66° - 45°$$

$$\text{Glass solar azimuth} = 21°$$

$$I_{DV} = I_D \cos A \cos D$$

$$I_{DV} = 832 \times \cos 43.5 \cos 21 \frac{\text{W}}{\text{m}^2}$$

$$I_{DV} = 563.4 \frac{\text{W}}{\text{m}^2}$$

$$\text{Direct transmitted irradiance, } Q_D = 0.85 \times I_{DV} \frac{\text{W}}{\text{m}^2} \times \text{sunlit area m}^2$$

$$\text{Direct transmitted irradiance, } Q_D = 0.85 \times 563.4 \frac{\text{W}}{\text{m}^2} \times ((3 \times 2.75) - 0.2)\ \text{m}^2$$

$$\text{Direct transmitted irradiance, } Q_D = 3855.1\ \text{W}$$

$$\text{Diffuse transmitted irradiance, } Q_d = 0.8 \times 43 \frac{\text{W}}{\text{m}^2} \times \text{glass area m}^2$$

$$\text{Diffuse transmitted irradiance, } Q_d = 0.8 \times 43 \frac{\text{W}}{\text{m}^2} \times 3 \times 2.75\ \text{m}^2$$

$$\text{Diffuse transmitted irradiance, } Q_d = 283.8 W$$

$$\text{Total transmitted heat gain} = Q_D + Q_d\ \text{W}$$

$$\text{Total transmitted heat gain} = 3855.1 + 283.8\ \text{W}$$

$$\text{Total transmitted heat gain} = 4138.9\ \text{W}$$

Absorbed solar irradiance causes the glass to rise in temperature above that of its surroundings. Air convection currents either side of the glazing remove the heat gained and under steady state conditions the glass remains at a constant temperature. However, as the irradiance is cyclic, stable conditions may be brief. Figure 3.15 shows the temperature gradient through glazing and a heat balance can be made if it is assumed that the glass is at a higher temperature than the adjacent air.

Consider 1 m^2 of glazing,

$$\text{Absorbed heat} = \text{heat lost to surroundings}$$

$$A \times I_{TV} = h_{so} \times (t_g - t_{ao}) + h_{si} \times (t_g - t_{ai})$$

Typically,

$$R_{so}\ 0.06\ \text{m}^2\text{K/W}, \text{ so } h_{so}\ 16.7\ \text{W/m}^2\ \text{K}$$

$$R_{si}\ 0.12\ \text{m}^2\text{K/W}, \text{ so } h_{si}\ 8.3\ \text{W/m}^2\ \text{K}$$

(CIBSE, 2006, Section A3)

$$A \times I_{TV} = 16.7 \times (t_g - t_{ao}) + 8.3 \times (t_g - t_{ai})$$

Rearrange the equation to find the glass temperature,

$$A \times I_{TV} = 16.7 \times t_g - 16.7 \times t_{ao} + 8.3 \times t_g - 8.3 \times t_{ai}$$

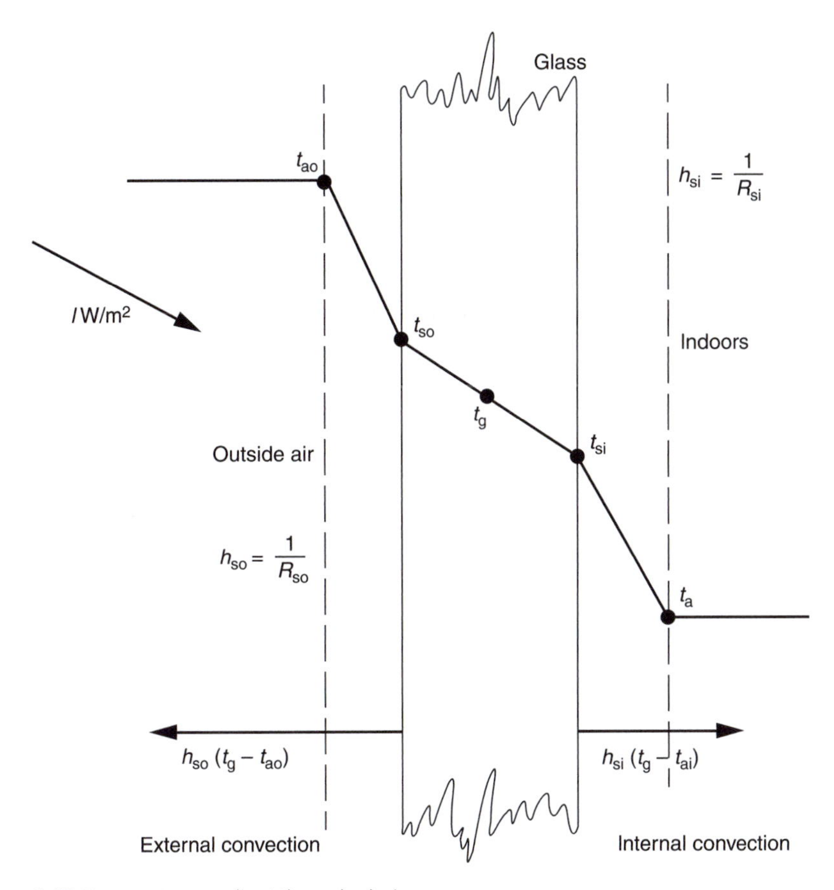

3.15 Temperature gradient through glazing.

Gather the t_g terms together,

$$A \times I_{TV} = 16.7 \times t_g + 8.3 \times t_g - 16.7 \times t_{ao} - 8.3 \times t_{ai}$$

$$A \times I_{TV} = (16.7 + 8.3)t_g - 16.7 \times t_{ao} - 8.3 \times t_{ai}$$

$$A \times I_{TV} = 25t_g - 16.7 \times t_{ao} - 8.3 \times t_{ai}$$

$$25t_g = A \times I_{TV} + 16.7 \times t_{ao} + 8.3 \times t_{ai}$$

$$t_g = \frac{A \times I_{TV} + 16.7 \times t_{ao} + 8.3 \times t_{ai}}{25}$$

EXAMPLE 3.15

Find t_g for 6 mm clear float glass that is in a total solar irradiance of 640 W/m^2. External and internal air temperatures are 29°C and 22°C. Absorptivity A of the glass is 0.15 and the transmissibility T is 0.78. Calculate the heat flow into the room.

$$t_g = \frac{A \times I_{TV} + 16.7 \times t_{ao} + 8.3 \times t_{ai}}{25}$$

$$t_g = \frac{0.15 \times 640 + 16.7 \times 29 + 8.3 \times 22}{25}$$

$$t_g = 30.5°C$$

It is worth noting that the glass temperature is higher than the mean of the air temperatures $(29+22)/2 = 25.5°C$ due to the absorbed irradiance. Highly heat absorbing 10 mm grey glass, where A is 0.71, would have a t_g of 44.9°C. When the glass temperature is greater than that of the external air, there can be no U value convection and conduction heat gain into the room, only direct and diffuse radiation heat flow.

The heat flow rate into the room through the glass Q is:

$$Q = h_{si}(t_g - t_{ai}) + T\, I_{TV}$$

$$Q = 8.3(30.5 - 22) + 0.78 \times 640 \frac{W}{m^2}$$

$$Q = 570 \frac{W}{m^2}$$

Q is a fraction of the available irradiance $(570/640) = 0.89$.
Glass manufacturers may describe this as the total transmittance.

Further data on glass characteristics, combinations of glasses and the effect of internal and exterior shading devices is given in CIBSE (2006) page A5-16. Tabulated values of the plant cooling load due to vertical glazing to maintain a constant internal dry resultant temperature in the UK are given in the CIBSE, (2006) pages A5-37 to A5-48. These are for lightweight buildings, which are the most common modern design, and 10 hour cooling plant operation in each 24 hour period, used in commercial applications. The tables are for unshaded 6 mm clear glazing and the same with the intermittent use of light coloured slatted blinds. Applications such as hospitals, manufacturing and containment buildings, where continuous air conditioning is used, require other data.

Heat gains through the opaque structure

The opaque structural elements of the building, walls and roof store solar heat gains for periods of up to 12 hours prior to having a significant influence upon the resultant internal temperature. Their brick, concrete, steel and masonry gradually rise in temperature during prolonged periods of hot sunny weather and remain warm. The time difference between the solar irradiance causing a heat gain to the outside surface of the structure and that same heat flow into storage, producing a corresponding heat flow into the interior, is important to the designer. The effect of cool evenings can be to produce an outflow of heat from the room, into the wall storage, at the time of peak solar heat gain into the room. The heat exchange between the room air and a dense wall at noon will be the result of heat loss from the outside wall surface to the external air at, perhaps, 9 h earlier, 03.00 h.

The CIBSE method is to calculate the 24 hour average heat flow through the structure and then the cyclic variation from the mean. The cyclic variation is calculated hourly and added to, or subtracted from, the mean heat flow. Thermal transmittance U, $W/m^2\,K$, is used to calculate the mean flows and admittance,

Y, W/m^2 K, is used for the periodic oscillations. The mean conduction heat gains, or losses through an opaque structure Q_u are:

$$Q_u = \frac{F_u}{F_v} AU(t_{eo} - t_c)$$

$$F_u = \frac{18\sum A}{18\sum A + \sum(AU)}$$

$$F_v = \frac{6\sum A}{6\sum A + 0.33NV}$$

t_{eo} = 24 hour mean sol-air temperature

t_c = 24 hour mean internal resultant temperature

A = surface area, m^2

U = thermal transmittance, W/m^2 K

$\sum A$ = summation of the room surface areas, m^2

N = room air infiltration ventilation rate, air changes per hour

V = room volume, m^3

The cyclic variation from the mean heat flow $\tilde{Q}_u$ is,

$$\tilde{Q}_u = \frac{F_y}{F_v} AUf\tilde{t}_{eo}$$

$$F_y = \frac{18\sum A}{18\sum A + \sum(AY)}$$

Where f = decrement factor, the ratio of the heat flow through the structure to the 24 hour mean heat flow. Glazing f is unity while dense concrete walls and roofs may be as low as 0.2 due to their energy storage and release to outdoors and indoors.
$\tilde{t}_{eo}$ = swing in the environmental temperature between that when the heat flow occurs at the external surface, and the 24 hour mean environmental temperature. The swing will be negative when t_{eo} at the outdoor time of the heat gain is less than the 24 hour mean t_{eo} and shows a loss of heat from the room. The time difference between these events is the time lag for the structure and this corresponds to the decrement factor reduction in the energy flows.
Y = thermal admittance, W/m^2 K, is the rate of heat flow into the structure. It is the same as the U value for glass but higher than the U value for insulated, dense structures.
F = surface factor that is the ratio used for internal walls, floors and ceilings, and it relates to the variation in heat flows between rooms.

The net heat flow through the structure is the sum of the 24 hour mean heat flow and the cyclic variation,

$$Q = Q_u + \tilde{Q}_u$$

EXAMPLE 3.16

An existing flat roof of 10 m × 8 m is over an office that is to be maintained at a resultant temperature of 21°C by an air conditioning system. The thermal transmittance of the roof is 2.5 W/m^2 K and the admittance is 7 W/m^2 K. The 24 hour average outdoor environmental temperature on 21 June in Basingstoke is 26°C. The roof decrement factor is 0.4 and its time lag is 7 hours. The outdoor environmental temperature at 05.00 h is 11.5°C. The room height is 3 m. There are 1.5 air changes per hour due to the infiltration of outdoor air. Calculate the room cooling load through the roof for noon.

$$\text{Roof area } \sum A = 10\ \text{m} \times 8\ \text{m}$$

$$\text{Roof area } \sum A = 80\ \text{m}^2$$

$$\text{Room volume } V = 10\ \text{m} \times 8\ \text{m} \times 3\ \text{m}$$

$$\text{Room volume } V = 240\ \text{m}^3$$

$$Q_u = \frac{F_u}{F_v} AU(t_{eo} - t_c)$$

$$F_u = \frac{18\sum A}{18\sum A + \sum(AU)}$$

$$F_u = \frac{18 \times 80}{18 \times 80 + 18 \times 2.5}$$

$$F_u = 0.88$$

$$F_v = \frac{6\sum A}{6\sum A + 0.33NV}$$

$$F_v = \frac{6 \times 80}{6 \times 80 + 0.33 \times 1.5 \times 240}$$

$$F_v = 0.8$$

Mean 24 hour cooling load Q_u:

$$Q_u = \frac{F_u}{F_v} AU(t_{eo} - t_c)$$

$$Q_u = \frac{0.88}{0.8} \times 80 \times 2.5 \times (26 - 21)\ \text{W}$$

$$Q_u = 1100\ \text{W}$$

Heat flow into the office at noon is due to the conditions 7 hours earlier, at 05.00 h. Swing in the environmental temperature between 05.00 h and noon is:

$$\tilde{t}_{eo} = (11.5 - 26)°\text{C}$$

$$\tilde{t}_{eo} = -14.5°\text{C}$$

$$\tilde{Q}_u = \frac{F_y}{F_v} A U f \tilde{t}_{eo}$$

$$F_y = \frac{18 \sum A}{18 \sum A + \sum (AY)}$$

$$F_y = \frac{18 \times 80}{18 \times 80 + 80 \times 7}$$

$$F_y = 0.72$$

$$\tilde{Q}_u = \frac{0.72}{0.8} \times 80 \times 2.5 \times 0.4 \times (-14.5)$$

$$\tilde{Q}_u = -1044W$$

Net heat flow through roof, $Q = Q_u + \tilde{Q}_u$

Net heat flow through roof, $Q = (1100 - 1044)$ W

Net heat flow through roof, $Q = 56$ W

Plant cooling load

Cooling equipment in each room or building module is designed to remove the peak heat gain that occurs in that location. This peak load is found from a total of:

1. transmitted solar heat gain through the glazing;
2. U value convection and conduction heat gains through the glazing when appropriate;
3. infiltration of warmer outdoor air directly into the room;
4. net heat flows into the room through the opaque structure;
5. heat output from the occupants;
6. electrical power input to lighting systems;
7. electrical power input to office equipment;
8. power input to electrical motors;
9. heat output from machinery or industrial process equipment;
10. heat content of high temperature items that are transported into the air conditioned room;
11. heat gains through the structure from adjacent warmer rooms;
12. ventilation air that enters from adjacent warmer rooms.

Transmitted heat gains from the direct and diffuse solar irradiances through the glazing are analysed as described earlier. Additional heat gains due to conduction and convection through the windows from the warmer outdoor air occur when the glass temperature is sufficiently low. The heat gain through the glazing is found from:

$$Q_u = \frac{F_u}{F_v} A U (t_{ao} - t_c)$$

t_{ao} = outdoor air temperature °C

Decrement factor for glass f is 1, and the time lag is zero.

Infiltration of outdoor air can be mainly avoided if the ducted air conditioning system supplies more fresh air into the building than it mechanically extracts. This results in positive pressurisation of the air within the building and causes leakage of air outwards. Under this circumstance, all the fresh outdoor air that is required for ventilation, is supplied through the air handling plant and is a cooling load upon the chilled water coil there. This cooling duty is found from the psychrometric chart calculations for the air handling plant. There is no fresh air cooling load within the room or its air conditioning unit. The alternative is to extract more air from the building than is mechanically supplied to create a negative static air pressure within it. Warm outdoor air will leak inwards and create a ventilation cooling load on the room terminal. Such depressurisation is used in containment buildings where it is essential to avoid the escape of contaminated air or particles and all the extract air is filtered to trap the dangerous or expensive matter. This is unlikely to be desirable for human comfort air conditioning, as untreated cold and hot air are continually drawn inwards. Where outdoor air is allowed to infiltrate the conditioned space, the instantaneous ventilation heat gain is calculated from:

$$Q_v = 0.33F_2NV(t_{ao} - t_c)$$

$$F_2 = 1 + \frac{F_u \sum AU}{6\sum A}$$

Heat output from room occupants is 90 W of sensible heat from an adult office worker in an air temperature of 22°C plus 50 W that is the latent heat equivalent of the evaporated moisture per person. Only the sensible heat is added to the cumulative gains for the room. The latent component is just moisture and does not influence the temperature of the room, only the person who is evaporating it. The designer needs to know how many occupants there will be and what their activity levels will be, as sensible heat output varies strongly with metabolic rate. Heavy manual work produces up to 220 W sensible and 220 W latent heat outputs but these are likely to be for short periods of time. Hourly and daily schedules of occupancy heat gains are needed.

The electrical power input to lighting systems, photocopiers, computers, facsimile machines, vending machines and any other items is emitted into the conditioned room as an equal amount of heat. This includes the visible light energy, that is electro-magnetic radiation. Visible radiation is absorbed by the room surfaces by multiple reflections and an amount of energy is absorbed on each reflection depending upon the surface absorptivity. A proportion of the light energy may be radiated out of the building through windows and doorways. The control equipment of each luminaire emits heat and so does the wiring. Some or all of the lighting system heat output might be carried away from the room with ventilated luminaires that have room air directly extracted through them.

Fans, pumps, lift motors and other electric motor applications all dissipate heat into their surroundings and into the fluid being blown or pumped. Their efficiency is usually less than 50% and this proportion of their input kVA power is released into the conditioned room. Other machinery or industrial process equipment is considered similarly.

High temperature items may be transported into the air conditioned room and their heat content that is above that of the room air temperature can be calculated from their mass and temperature. Other fuel burning appliances such as gas-fired furnaces, water heaters or incinerators will have some of their input energy dissipated into the room, though most of the waste heat will be flued.

Adjacent rooms that are at higher temperature provide heat flows through the structure into the conditioned room and these are *U* value, area and temperature difference calculations hourly. Openings into adjacent rooms and corridors cause ventilation air movements that can lead to heat exchanges with the conditioned space. When static pressure differences exist across doorways and openings, the air flow rate is calculated from the area of opening and its flow coefficient depending upon its shape. The time duration of the opening allows calculation of the total quantity of air transferred. Otherwise, flow through the doorway is created by natural convection and an estimate can be made. Some flows between rooms will be deliberately created, such as from occupied rooms, through corridors, into toilet accommodation or kitchens and then extracted through ductwork.

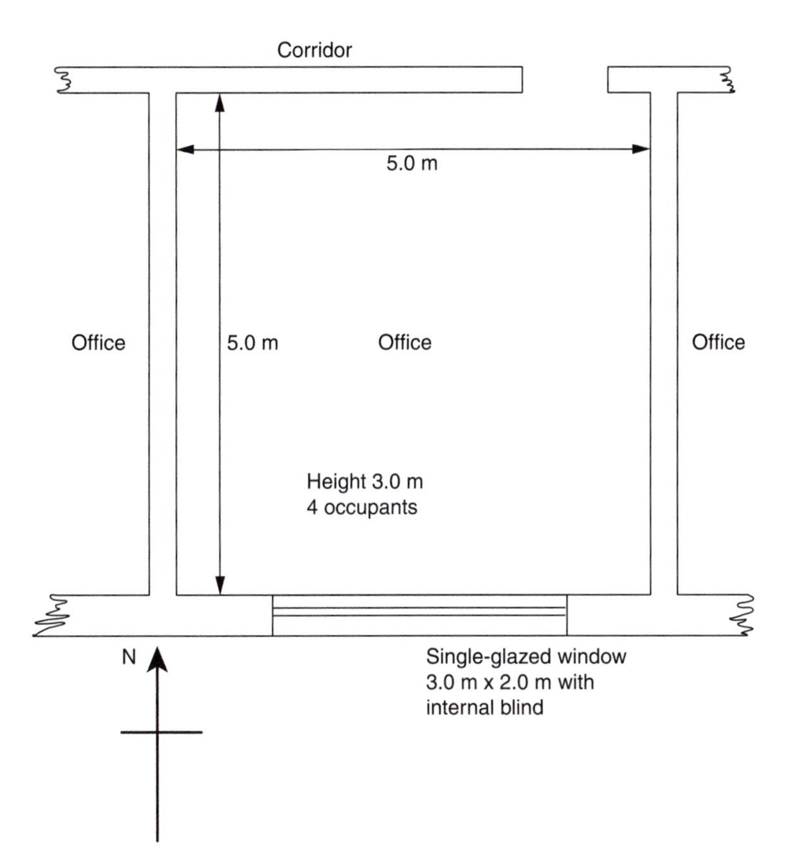

3.16 Plan of the office in example 3.16.

EXAMPLE 3.17

A south facing top floor Southampton office is shown in Figure 3.16. The exposure is normal. The office is to be maintained at a resultant temperature of 21°C. There are four sedentary occupants and two continuously used computers having power consumptions of 200 W each. The air conditioning plant operates for 10 hours per day. There are 1.25 air changes per hour due to natural infiltration of outdoor air. The adjacent offices, corridor and office below are maintained at the same temperature as the example office. The constructional details are:

(a) Double glazed window, clear float glass in an aluminium frame with thermal break and internal white coloured venetian blinds;
(b) External wall 100 mm heavyweight concrete block, 75 mm glass fibre, 100 mm lightweight concrete block, 13 mm lightweight plaster, dark exterior colour;
(c) Internal walls 100 mm medium weight concrete block, 13 mm lightweight plaster both sides;
(d) Floor 50 mm screed, 150 mm cast concrete, 25 mm wood block floor;
(e) Roof 19 mm asphalt, 13 mm fibreboard, 25 mm air gap, 75 mm glass fibre, 10 mm plasterboard, dark exterior colour.

Table 3.5 Heat transfer data for example 3.17

Surface	*A*, m^2	*U* W/m^2K	*AU*	*Y* W/m^2K	*AY*	*f*	*Lag, h*
Glass	6	3.3	19.8	3.3	19.8	1	0
External wall	9	0.33	3	2.4	21.6	0.35	9
Internal wall	45	1.7	0	3.5	157.5	0.72	1
Floor	25	1.5	0	2.9	72.5	0.7	1
Roof	25	0.4	10	0.7	17.5	0.99	1
	$\sum A = 110$		$\sum AU = 32.8$		$\sum AY = 288.9$		

Calculate the peak cooling load for the room air conditioning unit.

Suitable values for thermal data for each surface are shown in Table 3.5.

Note that the internal walls and floor have no heat exchange with their adjacent rooms, so *AU* is zero. These internal surfaces do not have solar radiation decrement factors *f* or time lags and surface factor *F* is used.

$$\text{Office volume} = 5\ \text{m} \times 5\ \text{m} \times 3\ \text{m}$$

$$\text{Office volume} = 75\ \text{m}^3$$

Find the cooling load due to the south facing double clear float glazing for a lightweight building at 51.7°N latitude with the intermittent use of internal slatted blinds. South facing glazing always has a peak cooling load at noon sun time and the peak for the year occurs on 22 September and 21 March at 333 W/m^2. A correction factor of 0.74 applies when the blinds are closed. Notice the effects of different glasses, single glazing and the use and position of the blinds from the correction factors.

$$\text{Solar gain through window} = 333 \frac{\text{W}}{\text{m}^2} \times 0.74 \times 6\ \text{m}^2$$

$$\text{Solar gain through window} = 1479\ \text{W}$$

Find the 24 hour mean sol-air temperatures and the sol-air temperatures at the time of the occurrence of the solar irradiance to the external surfaces. Table 3.6 shows the results. The time of irradiation of the wall is 9 h before noon, 03.00 h and the t_{eo} then is 10.5°C. The swing in the t_{eo} between the 24 hour mean and the 03.00 h value is −14.5°C. Similarly, at 11.00 h for the roof, the t_{eo} is 38°C and this is 18°C above the 24 hour mean. The window is subject to air temperature heat transfers not sol-air temperatures and it has zero time lag. The outdoor air is at 18.5°C at noon that is a swing of 3°C above the 24 hour mean.

$$F_u = \frac{18\sum A}{18\sum A + \sum (AU)}$$

Table 3.6 Sol air-data for example 3.16

Surface	*lag, h*	*24 h* t_{eo}	*24 h* t_{ao}	*time, h*	t_{eo}	*swing* $\tilde{t}_{eo}$
External wall	9	25	15.5	03.00	10.5	−14.5
Roof	1	20	15.5	11.00	38	18
					t_{ao}	$\tilde{t}_{ao}$
Window	0	—	15.5	12.00	18.5	3

$$F_u = \frac{18 \times 110}{18 \times 110 + 32.8}$$

$$F_u = 0.98$$

$$F_v = \frac{6\sum A}{6\sum A + 0.33NV}$$

$$F_v = \frac{6 \times 110}{6 \times 110 + 0.33 \times 1.25 \times 75}$$

$$F_v = 0.896$$

$$F_y = \frac{18\sum A}{18\sum A + \sum(AY)}$$

$$F_y = \frac{18 \times 110}{18 \times 110 + 288.9}$$

$$F_y = 0.97$$

$$F_2 = 1 + \frac{F_u \sum AU}{6\sum A}$$

$$F_2 = 1 + \frac{0.98 \times 32.8}{6 \times 110}$$

$$F_2 = 1.05$$

Calculate the 24 hour mean conduction heat gains with:

$$Q_u = \frac{F_u}{F_v} AU(t_{eo} - t_c)$$

For the external wall,

$$Q_u = \frac{0.98}{0.96} \times 9 \times 0.33 \times (25 - 21)$$

$$Q_u = 12 \text{ W}$$

For the roof,

$$Q_u = \frac{0.98}{0.96} \times 25 \times 0.4 \times (20 - 21)$$

$$Q_u = -10 \text{ W}$$

For the window, use $(t_{ao} - t_c)$,

$$Q_u = \frac{0.98}{0.96} \times 6 \times 3.3 \times (15.5 - 21)$$

$$Q_u = -111 \text{ W}$$

$$\text{Mean 24 h conduction gain} = (12 - 10 - 111) \text{ W}$$

$$\text{Mean 24 h conduction gain} = -109 \text{ W, a net loss.}$$

Cyclic variation from the mean heat flow is,

$$\tilde{Q}_u = \frac{F_y}{F_v} AUf\tilde{t}_{eo}$$

For the external wall,

$$\tilde{Q}_u = \frac{0.87}{0.96} \times 9 \times 0.33 \times 0.35 \times (-14.5) \text{ W}$$

$$\tilde{Q}_u = -14 \text{ W}$$

For the roof,

$$\tilde{Q}_u = 0.91 \times 25 \times 0.4 \times 0.99 \times 18 \text{ W}$$

$$\tilde{Q}_u = 162 \text{ W}$$

For the window, use $\tilde{t}_{ao}$.

$$\tilde{Q}_u = 0.91 \times 6 \times 3.3 \times 1 \times 3 \text{ W}$$

$$\tilde{Q}_u = 54 \text{ W}$$

Net swing in conduction gain $= (-14 + 162 + 54)$ W

Net swing in conduction gain $= 202$ W

The outdoor air that is at 18.5°C at noon infiltrates the conditioned office causing a heat gain of,

$$Q_v = 0.33 F_2 NV (t_{ao} - t_c)$$

$$Q_v = 0.33 \times 1.05 \times 1.25 \times 75 \times (18.5 - 21) \text{ W}$$

$$Q_v = -81 \text{ W}$$

The four occupants each emit 90 W,

$$Q = 4 \times 90 \text{ W}$$

$$Q = 360 \text{ W}$$

Two computers each emit 200 W,

$$Q = 2 \times 200 \text{ W}$$

$$Q = 400 \text{ W}$$

Total net heat gain Q to the office at noon 22 September and 21 March is the sum of:

1. Solar gain through the glazing, 1479 W;
2. 24 hour mean conduction through the structure, −109 W;
3. Net swing in the conduction gain, 202 W;

4. Ventilation air infiltration, −81 W;
5. Occupancy gain, 360 W;
6. Electrical equipment emission, 400 W;

$$Q = (1479 - 109 + 202 - 81 + 360 + 400)\ \text{W}$$

$$Q = 2251\ \text{W}$$

A close approximation to this answer could have been acquired by ignoring the conduction gains through the structure and the cooling effect of the infiltration, 2239 W; however, this may not be appropriate in every case.

Energy used by an air conditioning system

Our propensity to build glass towers for work, leisure and living accommodation in almost any climate of the world brings with it the need for mechanical ventilation and refrigeration for cooling. As we discussed in relation to peak summertime temperature, we expect air conditioned travel, work places, shops, restaurants, hotels and, increasingly, homes. In hot to tropical climates, where outdoor air is often in the 28°C to 40°C range, along with intense solar radiation, we change the interior climate with powered systems that are said to be changing the external climate to our detriment. Small packaged air conditioning units, multi-zone direct expansion split systems and all the way up to central chilled water plant, all consume large amounts of power at times of peak electrical demand. Air conditioning uses large amounts of peak electrical energy. There are ways to mitigate peak hours energy kWh, peak demand kW and costs, but the fact of high energy use remains.

What are the consumers of energy in a system? An energy auditor lists all the fan, pump and refrigeration compressor plant data, as these are the users of energy. Automatic control systems do use some energy to drive their computers, valves and damper motors, but these are very small amounts and do not concern the energy auditor. Refrigeration, and evaporative cooling, is a power demand from electric motors. When gas-fired absorption refrigeration plant is used, electrical demand is reduced. Plant operation hours depend on the application. Hospitals and other 24 hour use buildings have plant running all the time. Commercial buildings may have plant switched off outside working hours. Homes, entertainment theatres, hotels and shops have variable running times, while large shopping supermarkets and malls run continuously. Fans often run at full speed and power for the duration of the system operation. These include the fans on cooling towers and also the condenser water pump that often circulates 35°C water from the basement to the roof of a tower building. Fans on dry air condensing units or heat exchangers may have multiple fans that only run when needed. Variable frequency controllers moderate speed and power use in accordance with the demand for air flow and temperature control. Demand controlled ventilation, DCV, may save energy by measurement of CO_2, or other parameters, to minimise power use in response to occupancy fluctuations.

Around 60% of the total annual energy used by hospitals, universities and schools in warm climates is from the air conditioning cooling system, while the energy used by air conditioning heating is around 80%. Energy use, peak electrical demand and the annual cost of running air conditioning systems are of major concern. That is, of major concern to the building owner, public electrical distribution system, national power station capacity providers and HM Government Carbon Plan (2011) objectives.

Bourke Street case study

A 14-storey mixed academic and commercial building, known to the author, in the warm climate of Melbourne, has 28 air handling units (AHU) delivering 145 m^3/s of conditioned air *SA* for 16 hours/day,

7 days/week and 365 days/year. Observe 235 Bourke Street from above and the street in Google Earth. Notice three cooling towers, an air cooled condenser water heat exchanger, lift motor room and stairway pressurisation fan room on the roof. A large centrifugal water chilling compressor is in the plant room on the roof, as well as gas-fired heating boilers, a hot water service boiler and the building management system control panels. Data provided here are fictionalised. Each floor has an area of 2200 m^2 and a room height of 2.8 m. 200 people occupy each floor, emitting 90 W each. There are 50 computer workstations of 200 W each on each floor. There is an average of 10 kW of lighting running on each floor. Total nameplate motor power for all the AHU fans is 300 kW. Motors are loaded at an expected 80% of their rated capacity. The minimum outside air intake quantity *OA* is 30 m^3/s. Leaks from the building Q_L amount to 5 m^3/s. Each floor has a single duct system with outside air, exhaust air and return air modulating dampers as shown in Figure 7.8. All air temperatures in these calculations are dry bulb. The system design supply air temperature from the air handling units in summer has a lower limit of 15°C and a winter maximum of 28°C in order to maintain room air temperature in the range 22°C to 24°C.

An economy air controller makes the maximum use of outside air to heat or cool the building as appropriate to the weather and internal air conditions. Outdoor winter design air temperature is −1°C. Outdoor summer design air temperature is 34.6°C. Outdoor air is admitted by the economy controller in variable quantities without any heating or cooling applied when the outdoor air is in the range 18°C to 22°C.

For 24 hour, 365 day building running time, there are 6677 annual hours when the outside air is in the range −1°C to 18°C, 1185 hours when the outside air is in the range 18°C to 22°C, and 898 hours when the outside air is in the range 22°C to 34.6°C; a total of 8760 hours in the year. There are several days in a typical year when the outside air rises to 45°C.

Assess the mean value of the sinusoidal variation in outdoor air temperature as 70.7% of the maximum temperature difference, in the same way as with an alternating current. Specific heat capacity of air is taken as 1.22 kJ/m^3 K. The gas-fired hot water heating system has an overall efficiency of 70%. The overall coefficient of performance, COP, of the chilled water refrigeration system, including water chillers, pumps and cooling towers, is 2.5. Ignore changes of humidity for the plant loads in this example. Natural gas costs 5 p/kWh, electricity costs 10 p/kWh plus, £10/kW per month chargeable demand.

The downloadable workbook *AC Energy Use.xls* is provided for this example and other questions.

Calculate peak heating and cooling capacity required for the plant, the annual energy consumption for the air conditioning system and how much the energy costs. Comment on what happens when the external air temperature rises to its occasional 45°C.

The peak design cooling load occurs at t_{ao} of 34.6°C and t_{ai} of 24°C where the intake of outdoor air *OA* is set at its minimum. Heating plant load is determined from t_{ao} of −1°C and t_{ai} of 22°C.

$$\text{Recycled air to the AHUs, } RA = SA - OA$$

$$\text{Recycled air to the AHUs, } RA = (145 - 30)\,\frac{m^3}{s}$$

$$\text{Recycled air to the AHUs, } RA = 115\,\frac{m^3}{s}$$

$$\text{Mixed air } = \frac{OA \times t_{ao} + RA \times t_{ai}}{SA}\,^{\circ}\text{C}$$

$$\text{Winter mixed air } = \frac{30 \times (-1) + 115 \times 22}{145}\,^{\circ}\text{C}$$

$$\text{Winter mixed air } = 17.2^{\circ}\text{C}$$

$$\text{AHU heating load} = 145\,\frac{m^3}{s} \times 1.22\,\frac{kJ}{m^3 K} \times (28 - 17.2)\,K \times \frac{1\ kWs}{1\ kJ}$$

$$\text{Peak AHU heating load} = 1911 \text{ kW}$$

$$\text{Summer mixed air } = \frac{30 \times 34.6 + 115 \times 24}{145} \text{°C}$$

$$\text{Summer mixed air } = 26.2\text{°C}$$

$$\text{AHU cooling load} = 145\frac{\text{m}^3}{\text{s}} \times 1.22\frac{\text{kJ}}{\text{m}^3\text{K}} \times (26.2 - 15)\text{K} \times \frac{1 \text{ kWs}}{1 \text{ kJ}}$$

$$\text{Peak AHU cooling load} = 1981 \text{ kW}$$

$$\text{Electrical input to refrigeration plant} = 1981 \text{ kW} \times \frac{1}{2.5}$$

$$\text{Electrical input to refrigeration plant} = 792 \text{ kW}$$

This electrical input includes the chilled water refrigeration compressors, cooling tower fans, chilled water pumps and condenser cooling water pumps.

$$\text{AHU fans electrical demand} = 80\% \times 300 \text{ kW}$$

$$\text{AHU fans electrical demand} = 240 \text{ kW}$$

There are 50 computer workstations of 200 W each, 10 kW of lighting and 2 kW of office machines, running on each floor.

$$\text{Internal electrical demand} = 14 \times \left(50 \times \frac{200}{10^3} + 2 + 10\right) \text{kW}$$

$$\text{Internal electrical demand} = 308 \text{ kW}$$

$$\text{Calculated peak demand} = (792 + 240 + 308) \text{ kW}$$

$$\text{Calculated peak demand} = 1340 \text{ kW}$$

This does not include lift motors, lift motor room ventilation fan, basement car park lighting and ventilation, the window cleaning crane, communications power, stairways fans, fire pumps, water service pressurisation pumps, refrigeration, catering equipment, cleaning equipment, general purpose power, external lighting, security systems, powered doors, ground floor retail or other ancillary items. The metered maximum electrical demand for this building was 1600 kW in February, so our assessment by calculation is sufficiently close for the purpose of this example.

$$\text{Building volume } V = 14 \times 2200 \text{ m}^2 \times 2.8 \text{ m}$$

$$\text{Building volume } V = 86240 \text{ m}^3$$

$$\text{Air change rate } N = 145\frac{\text{m}^3}{\text{s}} \times \frac{1 \text{ air change}}{86240 \text{ m}^3} \times \frac{3600 \text{ s}}{1 \text{ h}}$$

$$\text{Air change rate } N = 6\frac{\text{air changes}}{\text{h}}$$

This is an expected minimum air change rate for a mechanically ventilated building.

When the supply air reaches the occupied spaces, it is capable of matching a room heat requirement of:

$$\text{Room heating load} = 145\frac{\text{m}^3}{\text{s}} \times 1.22\frac{\text{kJ}}{\text{m}^3\text{K}} \times (28 - 22)\text{K} \times \frac{1\ \text{kWs}}{1\ \text{kJ}}$$

$$\text{Room heating load} = 1061\ \text{kW}$$

This is the figure produced from calculation of the room heat loss through the external surfaces and does not include the heat load of the outside air intake and what happens in the AHUs.

When the supply air reaches the occupied spaces, it is capable of matching a room cooling requirement of:

$$\text{Room cooling load} = 145\frac{\text{m}^3}{\text{s}} \times 1.22\frac{\text{kJ}}{\text{m}^3\text{K}} \times (24 - 15)\text{K} \times \frac{1\ \text{kWs}}{1\ \text{kJ}}$$

$$\text{Room cooling load} = 1592\ \text{kW}$$

This is the figure produced from calculation of the room heat gains through the external surfaces, occupants, lighting and office electronic equipment. It does not include the heat load of the outside air intake and what happens in the AHUs.

Annual heating energy is manually estimated from the hours of the year when the outside air temperature is below 18°C. At warmer temperatures, the mixture of outside and return air increases from 21°C. Such days also have solar heat gains and do not need the heating plant to operate. In fact, internal heat gains from people, lights and computers will cause the heating plant to reduce output at much lower outdoor temperatures and the economy air cycle should be able to provide free cooling if the automatic control system allows it. In practice, economy air cycle dampers may not always be fully functional due to lack of maintenance, lubrication, bearing failures and movement verification. They seize into a fixed position, causing the refrigeration plant to operate all year in some such buildings.

Heating energy is used to raise supply air from the minimum mixed air temperature of 17.2°C to the supply air of 28°C during the 6677 hours of winter, with an RMS mean value of 70.7% of this range. The annual fuel input to the heating plant includes the 70% overall system efficiency.

$$\text{Heating energy} = 145\frac{\text{m}^3}{\text{s}} \times 1.22\frac{\text{kJ}}{\text{m}^3\text{K}} \times \frac{70.7}{100} \times (28 - 17.2)\text{K} \times 6677\ \text{h} \times \frac{1\ \text{kWs}}{1\ \text{kJ}} \times \frac{100}{70} \times \frac{1\ \text{MW}}{10^3\ \text{kW}}$$

$$\text{Heating energy} = 12884\ \text{MWh}$$

This will be an overestimate due to the internal heat gains from adventitious solar gains, people, lights and office equipment but that does not concern us for the purpose of this example.

During the 1185 annual hours when the outside air is in the range 18° to 22°, no heating or cooling energy is used as the economy cycle controls the indoor conditions.

Refrigeration system energy is used to cool supply air from the maximum mixed air temperature of 26.2°C to the supply air of 15°C during the 898 hours of summer, with an RMS mean value of 70.7% of this range. The annual electrical energy input to the cooling plant includes the 2.5 overall coefficient of performance.

$$\text{Cooling energy} = 145\frac{\text{m}^3}{\text{s}} \times 1.22\frac{\text{kJ}}{\text{m}^3\text{K}} \times \frac{70.7}{100} \times (26.2 - 15)\text{K} \times 898\ \text{h} \times \frac{1\ \text{kWs}}{1\ \text{kJ}} \times \frac{1}{2.5} \times \frac{1\ \text{MW}}{10^3\ \text{kW}}$$

$$\text{Cooling energy} = 503\ \text{MWh}$$

Energy used by continuously running the AHU fans during the year is found from,

$$\text{Fan energy} = 240\ \text{kW} \times 8760\ \text{h} \times \frac{1\ \text{MW}}{10^3\ \text{kW}}$$

$$\text{Fan energy} = 2102\ \text{MWh}$$

This is a remarkable amount of energy to move air around, greatly exceeding that consumed by the whole of the refrigeration system. However, in reality, refrigeration compressors, water pumps and cooling tower fans are not always switched off outside of the summer season. They continue to consume energy in standby mode and run whenever internal air temperatures rise, for whatever reason; particularly so in warm climates. Plant operators do not want complaints of overheating during mild weather. In tropical climates, year-round dehumidification and cooling requires the refrigeration plant to be operational.

Annual energy cost for the air conditioning system:

$$\text{Electrical cost} = \text{cooling energy} + \text{fan energy} + \text{demand charge}$$

$$\text{Electrical energy cost} = (503 + 2102)\frac{\text{MWh}}{\text{yr}} \times 10\frac{\text{p}}{\text{kWh}} \times \frac{10^3\ \text{kWh}}{1\ \text{MWh}} \times \frac{£1}{100\ \text{p}}$$

$$\text{Electrical energy cost} = £260500\ \text{p.a.}$$

$$\text{Electrical demand} = \text{peak system demand} + \text{fan demand}$$

$$\text{Electrical demand} = (792 + 240)\,\text{kW}$$

$$\text{Electrical demand} = 1032\ \text{kW}$$

$$\text{Electrical demand cost} = \frac{£10}{\text{kW}} \times \frac{12\ \text{months}}{1\ \text{year}} \times 1032\ \text{kW}$$

$$\text{Electrical demand cost} = £123840\ \text{p.a.}$$

$$\text{Gas heating energy cost} = 12884\ \text{MWh} \times \frac{10^3\ \text{kWh}}{1\ \text{MWh}} \times \frac{5\ \text{p}}{\text{kWh}} \times \frac{£1}{100\ \text{p}}$$

$$\text{Gas heating energy cost} = £644200\ \text{p.a.}$$

$$\text{Air conditioning running cost} = (£260500 + £123840 + £644200)\ \text{p.a.}$$

$$\text{Air conditioning running cost} = £1028540\ \text{p.a.}$$

$$\text{Air conditioning running cost per m}^2 = \frac{£1028540}{14 \times 2200\ \text{m}^2\text{yr}}$$

$$\text{Air conditioning running cost per m}^2 = \frac{£33.4}{\text{m}^2\text{yr}}$$

$$\text{Air conditioning energy per m}^2 = (503 + 2102 + 12884)\frac{\text{MWh}}{\text{yr}}\ \frac{1}{14 \times 2200\ \text{m}^2} \times \frac{10^3\ \text{kWh}}{1\ \text{MWh}}$$

$$\text{Air conditioning energy per m}^2 = \frac{503\ \text{kWh}}{\text{m}^2\text{yr}}$$

Bear in mind these calculations were for the building operating fully for 24 hours, 365 days, 8760 h per year; an unusual use. If it ran for conventional office hours, 10 h/day, 5 days/week, 52 weeks, minus 9 public holidays, annual hours would normally be 2510 h per year, and expected energy use would be:

$$\text{Air conditioning energy per m}^2 = \frac{503 \text{ kWh}}{\text{m}^2\text{yr}} \times \frac{2510 \text{ h}}{8760 \text{ h}}$$

$$\text{Air conditioning energy per m}^2 = \frac{144 \text{ kWh}}{\text{m}^2\text{yr}}$$

This aligns with the expected range of usage in a warm, low humidity climate region.

At higher external air temperatures, internal conditions will not be maintained at their design temperature unless the refrigeration compressor, cooling tower and AHU cooling coils have additional capacity; this is unlikely. It might be possible to achieve a lower supply air temperature into the rooms. Reducing the intake quantity of outdoor air might be possible if the outside air intake damper can be further closed when occupancy is below the design number. The refrigeration compressors may have spare or redundant capacity and be able to increase chilled water flow rates if there is pump capacity. This carries the risk of increasing peak electrical demand on days of the year when everyone else is maximising their cooling output. The public supply may not cope with extreme demand. Increasing demand above the previously established chargeable demand for this building incurs additional charges that last all through the year.

During periods where the outside air temperatures is 30°C and above, along with intense solar irradiance, air conditioning is a necessity for almost any building. No amount of shading and ventilation maintains a comfortable working environment.

The Shard case study

How much energy will be used by the air conditioning system for The Shard, 4–6 London Bridge Street, London, SE1 9SG? A triple glazed, 95-storey, 309.6 m all-glass pyramid above London Bridge railway station on the south bank of the river Thames opened in February 2013, it makes a striking statement of architecture among traditional low rise brick and Portland stone solidity. Visit the-shard.com website, also in Wikipedia, and look at the features. Locate the corner of Borough High Street, St. Thomas Street and Bedale Street in Google Earth and stand there to view The Shard at a distance. Move around The Shard and its surroundings. View the many photographs taken and relate the design to its surroundings. Many views show construction underway. What do you think about the contrast between The Shard and older buildings in London? Opinion is always valued; preferably before someone else builds another one as a rival. Is this the way to comply with the intentions of the HM Government Carbon Plan (2011) and the Kyoto Protocol (1997)? Does anyone really have any concern?

The Shard approximates to a square plan pyramid, a frustum, whose sloping sides face north, south, east and west, with a glazed surface area of 14000 m^2 on each face. Assume that internal blinds are lowered on the south, east and west sides for the summer heat gain calculation, solar glazing correction factor 0.15 and alternating 0.14; north side 0.37 and 0.35 with no blinds used. Window U 3 W/m^2 K, Y 3 W/m^2 K, f 1 and time lag 0 h. Ground floor area is 2102 m^2, U 1 W/m^2 K, Y 1.5 W/m^2 K, f 0.9 and time lag 10 h, ignoring the basement. Total floor area is 110000 m^2. The highest occupied floor is the Observatory level 72 that has a roof area of 150 m^2, U 1 W/m^2 K, Y 1.5 W/m^2 K, f 0.5 and time lag 10 h. Volume of the building is approximately 213000 m^3. There are no opaque exterior surfaces. The upwardly sloping faces will have a higher solar heat gain than the vertical surfaces used in our sample data.

A combined heat and power plant (CHP) provides heat and power to the building and its surroundings from National Grid natural gas. The CHP produces 1.13 MW of electricity at an overall efficiency of 41%. Heat recovered from the engine amounts to 1.2 MW. Take natural gas as costing 5 p/kWh and purchased electricity 10 p/kWh for this site.

The downloadable file *Shard energy.xls* is an adaptation of the peak summertime temperature file and calculates the plant cooling load and design supply air flow rate to assess peak design demand. A conservative sample schedule of occupancy, lighting and power is entered in this workbook. The design peak heat gain is calculated to be 7973 kW at 15.00 h for the cooling plant to deliver at a supply air flow rate of 727 m^3/s, creating 12 air changes/h through the building. The cooling load is calculated to be 72.5 W/m^2 based on floor area, although this may be a lower figure than reality due to the simple assumptions made. For example, occupancy may be much higher, blinds may not always be used at peak heat gain times, lighting may be higher power, internal heat gains from computer and server systems and the 44 lifts of maybe 30 kW each, all contribute to demand that we have not included.

The downloadable file *Shard energy use AC system.xls* calculates the annual energy use and cost, not including the ancillary demands that we have not calculated. Peak electrical demand from the air conditioning system amounts to 3748 kW. Additional ancillary demands plus the lifts increase this substantially; 44 lifts at 30 kW each would amount to 1320 kW demand. Peak demand for The Shard may be around 6 MW. The CHP output of 1.13 MW only meets a minor percentage that may be related to emergency power needs.

If the purchased natural gas costs 5 p/kWh, the 41% efficiency CHP generates at:

$$\text{CHP power cost} = \frac{5\ \text{p}}{\text{kWh}} \times \frac{100}{41}$$

$$\text{CHP power cost} = \frac{12.2\ \text{p}}{\text{kWh}}$$

Whether this is at a lower cost than purchased power we may not know. An advantage of the CHP is that recovered heat output reduces purchases of natural gas, when heat is needed, otherwise engine heat is ejected to the atmosphere with air cooling. Peak heat demand from the air conditioning system amounts to 6163 kW, 6.163 MW, while the CHP recovered heat output is 1.2 MW, and so is only a small proportion of the total load.

The gas energy consumed during the predicted year weather cycle is 46695 MWh at a cost of £2334733. The heating load comes from the air handling unit performance where outside intake air is mixed with recycled room air and heated to 28°C from the mixed air of 21.1°C. The Shard is not an intermittently heated application that is allowed to go cold at night and weekends. The hotel, apartments, extensive public viewing access, restaurants, retail and circulation requirements, plus many offices probably needing continuous service, rule that out.

Electrical energy consumed by the refrigeration compressors, chilled water pumps, condenser cooling water pumps and the cooling tower fans during the predicted year weather cycle is 1089 MWh at a cost of £108905. The cooling load comes from the air handling unit performance where outside intake air is mixed with recycled room air and cooled to 15°C from the mixed air of 24.4°C. Air conditions are maintained continuously in The Shard.

$$\text{Electrical energy} = (\text{cooling plant} + \text{fans})\ \text{MWh/yr}$$

$$\text{Electrical energy} = (1089 + 3504)\ \text{MWh/yr}$$

$$\text{AC electrical system energy} = 4593\ \text{MWh/yr}$$

$$\text{Electrical cost} = \pounds\,(\text{cooling energy} + \text{fan energy} + \text{demand})\ /\text{yr}$$

$$\text{Electrical cost} = (\pounds 108905 + \pounds 350400 + \pounds 449780)\ /\text{yr}$$

$$\text{AC electrical system cost} = \pounds 909085/\text{yr}$$

Air conditioning energy and demand cost = gas + electrical

Air conditioning energy and demand cost = £2334733 + £909805

Air conditioning energy and demand cost = £3243818 per year

Balance temperature

The balance temperature of a building is that outdoor air temperature at which there is no net heat loss or gain between the internal and external environments. At this condition, the heat losses to the cooler outdoor air are balanced by the solar and internal heat gains. A low balance temperature either means that the building is highly insulated and makes efficient use of the available heat gains, or that it has very high internal heat generation from people and equipment. The heat flows into and out from the building need to be known for both the peak summer and winter conditions and at intermediate outdoor temperatures. If only the peak values are known, it may be assumed that the gains and losses are straight line graphs between these points. Some inaccuracy will result as solar gains have a sinusoidal variation.

Figure 3.17 shows the variation of heat losses and gains for an office building in London where:

1. there is a constant heat gain of 200 kW from the occupants and equipment during the working day;
2. thermal transmittance and ventilation heat losses occur up to an outdoor air temperature of 20°C; above 20°C these turn into heat gains;
3. solar heat gains follow the curves shown.

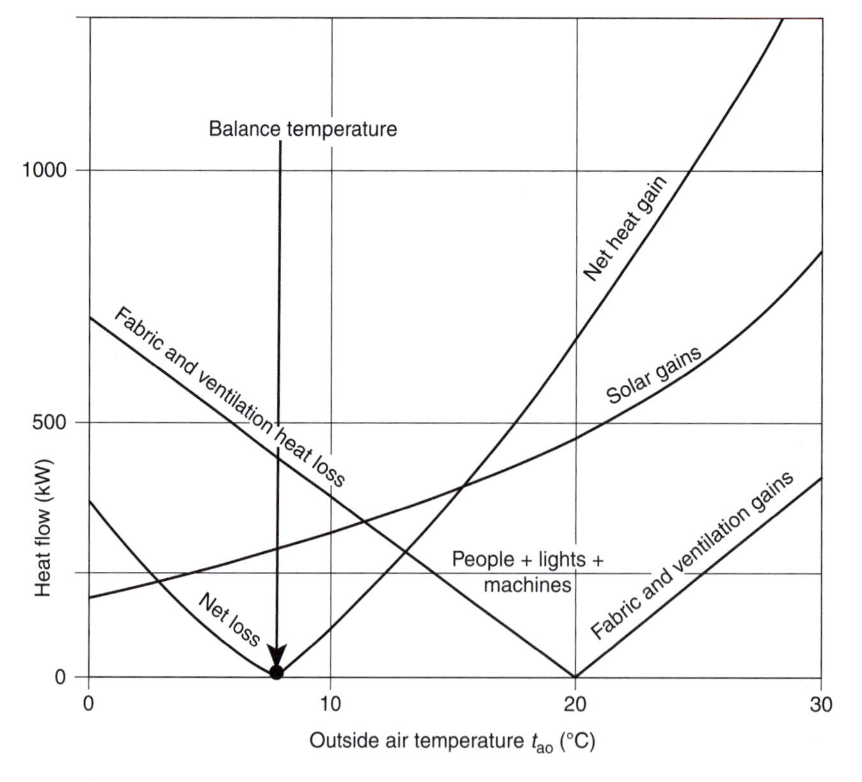

3.17 Balance temperature for an air conditioned building.

Addition of the heat gains and losses at 2°C intervals produce the net heat loss and net heat gain curve that balances to zero at an outdoor air temperature of 8°C. This is the building's balance temperature.

Questions

Data are provided for these questions and the reader should verify the information from the references to become familiar with its use.

1. State the sources of heat gain that will affect the internal thermal environment in residences, offices, retail premises, the atrium in a shopping concourse, a pharmaceutical manufacturing building, an air conditioning plant room, a railway passenger carriage, a motor car, an aeroplane and an entertainment theatre.
2. Explain how heat gains to an occupied building have intermittent characteristics.
3. Describe the angular position of the sun in relation to the four walls and the roof of a building. Define each term used.
4. Explain what is meant by the incidence of solar radiation upon a surface and define how it is found for vertical walls, horizontal roofs and sloping surfaces. State the reason for needing to know the incidence in the calculation of solar radiation heat gains to a surface.
5. A surveyor needed to assess the height of a tower and measured an angle of 52° from the ground to the top of the tower at a distance of 35 m from it. Calculate the building height.
6. A 19-storey office building was being built on a hillside and the surveyor needed to check the height of the structural steelwork during construction. The nearest point of the top of the structure was at an elevation of 40° from the surveyor's position and 55 m horizontally away. The furthest point of the steel frame down the hill was 75 m from the surveyor when measured down the slope of the hill and at decline angle of 12°. Calculate the overall height of the frame and the average floor to floor height.
7. A building has a side 30 m long that slopes forwards from ground level at 78° from the horizontal. The roof line is at 66° from the ground at a distance of 32 m from the foot of the wall. Calculate the vertical height of the building, the length of the sloping face and the sloping face area.
8. A vertical wall in London faces south. At 10.00 hours sun time on 22 June the solar altitude is 55° and solar azimuth 128°. Calculate the wall solar azimuth and incidence angles.
9. Atrium glazing in Glasgow, latitude 55° 52′N, faces west and slopes 12° forwards from the vertical. The sun time is 13.00 hours on 21 August, the solar altitude is 45° and the solar azimuth is 201°. The intensity of the direct solar radiation normal to the sun is 840 W/m^2. Calculate the incidence angle and the solar intensity upon the glazing.
10. A flat plate solar collector near Sidney at latitude of 35° S slopes at 45° to the horizontal and faces north. At 12.00 hours sun time on 21 December the solar altitude is 78°, azimuth 0° and direct intensity normal to the sun is 918 W/m^2. Find the incidence angle by sketching the angles and not applying the general equation, then calculate the solar intensity that is normal to the collector surface.
11. A solar collector in south east England faces due south at an angle of 40° to the horizontal. Calculate the maximum solar irradiance and state when this will occur.
12. A window facing south is 4 m long and 3 m high and has a shaded area of 5 m^2. Solar altitude is 55°, solar azimuth is 12° east of south, direct solar irradiance normal to the sun is 540 W/m^2 and diffuse irradiance is 210 W/m^2. The glazing transmissibilities of direct and diffuse irradiances are 0.74 and 0.69. Calculate the solar heat gain transmitted through the glass.
13. The upper glazing on an atrium slopes at 60° to the horizontal and faces south east. It is 2 m long and 2 m high and has a shaded area of 1 m^2. At 10.00 h on 22 September the design direct solar

irradiance on a horizontal surface I_{DH} is 355 W/m^2. The design direct solar irradiance on a vertical surface I_{DV} is 525 W/m^2. The design diffuse irradiance I_{dH} is 160 W/m^2. The glazing transmissibility of direct and diffuse irradiances are 0.46 and 0.4. Calculate the solar heat gain transmitted through the glass.

14. Find the glass temperature t_g, total heat flow into the room Q and the total transmittance T for 6 mm silver reflective float glass that is in a total solar irradiance of 625 W/m^2 at 08.00 h on 21 June facing east. The external and internal air temperatures are 31°C and 23°C. The absorptivity A of the glass is 0.29, the transmissibility T is 0.43 and reflectance R is 0.28. The glass is low emissivity and has values of R_{si} 0.3 m^2 K/W and R_{so} 0.07 m^2 K/W.
15. Find the glass temperature t_g, direction of heat flow Q and the total transmittance T for 10 mm bronze tinted heat absorbing float glass that is in a total solar irradiance of 175 W/m^2 at 13.00 h on 21 December facing west. The external and internal air temperatures are −2°C and 18°C. The absorptivity A of the glass is 0.67, the transmissibility T is 0.29 and reflectance R is 0.04. The glass is high emissivity and has values of R_{si} 0.12 m^2 K/W and R_{so} 0.03 m^2 K/W due to its exposed location.
16. A new light coloured flat roof of 25 m × 15 m is over a production area that is to be maintained at a resultant temperature of 19°C by an air conditioning system. The thermal transmittance of the roof is 0.4 W/m^2 K and the admittance is 0.7 W/m^2 K. The 24 hour average outdoor environmental temperature for a roof on 23 July in Southampton is 22°C. The roof decrement factor is 0.99 and its time lag is 1 h. The outdoor environmental temperature at 13.00 h is 37.5°C. The room height is 3 m and it has 0.5 air changes per hour due to the infiltration of outdoor air. Calculate the room cooling load through the roof for 14.00 h.
17. A south facing office wall is 30 m long and 4 m high. The thermal transmittance of the wall is 0.33 W/m^2 K and the admittance is 2.4 W/m^2 K. The office has a volume of 1440 m^3 and is to be maintained at a resultant temperature of 20°C by an air conditioning system. The 24 hour average outdoor environmental temperature for a dark coloured wall on 23 July in Bournemouth is 26°C. The wall decrement factor is 0.35 and its time lag is 9 h. The outdoor environmental temperature at 08.00 h is 25.5°C. The room has 1.5 air changes per hour due to the infiltration of outdoor air. Calculate the room cooling load through the wall for 17.00 h.
18. A south east facing top floor London office is similar to that shown in Figure 3.16. The exposure is normal. The office is to be maintained at a resultant temperature of 19°C. There are two sedentary occupants and two continuously used computers having power consumptions of 250 W each. The air conditioning plant operates for 10 hours per day. There are 1.5 air changes per hour due to natural infiltration of outdoor air. The adjacent offices, corridor and office below, are maintained at the same temperature as the example office. Room volume V is 600 m^3. The glazing cooling load peak is 313 W/m^2 at 11.00 h on 22 September and the correction factor for glass and shading types is 0.55. Use the data provided and calculate the peak cooling load for the room air conditioning unit. The thermal data are given in Tables 3.7 and 3.8.

Table 3.7 Heat transfer data for question 18

Surface	*A, m^2*	*U W/m^2 K (AU)*	*Y W/m^2 K (AY)*	*f*	*Lag, h*
Glass	25	3.3	3.3	1	0
External wall	60	0.3	2.7	0.28	8
Internal wall	100	1.5	3	0.72	1
Floor	200	1.2	2.4	0.7	1
Roof	200	0.25	0.8	0.99	1
	$\sum A =$	$\sum AU =$	$\sum AY =$		

Table 3.8 Sol-air data for question 18

Surface	*Lag, h*	*24 h t_{eo}*	*24 h t_{ao}*	*Time, h*	t_{eo}	*Swing $\tilde{t}_{eo}$*
External wall	8	23.5	15.5	03.00	10.5	
Roof	1	20	15.5	10.00	33.5	
					t_{ao}	$\tilde{t}_{ao}$
Window	0	—	15.5	11.00	17	

19. The top floor of an office building in Bristol is to be maintained at a resultant temperature of 20°C. Use the data provided to calculate the peak cooling load for the top floor office.

    The air conditioning plant operates for 10 hours per day. There is 0.75 of an air change per hour due to natural infiltration of outdoor air and 0.25 of an air change per hour from uncooled air entering from the rooms on the floors below. Only the top floor of the building is cooled. The lower floors are expected to be at an air temperature of 26°C at times of peak cooling load. The top floor is a rectangular open plan general office 40 m long, 15 m wide and 3.5 m high with one long side facing south. There is 40 m^2 of glass on the north and south sides but no glazing on the east and west walls. There are 40 occupants each emitting 90 W, 20 computers of 200 W each, one photocopier of 400 W and 10 fluorescent lamps permanently used of 65 W each. The thermal data are given in Tables 3.9 and 3.10.

Table 3.9 Heat transfer data for question 19

Surface	*A, m^2*	*U W/m^2 K (AU)*	*Y W/m^2 K (AY)*	*f*	*Lag, h*
South glass		5.7	5.7	1	0
North glass		3.3	3.3	1	0
South wall		0.3	2.5	0.3	8
East wall		0.4	3	0.7	9
West wall		0.4	3	0.7	9
North wall		0.3	2.5	0.3	8
Floor		1.8	2.8	0.7	1
Roof		0.25	0.9	0.9	2
	$\sum A =$	$\sum AU =$	$\sum AY =$		

Table 3.10 Sol-air data for question 19

Surface	*Lag, h*	*24 h t_{eo}*	*24 h t_{ao}*	*Time, h*	t_{eo}	*Swing $\tilde{t}_{eo}$*
South wall	8	25	15.5	04.00	10.5	
East wall	9	20.5	15.5	03.00	10.5	
West wall	9	20.5	15.5	03.00	10.5	
North wall	8	17	15.5	04.00	10.5	
Roof	2	20	15.5	10.00	33.5	
					t_{ao}	Swing $\tilde{t}_{ao}$
South glass	0	—	15.5	12.00	18.5	
North glass	0	—	15.5	12.00	18.5	

It is expected that the peak cooling load for the top floor will occur at the time and date of the peak irradiance on the south glazing, that is noon sun time on 22 September and 21 March. This might not be the case and other times and dates could be analysed for comparison. South single glazing has a cooling load of 333 W/m^2 and a correction factor of 0.77, the north double glazing has a cooling load of 104 W/m^2 and a correction factor of 0.95.

20. Calculate the net heat transfer through a west facing single glazed window for 22 April at 14.00 h when the cooling load is 207 W/m^2 with a shading correction factor of 0.77, $\frac{F_u}{F_v}$ is 0.9, $\frac{F_u}{F_y}$ is 0.95, outside air temperature is 14.8°C, room resultant temperature is 21°C, 24 hour mean external air temperature is 9°C, window U value is 5.7 W/m^2 K and dimensions are 2.5 m × 1.5 m.
21. List the ways in which the south facing office in example 3.17 could be made comfortable with passive architectural changes and mechanical cooling methods.
22. A west facing Plymouth office of 15 m × 5 m × 3 m high has double glazed gold coloured heat reflecting window openings of 25 m^2. The surrounding rooms are all similar. There are six occupants emitting 90 W and five electrical items of 150 W each. The office is used for 8 h in each 24 hours. Windows and door are shut at other times and the ventilation rate is 1.5 air changes per hour. Use the data provided to estimate the 24 hour mean and peak internal environmental temperatures. The peak solar irradiance on a west facing vertical window is 625 W/m^2 at 16.00 h on 21 June in south east England and the daily mean is 185 W/m^2. The mean solar gain correction factor for the glazing without blinds is 0.25 and the alternating factor is 0.2. The thermal data are given in Tables 3.11 and 3.12.

Table 3.11 Heat transfer data for question 22

Surface	*A, m^2*	*U W/m^2 K (AU)*	*Y W/m^{2}K(AY)*	*f*	*Lag, h*
Glass		3.3	3.3	1	0
External wall		0.57	3.6	0.31	9
Internal wall		1	3.6	0.62	1
Floor		2	4.3	0.59	2
Ceiling		2	6	0.46	3
	$\sum A =$	$\sum AU =$	$\sum AY =$		

Table 3.12 Sol-air data for question 22

Surface	*Lag, h*	*24 h t_{eo}*	*24 h t_{ao}*	*Time, h*	*t_{eo}*	*Swing $\tilde{t}_{eo}$*
External wall	9	24.5	16.5	07.00	15.5	
					t_{ao}	Swing $\tilde{t}_{ao}$
Window	0	—	16.5	16.00	22	

23. A vehicle production factory that was constructed in 1960 at Southampton is 100 m × 100 m × 10 m high and has no glazing. The structural steel frame is clad in corrugated metal sheet with no insulation for the walls and a lightweight flat roof, all painted white externally. The workforce of 200 occupies two 8 h shifts per 24 hours, emitting 110 W each. Heat producing motors, lights, tools and processes generate 50 kW during the working periods. The 100% outdoor air mechanical ventilation systems operate for 24 hours per day and 7 days per week and produce 1 air change per hour. The thermal data are given in Tables 3.13 and 3.14. Use the data provided to estimate the peak internal environmental temperature.

Table 3.13 Heat transfer data for question 23

Surface	*A*, m^2	*U W/m^2 K (AU)*	*Y W/m^2 K (AY)*	*f*	*Lag, h*
South wall		5.7	5.7	1	0
East wall		5.7	5.7	1	0
West wall		5.7	5.7	1	0
North wall		5.7	5.7	1	0
Floor		1.7	5.2	0.72	3
Roof		1.1	1.2	0.99	1
	$\sum A =$	$\sum AU =$	$\sum AY =$		

Table 3.14 Sol-air data for question 23

Surface	*Lag, h*	*24 h t_{eo}*	*24 h t_{ao}*	*Time, h*	t_{eo}	*Swing $\tilde{t}_{eo}$*
South wall	0	22.5	19	12.00	36	
East wall	0	22.5	19	12.00	26.5	
West wall	0	22.5	19	12.00	26.5	
North wall	0	20	19	12.00	26	
Roof	1	22	19	11.00	34.5	
					t_{ao}	*Swing $\tilde{t}_{ao}$*
Air	0	—	19	12.00	21.5	

24. Describe, with the aid of sketches, how shading devices and glass types are used to assist in the provision of thermal comfort within buildings and how they affect the natural illumination and cooling plant loads.
25. A window 3 m long and 2 m high is recessed 200 mm from the face of a building. The wall azimuth is 170° and the solar azimuth at 13.00 h sun time is 190°. Calculate the shade width on the window.
26. A window 3.5 m long and 1.75 m high is recessed 150 mm from the face of a building. The solar altitude at 15.00 h on 24 August is 37° and the solar azimuth is 240°. The wall azimuth is 275°. Calculate the unshaded area of the glass.
27. State the ways in that the shaded areas of buildings can be predicted and the uses for this information.
28. Calculate the net heat transfer through a south facing single glazed window on 21 December at latitude of 51.7°N at noon when the solar irradiance cooling load is 273 W/m^2, the outside air temperature is 5°C and the room resultant temperature is 20°C. The 24 hour mean outdoor air temperature is 2°C. The window is 2.5 m × 2.5 m and its thermal transmittance is 5.7 W/m^2 K.
29. Explain the use of the peak summertime internal environmental temperature in the analysis of the thermal comfort conditions. Include in your explanation when needs to be calculated, what importance it has to the building owner and the design engineer and what part it plays in the decisions to be made on the choice of air conditioning system.
30. Explain why thermal admittance, decrement factor, surface factor, structural time lag and environmental temperature swing are used in preference to thermal transmittance in the assessment of summer internal conditions.
31. List the advice that can be given to the owner of a building that suffers from summer overheating. Explain why the use of additional outdoor air mechanical ventilation may be an unsuitable solution for some applications but correct for some buildings.

32. Draw a graph showing the balance temperature for a building from the following data.
    (a) Constant heat gain from the occupants, lights and electrical equipment of 10 kW
    (b) Solar gains of 20 kW at an outdoor air temperature of 0°C, 30 kW at 10°C, 50 kW at 20°C and 70 kW at 30°C
    (c) Conduction and ventilation heat loss of 90 kW at 0°C zero at 21°C and a heat gain of 30 kW at 30°C.
33. Explain how condensation can occur in both cool climates such as the UK and in hot humid tropical climates, causing damage to buildings, plant and equipment. Sketch examples of how such risks are combated.
34. Provide examples of comfort conditions for sedentary workers within buildings in different climate regions, namely, Arctic, northern Europe, Middle East, desert, Mediterranean, equatorial and sub-tropical. Identify each climate location being described. State which form of air conditioning is used in each location and give reasons for their suitability.
35. List the eight climate regions found within the tropics. State a typical location for each and what types of air conditioning system are used for commercial and domestic buildings. What do people without access to air conditioning do in each of the eight regions to maintain their living conditions? Would you abandon your present lifestyle to live in any or all of these eight regions if you had no access to air conditioning, and why?
36. Use the file *Around the world.xls* to compare air conditioning cooling loads for different locations and climates. Add further locations and data as needed. Copy the file *Shard in Dubai.xls* using another name and use it to estimate air conditioning loads and costs around the world as you decide.
37. Use the file *Around the world.xls* and copy the file *Shard in Dubai.xls* using another name and use it to estimate air conditioning loads and costs as if it were built in Doha, Qatar. Find the climate data for Doha. Comment on what you find.

4 Psychrometric design

Learning objectives

Study of this chapter will enable the reader to:

1. state the constituents of humid air;
2. understand the composition of atmospheric pressure;
3. use a sketch of the CIBSE psychrometric chart;
4. read a CIBSE psychrometric chart;
5. understand, use, calculate and read from tables, charts and an App, the properties of humid air – dry and wet bulb air temperatures, moisture content, percentage saturation, specific enthalpy, specific volume and density;
6. calculate and find the dew point temperature of air;
7. calculate the properties of humid air from formulae;
8. understand the meaning of saturation state;
9. calculate and use air vapour pressure;
10. find all the physical properties of humid air with readings taken from a sling psychrometer;
11. know what affects the density of humid air;
12. understand the psychrometric processes of heating, mixing, cooling, humidification and dehumidification;
13. calculate the air conditioning plant loads of heating, cooling and humidity control processes;
14. understand the meaning of 'cooling coil contact factor';
15. compare steam humidification with that by water sprays.

Key terms and concepts

Barometric pressure 100; dew point 116; dry bulb temperature 109; moisture content 109; percentage saturation 109; relative humidity 109; specific enthalpy 114; specific volume 113; vapour pressure 110; wet bulb temperature 110.

Introduction

The properties of humid air are introduced and the use of the CIBSE psychrometric chart is thoroughly described in logical and manageable steps. Worked examples are used to demonstrate the calculations needed to find the physical properties of humid air. Calculated properties are compared with tabulated values and those read from the psychrometric chart. A supply of CIBSE psychrometric charts and access to the CIBSE Guide A (1996) are needed for completion of the chapter. Psychrometric applications are available for mobile computers, some are free, and can easily be used for examples in this book and at work. The use of sketch charts is introduced for explanatory purposes and for use during preliminary design.

Sufficient accuracy can be achieved for most learning purposes by means of the calculation techniques used, but for precise data and professional use, either use published CIBSE data or commercial computer programs.

The psychrometric processes of heating, mixing, cooling, dehumidification and humidification are described. Example calculations of conditions and complete processes are detailed; these processes are fundamental to air conditioning systems.

Properties of humid air

Understanding the use of the thermal properties of humid air is fundamental to a study of air conditioning. Formulae, data and psychrometric charts from the CIBSE Guide A (2006) will be referred to for the normal atmospheric pressure of 101,325 Pa (1013.25 mb) at sea level. All calculations, examples and questions in this book will use normal atmospheric pressure and the sling psychrometer temperatures unless stated otherwise. The standard of measurement is the kg of dry air, so specific volume and enthalpy are both per kg dry air. The reader requires a supply of CIBSE psychrometric charts or an App to use during the worked

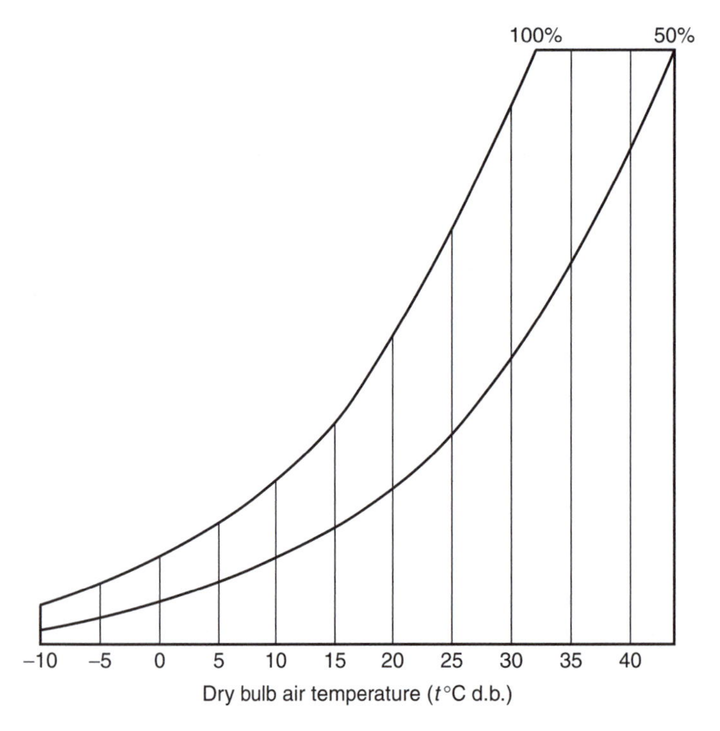

4.1 Sketch of a psychrometric chart.

examples and questions, and for practical design work. A sketch of the psychrometric chart will suffice for much of the work and Figure 4.1 can be reproduced for plotting approximate states and lines representing air conditioning processes.

In the early stages of design calculations, it is easier to use a freehand sketch than to be limited by the rigour of plotting exact locations. Precise data can be subsequently plotted onto a full size chart. There are computer programs to calculate and plot psychrometric states and lines. They have not replaced the need to handle the data and charts manually.

The uppermost curve of the psychrometric chart demonstrates the increasing capacity of air to hold moisture in suspension as the air temperature increases. This is the 100% saturation curve. Thermodynamic states to the left of the 100% curve represent fully saturated air (wet fog, plus a pool of water on the floor), while those to the right are conditions that can be plotted from dry and wet bulb air temperatures. The chart is plotted from two linear axes. The horizontal scale is that of dry bulb air temperature t°C d.b. with equally spaced increments of 1°C with 0.5°C subdivisions. The right hand side vertical scale is that of air moisture content g kg H_2O per kg of dry air, kg/kg. This may be called humidity ratio. At each value of dry bulb temperature, air can sustain maximum moisture content, in the form of water vapour, of g_s kg/kg. If the air is less than saturated, then its moisture content will be g kg/kg, and it will have a percentage saturation PS of:

$$PS = 100\frac{g}{g_s}\%$$

EXAMPLE 4.1

Air in an occupied room is at a temperature of 20°C d.b. and has a moisture content of 0.007376 kg/kg at 50 m altitude. When air at 20°C d.b. is fully saturated, it can hold 0.01475 kg/kg. Calculate the percentage saturation of the room air. Check the answer with a psychrometric chart, data tables of the properties of humid air and your psychrometric App.

$$g = 0.007376\ \frac{\text{kg}}{\text{kg}}$$

$$g_s = 0.01475\ \frac{\text{kg}}{\text{kg}}$$

$$PS = 100\ \frac{0.007376}{0.01475}\%$$

$$PS = 50\%$$

Relative humidity RH % is used extensively and is found from the ratio of vapour pressure to saturation vapour pressure at the dry bulb temperature.

$$RH = 100\frac{p_v}{p_s}\%$$

Wet bulb air temperature scale t_{sl}°C is linear but sloping at 35° downwards to the right and has 1°C increments. Wet and dry bulb temperatures coincide at 100% percentage saturation as shown in Figure 4.2.

Plotting an air condition with these two values allows all other data to be calculated or read from the chart or data tables.

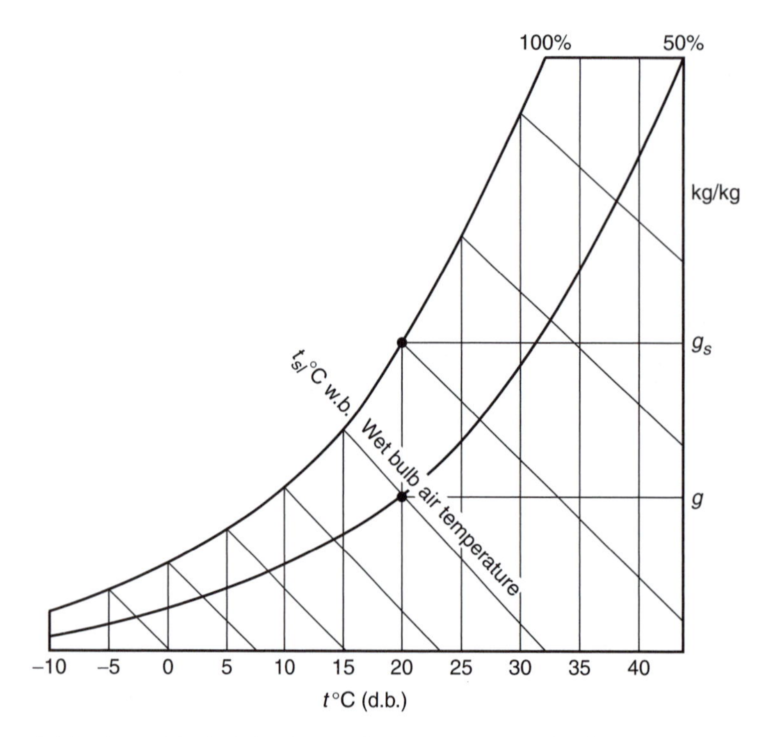

4.2 Psychrometric chart showing wet bulb temperature and moisture content.

Moisture content is calculated from the air vapour pressure p_v kPa and the air vapour pressure at saturation p_s kPa. The procedure is as follows.

1. Find the values of t°C and t_{sl}°C.
2. Saturation vapour pressure p_s kPa is found from;

$$\log_{10} p_s = 30.59051 - 8.2 \log_{10} T + \frac{2.4804}{10^3} T - \frac{3142.31}{T}$$

   Where, $T = (t^{\circ}\text{C} + 273.15)$ K

   Dry and wet bulb temperatures t are equal at saturation, but the air is less than saturated and the percentage saturation is to be found. The saturated vapour pressure corresponds to the dry bulb temperature, so, $t = t^{o}$C d.b.

   Vapour pressures will nearly always be between 0 mb and 70 mb, that is 0 kPa to 7 kPa.
3. Calculate the saturation moisture content g_s from p_s mb:

$$g_s = 0.62197 \frac{p_s}{(1013.25 - p_s)} \frac{\text{kg}}{\text{kg}}$$

   But use this when p_s kPa is entered:

$$g_s = 0.62197 \frac{p_s}{(101.325 - p_s)} \frac{\text{kg}}{\text{kg}}$$

4. Calculate the saturation vapour pressure p_{sl} mb at the wet bulb temperature from the formula above.
5. Calculate the actual vapour pressure p_v using the dry bulb t and wet bulb t_{sl} from p_{sl} mb:

$$p_v = p_{sl} - 1013.25 \times \frac{6.66}{10^4}(t - t_{sl})\,\text{mb}$$

But when p_{sl} kPa is used:

$$p_v = p_{sl} - 101.325 \times \frac{6.66}{10^4}(t - t_{sl})\,\text{kPa}$$

If t_{sl} is below 0°C, calculations relate to ice rather than water and other formulae are used. We will only concern ourselves with liquid and vapour states.

6. Actual moisture content;

$$g = \frac{0.62197 p_v}{(1013.25 - p_v)}\text{mb}$$

7. Percentage saturation;

$$PS = 100\frac{g}{g_s}\%$$

EXAMPLE 4.2

A sling psychrometer shows that the air condition in an occupied room is 22°C d.b. and 17°C w.b. at 50 m altitude. Calculate the percentage saturation and relative humidity. Check the values with published data and on your App.

At the dry bulb temperature,

$$\log_{10} p_s = 30.59051 - 8.2 \log_{10} T + \frac{2.4804}{10^3}T - \frac{3142.31}{T}$$

$$T = (22 + 273.15)\,\text{K}$$

$$T = 295.15\ \text{K}$$

$$\log_{10} p_s = 30.59051 - 8.2 \log_{10} 295.15 + \frac{2.4804}{10^3}295.15 - \frac{3142.31}{295.15}$$

$$\log_{10} p_s = 0.4214$$

$$\text{and, } p_s = 2.638\ \text{kPa}$$

As the equation consistently uses $\log_{10}$, the logarithm base will be assumed from here.

$$g_s = 0.62197\frac{p_s}{(1013.25 - p_s)}\frac{\text{kg}}{\text{kg}}$$

$$g_s = 0.62197\frac{2.638}{(101.325 - 2.638)}\frac{\text{kg}}{\text{kg}}$$

$$g_s = 0.01663 \frac{kg}{kg}$$

Using the wet bulb temperature, $t_{sl} = 17°C$;

$$T = (17 + 273.15)\,K$$

$$T = 290.15\ K$$

$$\log p_s = 30.59051 - 8.2 \log 290.15 + \frac{2.4804}{10^3} 290.15 - \frac{3142.31}{290.15}$$

$$\log p_s = 0.2867$$

$$p_s = 1.935\ kPa$$

Find the actual vapour pressure;

$$p_v = p_{sl} - 1013.25 \times \frac{6.66}{10^4}(t - t_{sl})\,mb$$

Convert this equation to a kPa pressure;

$$p_v = 1.935\ kPa - 1013.25 \times \frac{6.66}{10^4}(22 - 17)\,mb \times \frac{100\,kPa}{10^3\,mb}$$

$$p_v = 1.6\ kPa$$

Moisture content for this vapour pressure;

$$g = 0.62197 \frac{p_v}{(101.325 - p_v)} \frac{kg}{kg}$$

$$g = 0.62197 \frac{1.6}{(101.325 - 1.6)} \frac{kg}{kg}$$

$$g = 0.01 \frac{kg}{kg}$$

$$PS = 100 \frac{g}{g_s} \%$$

$$PS = 100 \frac{0.01}{0.01663} \%$$

$$PS = 60\%$$

$$RH = 100 \frac{p_v}{p_s} \%$$

$$RH = 100 \frac{1.6}{2.638} \%$$

$$RH = 60.7\%$$

Calculated data agree with those published. Check this with your App.

The density ρ kg/m^3 of humid air is the reciprocal of specific volume v m^3/kg and it depends upon the following factors.

Dry bulb temperature

This changes significantly throughout most air conditioning systems due to mixing, heating and cooling coil processes, the turbulent heating effect of the fan, heat generation by the fan motor and heat exchanges with the ambient environment through the duct wall.

Atmospheric pressure

Density varies linearly with total atmospheric pressure, increasing as pressure increases and reducing as pressure reduces. Height above sea level of a building determines the local atmospheric pressure. The weather can change the normal value by up to 10% due to high or low atmospheric pressure variations. Increases in wind velocity produce reductions in atmospheric, static, pressure according to Bernoulli's theorem. Prevailing wind force on the side of a building temporarily creates positive or negative pressure changes. These possible changes are usually ignored. Standard psychrometric data are usually employed.

Moisture content

Water vapour in air exists at its own partial pressure. Water vapour has a lighter molecular mass, 18.015 kg/kmol, than dry air, 29 kg/kmol. Thus the greater the amount of water vapour in dry air, the less the mixture weighs. As g kg H_2O/kg dry air, increases, so does the specific volume v m^3/kg and density ρ kg/m^3 reduces.

The Standard Condition for stating air density is 0°C d.b. and 101.325 kPa atmospheric pressure. 1.01325 bars, 1013.25 mb, 101325 Pa, 101325 $\frac{N}{m^2}$, 10.33 m H_2O and 760 mm Hg are also used. The standard density of dry air can be found from the general gas law:

$$pv = mRT$$

$$\text{Air density } \rho = \frac{m}{v}\frac{\text{kg}}{\text{m}^3}$$

$$\text{Air density } \rho = \frac{p}{RT}$$

$$\text{Air density } \rho = 101325\frac{\text{N}}{\text{m}^2} \times \frac{\text{kg K}}{287.1 \text{ kJ}} \times \frac{1}{273 \text{ K}}$$

$$\text{Air density } \rho = 1.293 \frac{\text{kg}}{\text{m}^3}$$

EXAMPLE 4.3

Find the densities of humid air at 25°C d.b. when it is at 20% saturation and then when it is at 23°C w.b. at 50 m altitude.

From tables, or an App, at 25°C d.b., 20% saturation:

$$\text{Specific volume } v = 0.8498\frac{\text{m}^3}{\text{kg}}$$

$$\text{Air density } \rho = \frac{1}{0.8498} \frac{kg}{m^3}$$

$$\text{Air density} \rho = 1.177 \frac{kg}{m^3}$$

From tables, or an App, at 25°C d.b., 23°C w.b.:

$$\text{Specific volume } v = 0.8672 \frac{m^3}{kg}$$

$$\text{Air density } \rho = \frac{1}{0.8672} \frac{kg}{m^3}$$

$$\text{Air density } \rho = 1.153 \frac{kg}{m^3}$$

Reading the psychrometric chart, shows specific volume lines at 73° sloping downwards to the right, spaced at 0.01 m^3/kg, of 0.85 and 0.87 m^3/kg for the air conditions in this example.

Specific enthalpy of humid air can be approximated from:

$$h = 1.0048t + g(2500.8 + 1.863t) \frac{kJ}{kg}$$

Where, 1.0048 is the specific heat of dry air at constant pressure at 293 K and 2500.8 is the specific enthalpy of dry saturated steam at 0°C. For the air condition in example 4.2:

$$h = 1.0048 \times 22 + 0.01 \times (2500.8 + 1.863 \times 22) \frac{kJ}{kg}$$

$$h = 47.52 \frac{kJ}{kg}$$

The data table shows 47.64 kJ/kg. The 0.3% error is due to the simplification and rounding of decimal places made. CIBSE data are to be used where accuracy has to be assured.

Specific volume of humid air can be found from the general gas law:

$$Pv = mRT$$

$$P = \text{absolute pressure of dry air, Pa}$$

$$P = \text{atmospheric pressure} - \text{vapour pressure}$$

$$P = 101325 - p_v Pa$$

$$v = \text{specific volume of 1 kg of dry air, } \frac{m^3}{kg}$$

$$m = 1 \text{ kg dry air}$$

$$R = \text{specific gas constant 0.2871, } \frac{kJ}{kg\ K}$$

$$T = \text{absolute temperature, K}$$

$$\text{So, } v = \frac{mRT}{P} \frac{m^3}{kg}$$

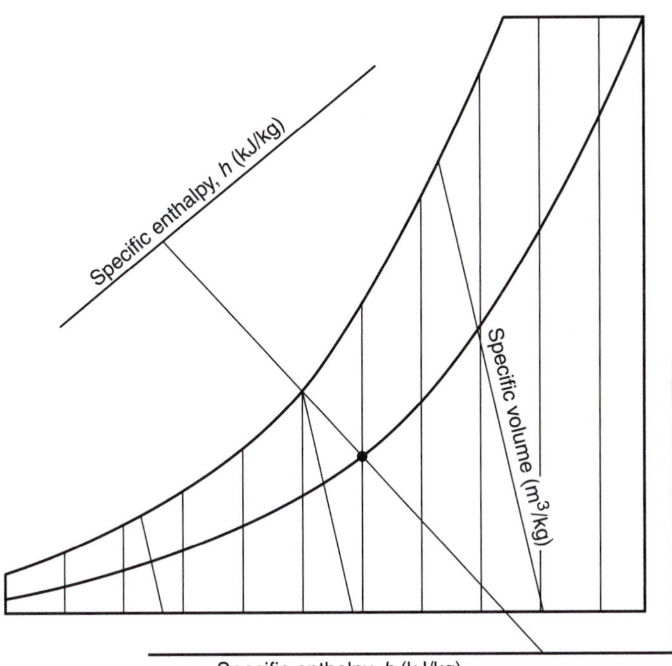

4.3 Psychrometric chart showing specific enthalpy and specific volume.

EXAMPLE 4.4

Calculate the specific volume of humid air at 22°C d.b., 17°C w.b. and vapour pressure 16 mb.

$$1\ \text{bar} = 100000\ \text{Pa}$$

$$1\ \text{mb} = 100\ \text{Pa}$$

$$1\ \text{Pa} = 1\ \frac{\text{N}}{\text{m}^2}$$

$$1\ \text{J} = 1\ \text{Nm}$$

$$p_s = 16\ \text{mb} \times \frac{100\ \text{Pa}}{1\ \text{mb}}$$

$$p_s = 1600\ \text{Pa}$$

$$\text{Dry air partial pressure } P = (101325 - 1600)\,\text{Pa}$$

$$\text{Dry air partial pressure } P = 99725\ \text{Pa}$$

$$\text{Dry air partial pressure } P = 99725\frac{\text{N}}{\text{m}^2}$$

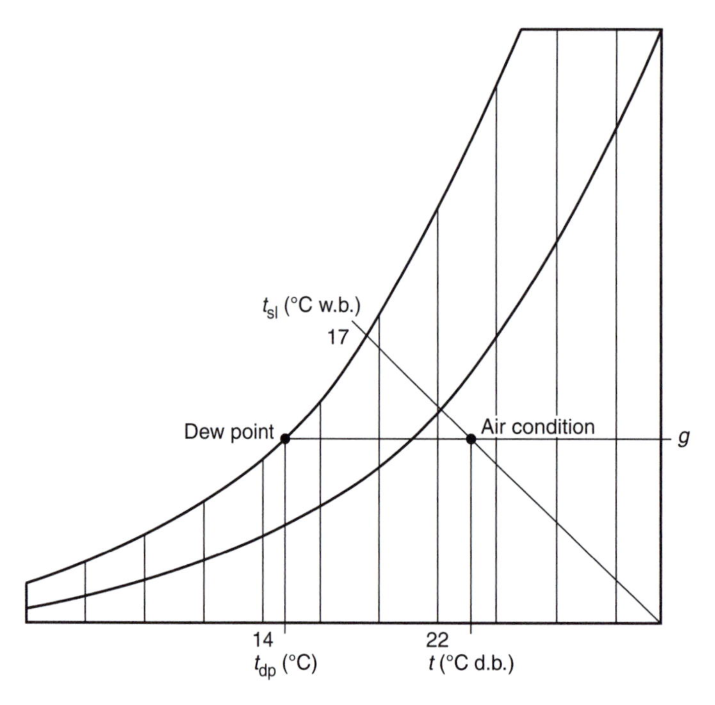

4.4 Psychrometric chart showing the location of dew point temperature.

$$v = 1\ \text{kg} \times 287.1 \frac{\text{J}}{\text{kgK}} \times 295\ \text{K} \times \frac{\text{m}^2}{99725\ \text{N}} \times \frac{1\ \text{Nm}}{1\ \text{J}}$$

$$v = 0.8493 \frac{\text{m}^3}{\text{kg dry air}}$$

This specific volume agrees with tabulated data and an App, and can be read from the psychrometric chart.
Dew point temperature t_{dp} of humid air can be found by:

1. Inspecting tabulated data or an App;
2. Moving horizontally leftwards from the air condition on the psychrometric chart until the 100% curve is reached (this means the temperature at which the moisture held in the air will begin to condense), as shown in Figure 4.4;
3. Using an equation having sufficiently close agreement with the 100% saturation curve that enables direct calculation of the dew point from the saturated air vapour pressure p_s Pa, such as:

$$t_{dp} = 14.62 \ln\left(\frac{p_s}{600.245}\right) ^\circ\text{C}$$

EXAMPLE 4.5

Find the dew point temperature of humid air at 22°C d.b., 17°C w.b. and vapour pressure 16 mb at an altitude of 50 m with the three methods described. A simplified psychrometric chart is shown in Figure 4.5.

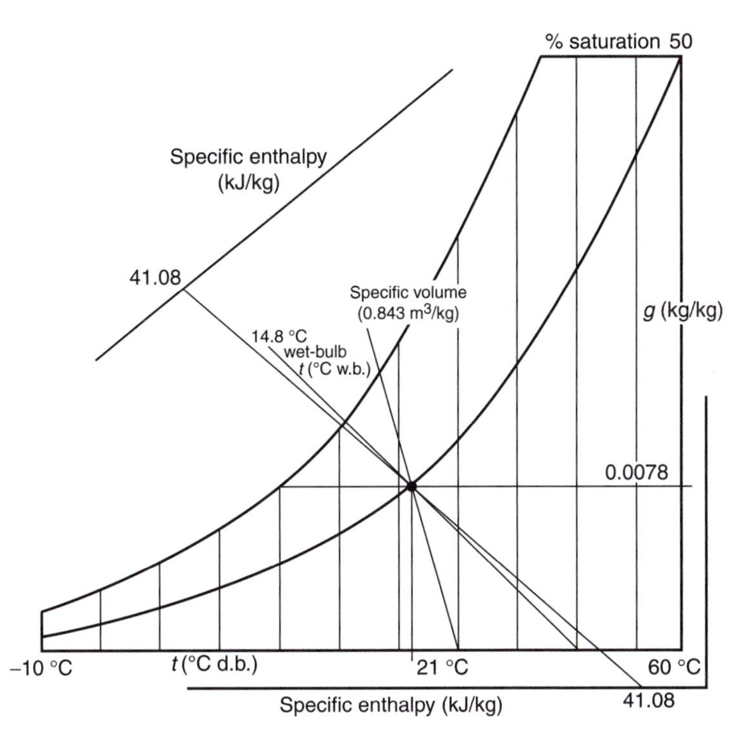

4.5 Simplified psychrometric chart.

Method 1

22°C d.b. and 17°C w.b., reference to table data and an App shows a dew point of 14°C.

Method 2

Figure 4.5 demonstrates that the dew point is 14°C.

Method 3

$$t_{dp} = 14.62 \ln\left(\frac{p_s}{600.245}\right) °C$$

The p_s of 1600 Pa becomes the saturated vapour pressure as the local air temperature has been lowered to the dew point that lies on the 100% curve:

$$p_s = p_v$$

$$p_s = 1600 \text{Pa}$$

$$t_{dp} = 14.62 \ln\left(\frac{1600}{600.245}\right) °C$$

$$t_{dp} = 14.3 °C$$

The three methods agree, bearing in mind calculation accuracy in both manual and App methods.

When the dew point is known, the air vapour pressure can be found from a transposition of the curve fit equation:

$$p_s = 600.245e^{(0.0684t_{dp})}\ \text{Pa}$$

For $t_{dp} = 14.3°C$

$$p_s = 600.245e^{(0.0684 \times 14.3)}\ \text{Pa}$$

$$p_s = 1596\ \text{Pa}$$

If the further decimal places in the dew point value are used, the original 1600 Pa is calculated.

Summary of psychrometric formulae

The quantities and formulae used are as follows.

1. $t°C$ dry bulb temperature from a mercury in glass thermometer.
2. $t^{\circ}_{sl}C$ wet bulb temperature from a mercury in glass thermometer covered by a saturated wick.
3. Sling psychrometer used for $t°C$ dry bulb and $t^{\circ}_{sl}C$ wet bulb.
4. Sea level atmospheric pressure 101325 Pa, 1013.25 mb.
5. Absolute temperature is $T = (t°C + 273.15)\,K$
6. Gas constant is $R = 287.1 \frac{J}{kgK}$
7. Air moisture content is g kg H_2O per kg of dry air, kg/kg.
8. Percentage saturation is $PS = 100 \frac{g}{g_s} \%$
9. Relative humidity is $RH = 100 \frac{p_v}{p_s} \%$
10. Saturation vapour pressure p_s kPa:

$$\log p_s = 30.59051 - 8.2 \log T + \frac{2.4804}{10^3} T - \frac{3142.31}{T}$$

11. Dew point temperature $t^{\circ}_{dp}C$ when p_s is in Pa:

$$t_{dp} = 14.62 \ln\left(\frac{p_s}{600.245}\right) °C$$

12. Moisture content when vapour pressure is in mb:

$$g_s = 0.62197 \frac{p_s}{(1013.25 - p_s)} \frac{kg}{kg}$$

13. Moisture content when vapour pressure is in kPa:

$$g = 0.62197 \frac{p_v}{(101.325 - p_v)} \frac{kg}{kg}$$

14. Vapour pressure when p_{sl} is in mb:

$$p_v = p_{sl} - 1013.25 \times \frac{6.66}{10^4} (t - t_{sl})\ \text{mb}$$

15. Vapour pressure when p_{sl} is in kPa:

$$p_v = p_{sl} - 101.325 \times \frac{6.66}{10^4}(t - t_{sl})\,\text{kPa}$$

16. Humid air density ρ:

$$\rho = \frac{p}{RT}\frac{\text{kg}}{\text{m}^3}$$

17. Specific volume v:

$$v = \frac{mRT}{P}\frac{\text{m}^3}{\text{kg}}$$

$$v = \frac{1}{\rho}\frac{\text{m}^3}{\text{kg}}$$

18. Specific enthalpy h:

$$h = 1.0048t + g(2500.8 + 1.863t)\frac{\text{kJ}}{\text{kg}}$$

19. Mixed air temperature:

$$t_3 = \frac{Q_1}{Q_3}t_1 + \frac{Q_2}{Q_3}t_2$$

20. Mass flow rate of air:

$$m_1\frac{\text{kg}}{\text{s}} = Q_1\frac{m^3}{s} \times \rho_1\frac{\text{kg}}{\text{m}^3}$$

$$m_1\frac{\text{kg}}{\text{s}} = Q_1\frac{m^3}{s} \times \frac{1}{v_1}\frac{\text{kg}}{\text{m}^3}$$

21. To find moisture content of mixed air:

$$m_3g_3 = (m_1g_1 + m_2g_2)\frac{\text{kg H}_2\text{O}}{\text{s}}$$

22. To find specific enthalpy of mixed air:

$$m_3h_3 = (m_1h_1 + m_2h_2)\,\text{kW}$$

23. To find water flow to a steam humidifier:

$$m_3 = m_1\frac{(g_2 - g_1)}{(1 - g_2)}\frac{\text{kg}}{\text{s}}$$

24. Heating or cooling input to a coil:

$$Q = m_1(h_2 - h_1)\,\text{kW}$$

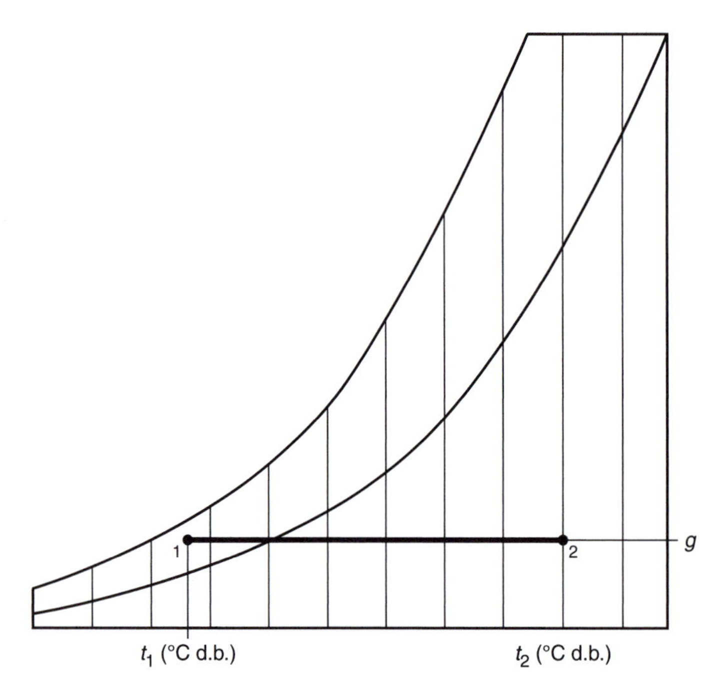

4.6 Psychrometric chart showing a heating process.

Psychrometric processes

Air conditions and processes of change are represented by coordinates and lines on the psychrometric chart shown in simplified form in Figure 4.5. A typical room condition of 21°C d.b. and 50% saturation is shown along with the other properties. When calculating the heating or cooling loads on coils, the inlet air specific volume v m^3/kg is used to convert the air volume flow rate into mass flow rate m kg/s. Changes in specific volume through the coil are dealt with by the coil manufacturer.

Heating

Heating air up to 40°C d.b. is performed by low or high pressure hot water finned pipe heater coil, electric resistance elements or fuel-fired heat exchangers. Figure 4.6 shows that air at condition 1 is being raised in dry bulb temperature and in specific enthalpy to condition 2, while the moisture content remains constant. The percentage saturation reduces because the saturation moisture content increases exponentially while the actual moisture content remains at the original level. The specific volume increases as the air expands, conversely, the air density reduces.

EXAMPLE 4.6

1 m^3/s of outdoor air at −2°C d.b., −3°C w.b. and 80% saturation enters a heating coil to be raised to 30°C d.b. and supplied to an office. Calculate the heating load in kW, sketch the process on a psychrometric chart and identify all the properties of the air before and after heating.

Data for entering air at condition 1 is found from tables, chart or App: −2°C d.b., 80% saturation, −3°C w.b., 0.002564 $\frac{kg}{kg}$, 4.392 $\frac{kJ}{kg}$, 0.7708 $\frac{m^3}{kg}$, 0.4141 kPa vapour pressure, −4.6°C dew point.

Similarly for the heated air at condition 2: 30°C d.b., the moisture content is known to be the same as at 1, so is 0.002564 kg/kg, the table for 30°C d.b. has air with this moisture content at around 10% saturation, linear interpolation between 8% and 12% produces 13.6°C w.b., 37.17 kJ/kg, 0.862 m^3/kg, 0.4141 mb vapour pressure, −4.1°C dew point that should be the same as at 1, −4.6°C. The erroneous dew point demonstrates that linear interpolation of exponential curve data suffers from inaccuracy.

The mass flow of air into the heater coil is:

$$m = 1\frac{m^3}{s} \times \frac{kg}{0.7708\ m^3}$$

$$m = 1.297\frac{kg}{s}$$

$$\text{Specific enthalpy rise} = h_2 - h_1$$

$$\text{Specific enthalpy rise} = (37.17 - 4.392)\frac{kJ}{kg}$$

$$\text{Specific enthalpy rise} = 32.778\ \frac{kJ}{kg}$$

$$\text{Heater input power} = 1.297\frac{kg}{s} \times 32.778\frac{kJ}{kg} \times \frac{kWs}{kJ}$$

$$\text{Heater input power} = 42.513\ kW$$

Values can also be calculated for condition 1:

$$T = (-2 + 273.15)K$$

$$T = 271.15\ K$$

$$\log p_s = 30.59051 - 8.2\ \log T + \frac{2.4804}{10^3}T - \frac{3142.31}{T}$$

$$\log p_s = 30.59051 - 8.2\ \log 271.15 + \frac{2.4804}{10^3}271.15 - \frac{3142.31}{271.15}$$

$$\log p_s = -0.2781$$

$$p_s = 0.527\ Pa$$

$$g = 0.62197\frac{p_v}{(101.325 - p_v)}\frac{kg}{kg}$$

$$g = 0.62197\frac{0.527}{(101.325 - 0.527)}\frac{kg}{kg}$$

$$g = 0.00325\frac{kg}{kg}$$

Using the wet bulb temperature, $t_{sl} = -3°C$:

$$T = (-3 + 273.15)K$$

$$T = 270.15\ K$$

$$\log p_s = 30.59051 - 8.2 \log T + \frac{2.4804}{10^3}T - \frac{3142.31}{T}$$

$$\log p_s = 30.59051 - 8.2 \log 270.15 + \frac{2.4804}{10^3}270.15 - \frac{3142.31}{270.15}$$

$$\log p_s = -0.31$$

$$p_s = 0.489 \text{ kPa}$$

$$p_v = p_{sl} - 101.325 \times \frac{6.66}{10^4}\left(t - t_{sl}\right) \text{kPa}$$

$$p_v = 0.489 - 101.325 \times \frac{6.66}{10^4}(-2-3) \text{kPa}$$

$$p_v = 0.422 \text{ kPa}$$

$$p_v = 422 \text{ Pa}$$

$$g = 0.62197 \frac{p_v}{(101.325 - p_v)} \frac{\text{kg}}{\text{kg}}$$

$$g = 0.62197 \times \frac{0.422}{(101.325 - 0.422)} \frac{\text{kg}}{\text{kg}}$$

$$g = 0.0026 \frac{\text{kg}}{\text{kg}}$$

$$PS = 100 \frac{g}{g_s} \%$$

$$PS = 100 \times \frac{0.0026}{0.00325} \%$$

$$PS = 80\%$$

$$h = 1.0048t + g(2500.8 + 1.863t) \frac{\text{kJ}}{\text{kg}}$$

$$h = 1.0048 \times (-2) + 0.0026 \times (2500.8 + 1.863 \times (-2)) \frac{\text{kJ}}{\text{kg}}$$

$$h = 4.48 \frac{\text{kJ}}{\text{kg}}$$

$$v = \frac{mRT}{P} \frac{\text{m}^3}{\text{kg}}$$

$$v = 1 \text{ kg} \times 287.1 \frac{\text{J}}{\text{kgK}} \times (273 - 2) \text{ K} \times \frac{\text{m}^2}{\left(101325 - 0.4223 \times 10^3\right)\text{N}} \times \frac{1 \text{ Nm}}{1 \text{ J}}$$

$$v = 0.771 \frac{\text{m}^3}{\text{kg dry air}}$$

At the saturation curve, $p_s = p_v = 422$ Pa

$$t_{dp} = 14.62 \ln\left(\frac{p_s}{600.245}\right) °\text{C}$$

$$t_{dp} = 14.62 \ln\left(\frac{422}{600.245}\right) °C$$

$$t_{dp} = -5.1 °C$$

Values calculated for condition 2 are:

$$T = (30 + 273.15) K$$

$$T = 303.15 \text{ K}$$

$$\log p_s = 30.59051 - 8.2 \log T + \frac{2.4804}{10^3} T - \frac{3142.31}{T}$$

$$\log p_s = 30.59051 - 8.2 \log 303.1 + \frac{2.4804}{10^3} \times 303.1 - \frac{3142.31}{303.1}$$

$$\log p_s = 0.627$$

$$p_s = 4.24 \text{Pa}$$

$$g = 0.62197 \frac{p_v}{(101.325 - p_v)} \frac{kg}{kg}$$

$$g = 0.62197 \frac{4.24}{(101.325 - 4.24)} \frac{kg}{kg}$$

$$g = 0.027 \frac{kg}{kg}$$

Using the wet bulb temperature, t_{sl} of $-3°C$ read from references:

$$T = (13.6 + 273.15) K$$

$$T = 286.75 \text{ K}$$

$$\log p_s = 30.59051 - 8.2 \log T + \frac{2.4804}{10^3} T - \frac{3142.31}{T}$$

$$\log p_s = 30.59051 - 8.2 \log 286.75 + \frac{2.4804}{10^3} \times 286.75 - \frac{3142.31}{286.75}$$

$$\log p_s = 0.192$$

$$p_s = 1.556 \text{kPa}$$

$$p_v = p_{sl} - 101.325 \times \frac{6.66}{10^4} (t - t_{sl}) \text{kPa}$$

$$p_v = 1.556 - 101.325 \times \frac{6.66}{10^4} (30 - 13.6) \text{kPa}$$

$$p_v = 0.449 \text{kPa}$$

$$p_v = 449 \text{Pa}$$

$$g = 0.62197 \frac{p_v}{(101.325 - p_v)} \frac{kg}{kg}$$

$$g = 0.62197 \times \frac{0.449}{(101.325 - 0.449)} \frac{\text{kg}}{\text{kg}}$$

$$g = 0.00277 \frac{\text{kg}}{\text{kg}}$$

$$PS = 100 \frac{g}{g_s} \%$$

$$PS = 100 \times \frac{0.00277}{0.02716} \%$$

$$PS = 10.2\%$$

$$h = 1.0048t + g(2500.8 + 1.863t) \frac{\text{kJ}}{\text{kg}}$$

$$h = 1.0048 \times 30 + 0.00277 \times (2500.8 + 1.863 \times 30) \frac{\text{kJ}}{\text{kg}}$$

$$h = 37.22 \frac{\text{kJ}}{\text{kg}}$$

$$v = \frac{mRT}{P} \frac{\text{m}^3}{\text{kg}}$$

$$v = 1\text{kg} \times 287.1 \frac{\text{J}}{\text{kgK}} \times (273 + 30)\text{K} \times \frac{\text{m}^2}{(101325 - 0.449 \times 10^3)\text{N}} \times \frac{1\ \text{Nm}}{1\ \text{J}}$$

$$v = 0.862 \frac{\text{m}^3}{\text{kg dry air}}$$

At the saturation curve, $p_s = p_v = 449\text{Pa}$

$$t_{dp} = 14.62 \ln\left(\frac{p_s}{600.245}\right) {}^\circ\text{C}$$

$$t_{dp} = 14.62 \ln\left(\frac{449}{600.245}\right) {}^\circ\text{C}$$

$$t_{dp} = -4.2^\circ\text{C}$$

The results show sufficiently good agreement.

Cooling

Chilled water, brine or refrigerant is passed through finned pipe cooling coils in the air stream. If the coolant is above the air dew point temperature, no condensation takes place, the process is sensible cooling only and the air remains at a constant moisture content. Figure 4.7 also shows the case when the coolant is below the air dew point and dehumidification takes place. This is the normal case for comfort air conditioning.

The precise path of the air being cooled from point 1 to point 2 is not easily defined as sensible and latent heat exchanges are taking place that involve the transfer of moisture mass from the humid air to the drip tray. Water is being squeezed out of the air quickly when particles of air contact the cold surfaces, but more slowly by forced convective cooling of the air that passes between the pipes and fins. The cooling and dehumidification line from 1 to 2 is more of a curve than straight. A dehumidifying coil operates wet and the exposed water surfaces increase the cooling area. There is little to be gained by attempting to

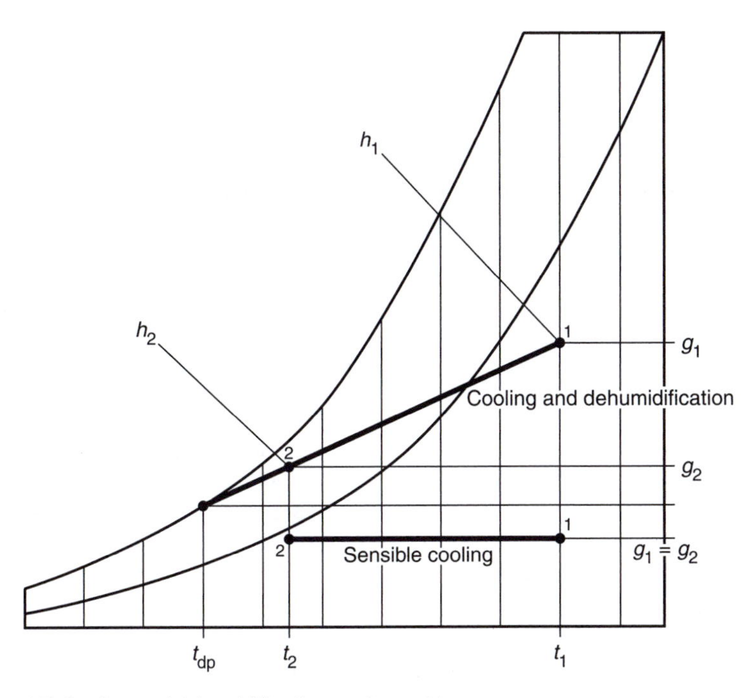

4.7 Cooling and dehumidification psychrometric processes.

plot detailed states between 1 and 2, unless a coil is being designed. The end conditions need only be considered. A slight curve is generated by the reducing performance of the second and subsequent rows of finned tubes in the cooling coil.

When designing the cooling and dehumidification process, the air condition that enters the coil, point 1, is known from the room and external air conditions, but the coil exit condition is not. The moisture content required upon leaving the coil, g_2, will be known, but an amount of variation may be allowable due to a tolerance in the room percentage saturation of upto 5% either side of the target 50%. For economy of refrigeration plant and operating costs of the air conditioning system, the refrigerant temperature needs to be as high as possible, consistent with the plant capacity to meet the cooling load.

Air entering the coil will be cooled towards the saturation curve. A straight line will be used to represent this process. The intersection of the line with the 100% curve will be the dew point temperature of the cooling coil. This dew point will be close to the lowest of the coolant temperatures and this should be found in the last row of finned pipes. There will be a small temperature rise between the coolant and coil dew point due to heat transfer through the pipe wall and along the fin. When refrigerant is within the pipes, it evaporates at constant temperature and pressure within the coil at around 4°C to 10°C for comfort systems but sub-zero for food chillers. Chilled water is often supplied to a coil at 6°C and leaves at up to 16°C. The coil dew point will be within 0.5°C of the lower coolant value. Incomplete mixing of the turbulent air flowing over the coil will mean that not all the air will be subjected to the lowest attainable temperature. The leaving condition, position 2, will be at about 90% of the distance between the entering and dew point states. This 90% is known as the coil contact factor whose value depends upon coil design in terms of pipe and fin spacing, surface corrugations, number of rows of coils, direction of flow of the cooling medium, staggering of pipes in the rows and the average face velocity of the air entering from the duct, around 2.5 m/s. Contact factor can be scaled along the three linear axes of dry bulb temperature, moisture content or specific enthalpy. Dry bulb temperature is normally used.

EXAMPLE 4.7

Summer outdoor air at 32°C d.b. and 22°C w.b. is cooled and dehumidified by a chilled water cooling coil before being passed into an office at a temperature of 16°C d.b. The cooling coil has a dew point of 5°C. Sketch the psychrometric cycle. Find the outside air and supply air properties. Calculate the contact factor for the coil and the amount of moisture removed per kg of dry air.

The procedure is as follows.

1. Sketch Figure 4.8 to show the cooling process.
2. Plot the data on a real chart and sketch.
3. Read the tabulated data for condition 1: 32°C d.b., 22°C w.b., 40% saturation, 0.01231 kg/kg, 63.7 kJ/kg, 0.8812 m^3/kg, 1.957 kPa vapour pressure, and 17.2°C dew point.
4. Read the properties for point 1 from the chart and App to compare for accuracy with the tabulated figures.
5. Plot the cooling coil dew point of 5°C on the 100% curve.
6. Use the cooling coil manufacturer's performance curve or draw a straight line from point 1 to the coil dew point in the first instance.
7. Plot the intersection of this cooling process with a vertical from 16°C d.b. This is point 2 and represents the air condition leaving the coil.
8. Read the wet bulb temperature of point 2 from the chart, 13.2°C w.b.

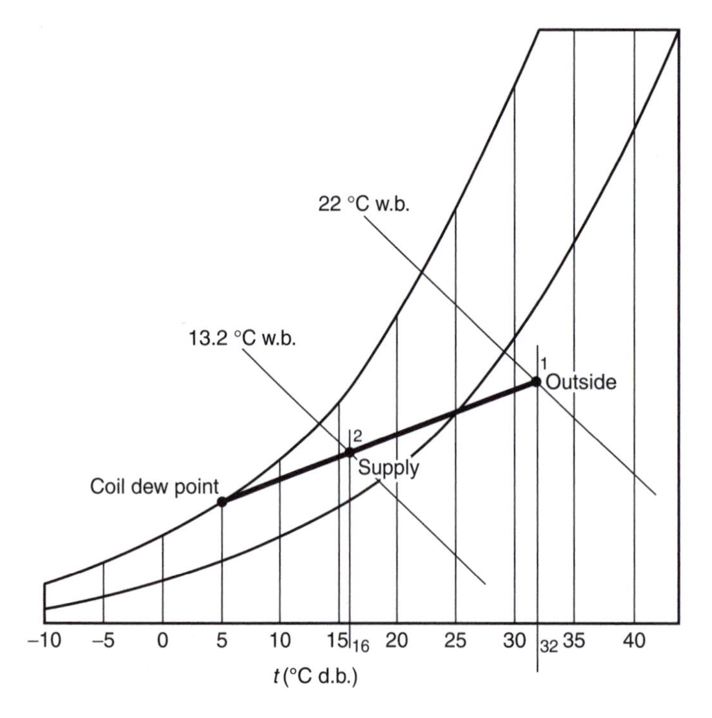

4.8 Solution for example 4.7.

9. Read the tabulated properties of point 2, interpolating as necessary, 16°C d.b., 13.2°C w.b., 73% saturation, 0.0083 kg/kg, 37.11 kJ/kg, 0.83 m^3/kg, 1.329 kPa and 11.2°C dew point.
10. Check the same values on the chart.
11. The contact factor of the cooling coil is found by ratio along the dry bulb scale:

$$\text{Contact factor} = 100 \times \frac{(32-16)}{(32-5)}\%$$

$$\text{Contact factor} = 59\%$$

12. The amount of moisture removed from the air as it is cooled:

$$\text{Condensate collected} = (g_1 - g_2)\frac{\text{kg H}_2\text{O}}{\text{kg dry air}}$$

$$\text{Condensate collected} = (0.01231 - 0.0083)\frac{\text{kg H}_2\text{O}}{\text{kg dry air}}$$

$$\text{Condensate collected} = 0.00401\frac{\text{kg H}_2\text{O}}{\text{kg dry air}}$$

Mixing

Mixing of air streams, shown in Figure 4.9 can take place when:

1. Fresh air drawn into an air handling unit is mixed with recirculated room air prior to conditioning and supply into the building;
2. Supply air is blown through supply grilles or diffusers into the room;
3. Exhaust air is discharged from the building into the external atmosphere;
4. Discharge air from an external dry or evaporative cooling tower is released into the atmosphere;
5. Steam from cooking equipment is released;
6. Opening doors between rooms connects areas held at different temperature, humidity or pressure states;
7. Air is extracted from different rooms and is mixed in the recirculation ductwork;
8. Air from different parts of the same room is drawn towards the extract grill and into ductwork.

Whenever such a mixing process takes place, changes occur in the air condition. Severe cases may produce condensation if highly humid warm vapour discharges from cooking ranges and is mixed with cool air from outdoors or another room. Deposition of grease, condensation or dust particles can occur. In comfort air conditioning, room air extracted into the return air duct may, or may not, accurately represent the room air condition around the occupants due to the relative mixing of air from different parts of the room. All examples are treated on their merits and importance.

The blending of two air streams is often an adiabatic mixing process. Little or no heat is transferred outwards from the process or inwards to it. Either the air handling unit has an insulated casing or the process is sufficiently rapid and is within a small space. Figure 4.9 shows a mixing process where there is continuity of mass of air and it is adiabatic.

$$\text{Mixed mass flow} = \text{mass flow of stream 1} + \text{mass flow of stream 2}$$

$$m_3 = (m_1 + m_2)\frac{\text{kg}}{\text{s}}$$

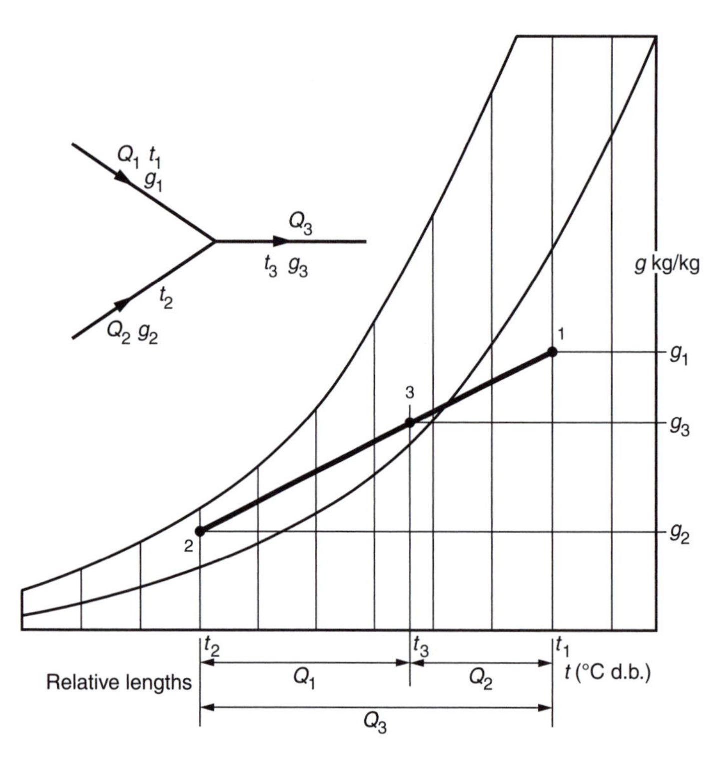

4.9 Mixing of two air streams.

It is important to summate mass flow rates for accurate results although an approximation can be made by adding volume flow rates. The inaccuracy here will be due to the changes in air density as the temperatures change during the mixing process.

$$Q_3 = (Q_1 + Q_2)\frac{\text{m}^3}{\text{s}}$$

Mass and volume flow rates are calculated from:

$$m\frac{\text{kg}}{\text{s}} = Q\frac{\text{m}^3}{\text{s}} \times \rho\frac{\text{kg}}{\text{m}^3}$$

Additive mass flows are:

$$Q_3\rho_3 = (Q_1\rho_1 + Q_2\rho_2)\frac{\text{kg}}{\text{s}}$$

By conservation of energy at the mixing, an enthalpy balance will be:

Total enthalpy in mixed stream 3 = enthapy stream 1 + enthalpy stream 2

$$m_3h_3 = (m_1h_1 + m_2h_2)\text{kW}$$

Similarly, a balance of moisture flows can be made:

$$m_3g_3 = (m_1g_1 + m_2g_2)\frac{\text{kg H}_2\text{O}}{\text{s}}$$

An approximation to the mixed air temperature can be made from the volume flow rates and dry bulb temperatures of the two streams, often with sufficient accuracy for plotting on the psychrometric chart; that is, to within 0.5°C d.b., by making the assumption that the air densities and specific heat capacities remain practically constant.

$$Q_3 t_3 = Q_1 t_1 + Q_2 t_2$$

Divide each side by Q_3:

$$t_3 = \frac{Q_1}{Q_3} t_1 + \frac{Q_2}{Q_3} t_2$$

This shows that the mixing process is a linear ratio of the dry bulb temperature scale using either volume or mass flow rates depending upon the accuracy required. These relative line lengths are indicated in Figure 4.9. It is similarly possible to scale the moisture content and specific humidity scales. The mixed air condition lies on a straight line connecting the two end states.

EXAMPLE 4.8

Outdoor air at −1°C d.b. and 100% saturation enters a mixing box at 1 m^3/s when measured at the outdoor air condition. Recirculated room air enters the mixing box at 22°C d.b. and 50% saturation at 2 m^3/s at the room condition. Calculate the mixed air condition using accurate and approximate methods and plot the data on a psychrometric chart.

The solution procedure is as follows:

1. Plot the room air condition, 1, and outdoor air, 2, on a sketch of the psychrometric chart.
2. Plot the conditions on a real chart.
3. Connect the two states with a straight line.
4. Estimate the mixed air volume flow rate, Q_3.

$$Q_3 = (Q_1 + Q_2) \frac{m^3}{s}$$

$$Q_3 = (2 + 1) \frac{m^3}{s}$$

$$Q_3 = 3 \frac{m^3}{s}$$

5. Carry out an approximation to the mixed air temperature:

$$t_3 = \frac{Q_1}{Q_3} t_1 + \frac{Q_2}{Q_3} t_2$$

$$t_3 = \frac{2}{3} \times 22 + \frac{1}{3} \times (-1)$$

$$t_3 = 14.3°C$$

6. Calculate the mass flow rate of air from the room:

$$m_1 \frac{kg}{s} = Q_1 \frac{m^3}{s} \times \rho_1 \frac{kg}{m^3}$$

$$m_1 \frac{kg}{s} = Q_1 \frac{m^3}{s} \times \frac{1}{v_1} \frac{kg}{m^3}$$

Room air at 22°C d.b. and 50% saturation:

$$v_1 = 0.847 \frac{m^3}{kg}$$

$$\rho_1 = 1.1806 \frac{kg}{m^3}$$

$$m_1 \frac{kg}{s} = 2 \frac{m^3}{s} \times 1.1806 \frac{kg}{m^3}$$

$$m_1 = 2.361 \frac{kg}{s}$$

7. Calculate the mass flow rate of air from outdoors:

$$v_2 = 0.7748 \frac{m^3}{kg}$$

$$m_2 = 1 \frac{m^3}{s} \times \frac{kg}{0.7748\ m^3}$$

$$m_2 = 1.291 \frac{kg}{s}$$

8. Find the total mass flow rate:

$$m_3 = (m_1 + m_2) \frac{kg}{s}$$

$$m_3 = (2.361 + 1.291) \frac{kg}{s}$$

$$m_3 = 3.652 \frac{kg}{s}$$

9. Read the moisture contents from tables, chart or App:

$$g_1 = 0.008366 \frac{kg}{kg}$$

$$g_2 = 0.003484 \frac{kg}{kg}$$

10. Calculate the moisture content of the mixed air,

$$m_3 g_3 = (m_1 g_1 + m_2 g_2) \frac{kg\ H_2O}{s}$$

$$3.652 \times g_3 = (2.361 \times 0.008366 + 1.291 \times 0.003484)\frac{\text{kg H}_2\text{O}}{\text{s}}$$

$$g_3 = \frac{0.02425}{3.652}\frac{\text{kg H}_2\text{O}}{\text{s}}$$

$$g_3 = 0.00664\frac{\text{kg H}_2\text{O}}{\text{s}}$$

11. Plot the intersection of the mixing process line previously drawn on the chart and the moisture content of the mixed air.
12. Read the mixed air temperature from the chart, 14°C d.b.
13. Read the percentage saturation from the chart, 66%.
14. Check that the tabulated value for air at 14°C has corresponding moisture content and percentage saturation. It does, to four significant decimal places.
15. Find the specific enthalpy of the two air streams:

$$h_1 = 43.39\frac{\text{kJ}}{\text{kg}}$$

$$h_2 = 7.702\frac{\text{kJ}}{\text{kg}}$$

16. Calculate the specific enthalpy of the mixture:

$$m_3 h_3 = (m_1 h_1 + m_2 h_2)\,\text{kW}$$

$$3.652 \times h_3 = (2.361 \times 43.39 + 1.291 \times 7.702)\,\text{kW}$$

$$h_3 = 30.77\,\text{kJ/kg}$$

17. Check h_3 with the table, chart and App: 30.77 kJ/kg
18. Read v_3 from the data resources: $v_3 = 0.8217\ \text{m}^3/\text{kg}$
19. Calculate the mixed air volume flow:

$$Q_3 = 3.652\frac{\text{kg}}{\text{s}} \times 0.8217\frac{\text{m}^3}{\text{kg}}$$

$$Q_3 = 3\frac{\text{m}^3}{\text{s}}$$

Steam humidification

Adding moisture to air is most hygienically done with steam injection from electric resistance heaters alongside the air duct. Potable cold water is taken from the mains drinking water pipework at around 10°C that is below the temperature at which micro-organisms in the water become active. Raising water to 100°C and then boiling it into steam at atmospheric pressure avoids bacterial contamination of the air if condensation of the injected moisture does not take place within the ducts. The supply of steam is carefully controlled for economy and health reasons. The specific enthalpy of steam is higher than that of the air being humidified. A little sensible heating takes place. The air moisture content increases. The dry bulb temperature of the air remains close to its lower moisture content value.

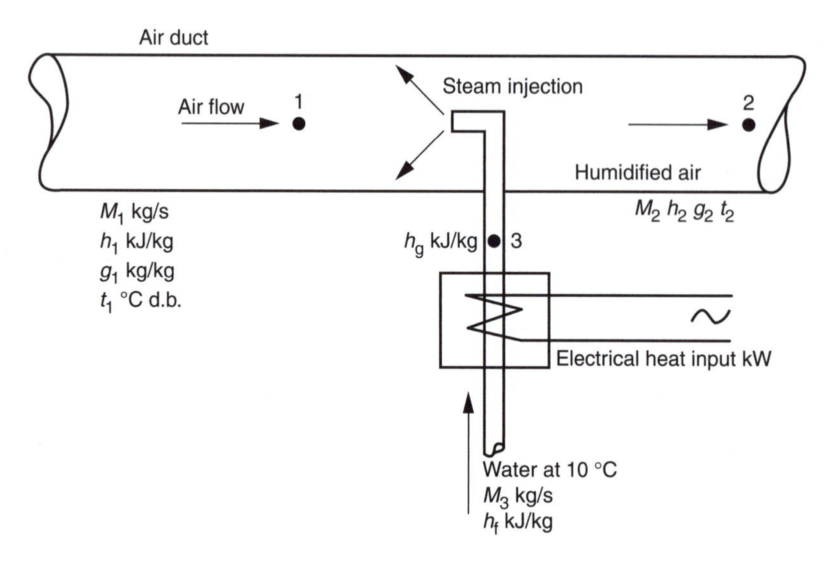

4.10 Steam humidification.

Consider the humidification to be an adiabatic process and there is no condensation of the steam. Figure 4.10 shows the general arrangement. Mass flow rate, enthalpy and moisture flow rate balances can be made:

$$\text{Mass flow rates, } m_1 + m_3 = m_2 \frac{\text{kg}}{\text{s}}$$

$$\text{Enthalpy equation, } m_1 h_1 + m_3 h_g = m_2 h_2 \text{ kW}$$

$$\text{Moisture mass flows, } m_1 g_1 + m_3 = m_2 g_2 \frac{\text{kg}}{\text{s}}$$

Moisture contents g_1 and g_2 are known from the required air conditions. The mass flow rate of steam, m_3 needs to be found. The mixed mass flow rate m_2 is unknown. To find m_3:

$$m_1 g_1 + m_3 = m_2 g_2 \frac{\text{kg}}{\text{s}}$$

$$m_1 g_1 + m_3 = (m_1 + m_3) g_2 \frac{\text{kg}}{\text{s}}$$

$$m_1 g_1 + m_3 = g_2 m_1 + g_2 m_3$$

By rearrangement:

$$m_3 - g_2 m_3 = g_2 m_1 - m_1 g_1$$

$$m_3 (1 - g_2) = m_1 (g_2 - g_1)$$

$$m_3 = m_1 \frac{(g_2 - g_1)}{(1 - g_2)} \frac{\text{kg}}{\text{s}}$$

Calculate the leaving air mass flow rate m_2 and enthalpy h_2:

$$m_2 = (m_1 + m_3)\frac{kg}{s}$$

$$m_2 h_2 = (m_1 h_1 + m_3 h_g)\, kW$$

h_1 is read from the chart or tables and h_g can be taken at 0°C, 2500.8 kJ/kg, although the water may normally be a few degrees warmer. h_2 can be found:

$$h_2 = \frac{m_1 h_1 + m_3 h_g}{m_2} \frac{kJ}{kg}$$

Dry bulb temperature of the humidified air can be calculated from the specific enthalpy formula:

$$h = 1.0048t + g(2500.8 + 1.863t)\frac{kJ}{kg}$$

EXAMPLE 4.9

Outdoor air at a flow rate of 0.5 m^3/s , 20°C d.b. and 10°C w.b. leaves a preheater coil and is humidified to a moisture content of 0.0074 kg/kg by steam injection. Find the steam flow rate, humidified air specific enthalpy, leaving air condition and humidifier heater input power.

The procedure adopted can be as follows.

1. The inlet air data from tables; 20°C d.b., 10°C (10.1 used) w.b, 0.00354 kg/kg, 29.1 kJ/kg, 0.8348 m^3/kg, 24% saturation.
2. $m_1 = 0.5\frac{m^3}{s} \times \frac{kg}{0.8348\ m^3}$

   $m_1 = 0.599\frac{kg}{s}$

3. After humidification, $g_2 = 0.0074\frac{kg}{kg}$
4. $m_3 = m_1\frac{(g_2 - g_1)}{(1 - g_2)}\frac{kg}{s}$

   $m_3 = 0.599 \times \frac{(0.0074 - 0.00354)}{(1 - 0.0074)}\frac{kg}{s}$

   $m_3 = \frac{2.329}{10^3}\frac{kg}{s}$

5. $m_2 = m_1 + m_3$

   $m_2 = \left(0.599 + \frac{2.329}{10^3}\right)\frac{kg}{s}$

   $m_2 = 0.601\frac{kg}{s}$

6. $h_2 = \dfrac{m_1 h_1 + m_3 h_g}{m_2} \dfrac{\text{kJ}}{\text{kg}}$

$$h_2 = \frac{0.599 \times 29.2 + \frac{2.329}{10^3} \times 2500.8}{0.601} \frac{\text{kJ}}{\text{kg}}$$

$$h_2 = 38.69 \frac{\text{kJ}}{\text{kg}}$$

7. Calculate the mixed air temperature:

$$h_2 = 1.0048 t_2 + g_2 (2500.8 + 1.863 t_2) \frac{\text{kJ}}{\text{kg}}$$

$$38.69 = 1.0048 \times t_2 + 0.0074 \times (2500.8 + 1.863 \times t_2) \frac{\text{kJ}}{\text{kg}}$$

$$t_2 = 19.82°\text{C d.b.}$$

   This is indistinguishable from the inlet air dry bulb temperature of 20°C. The air is humidified at virtually constant dry bulb temperature.

8. Plot the two air conditions on a chart, Figure 4.11, and connect them with a straight vertical line.
9. Calculate the heater input power Q,

$$Q = m_1 (h_2 - h_1) \text{kW}$$

$$Q = 0.599 (38.69 - 29.1) \text{kW}$$

$$Q = 5.744 \text{kW}$$

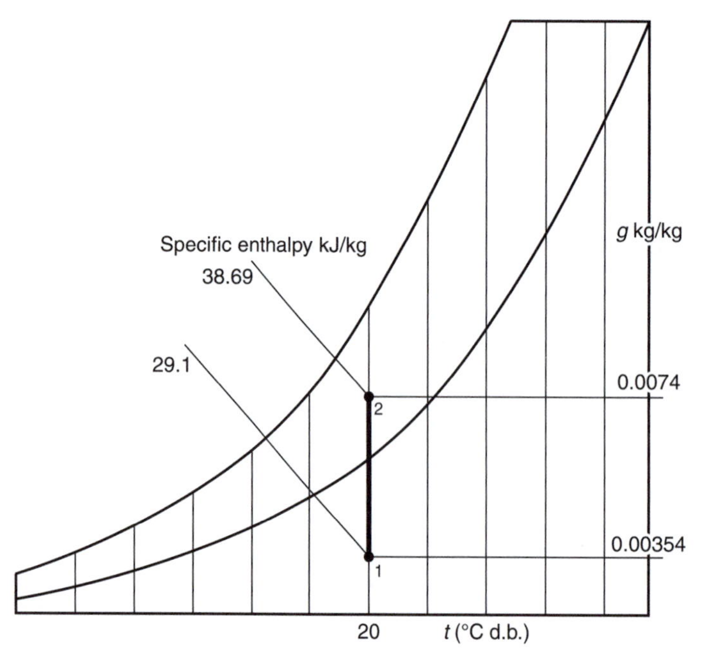

4.11 Steam humidification in example 4.9.

10. Heater power can be estimated from the heat input to the steam when there are no losses:

$$Q = m_3 h_g \text{ kW}$$

$$Q = \frac{2.329}{10^3} \times 2500.8 \text{ kW}$$

$$Q = 5.824 \text{ kW}$$

Direct injection humidification

Direct injection of potable water into the room air through atomising nozzles connected to overhead pipework can be used in industrial premises such as paper and cotton factories. The distribution system is separated from the drinking water system through a storage tank so that a pumped and pressurised supply can be assured. This avoids the use of water spray humidification within the air handling plant and the attendant need for hygiene maintenance. There is some evaporative cooling of the room air as the water turns from atomised liquid into vapour and the dry bulb air temperature will reduce. The calculations are similar to those for steam injection. An ultrasonic humidifier is one which drives water droplets from the surface of a water tank due to the action of a sound generator beneath the water surface. Sound pressure waves cause rapid oscillation of water molecules. Some water molecules are driven off from the surface and into the moving air stream. Electrical energy is needed to generate the ultra high frequency sound waves. There is no appreciable addition of heat to the water. The water does not change state into steam. The air that is being humidified is evaporatively cooled as the latent heat of vapourisation of the water is taken from the air. The air dry bulb temperature will reduce. The distribution of water into the air stream will depend upon the location of the humidifier. There is a risk that water droplets may condense on cool surfaces within the air handling plant, supply air duct or within the conditioned room. The on/off control of any humidifier can result in wide fluctuations in room percentage saturation.

EXAMPLE 4.10

Water at 4 b gauge pressure and 12°C is sprayed into a factory 50 m × 30 m and 5 m high where paper is being produced. The factory air saturation must not drop below 70% at a dry bulb of 20°C. The winter outdoor air temperature is expected to fall to −3°C d.b. at −6°C w.b. on the worst day. A mechanical ventilation system flushes the factory with 1.5 air changes per hour of heated fresh air. Calculate the air preheater coil duty and the expected water supply rate into the humidification pipe and nozzle system. Note the data for each air condition. Compare the results with those from a preheater and steam humidifier that are both in the air handling plant. Figure 4.12 shows the psychrometric process.

A similar procedure to the previous example is adopted.

1. The factory air data from tables; 20°C d.b., 70% saturation, 16.5°C w.b., 0.01033 kg/kg, 46.32 kJ/kg, 0.8438 m^3/kg.
2. Factory volume $V = 50 \text{ m} \times 30 \text{ m} \times 5 \text{ m}$

   Factory volume $V = 7500 \text{ m}^3$

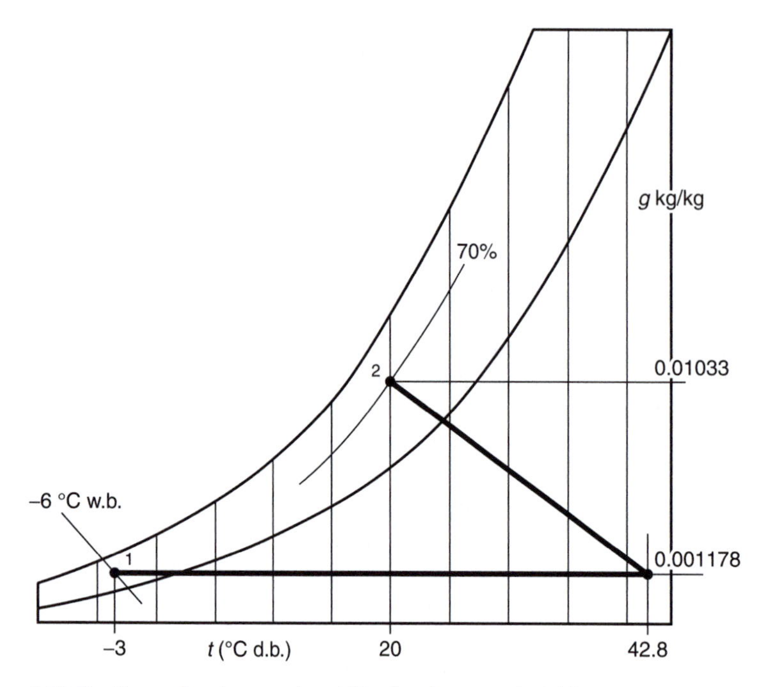

4.12 Heating and water spray humidification for example 4.10.

$$\text{Ventilation rate } Q = 7500\text{m}^3 \times 1.5\frac{\text{air changes}}{\text{h}} \times \frac{1\text{ h}}{3600\text{ s}}$$

$$\text{Ventilation rate } Q = 3.125\frac{\text{m}^3}{\text{s}} \text{ at factory conditions}$$

$$m_1 = 3.125\frac{\text{m}^3}{\text{s}} \times \frac{\text{kg}}{0.8438\text{ m}^3}$$

$$m_1 = 3.704\frac{\text{kg}}{\text{s}}$$

3. After humidification, $g_2 = 0.01033\frac{\text{kg}}{\text{kg}}$
4. Outdoor air from sources: −3°C d.b., −6°C w.b., 40% saturation, 0.001178 kg/kg, −0.076 kJ/kg, 0.7663 m^3/kg. Outdoor enters the factory at $g_2 = 0.001178$kg/kg
   Water spray mass flow rate m_3:

$$m_3 = m_1\frac{(g_2 - g_1)}{(1 - g_2)}\frac{\text{kg}}{\text{s}}$$

$$m_3 = 3.704 \times \frac{(0.01033 - 0.001178)}{(1 - 0.01033)}\frac{\text{kg}}{\text{s}}$$

$$m_3 = \frac{34.253}{10^3}\frac{\text{kg}}{\text{s}}$$

5. $m_2 = m_1 + m_3$

$$m_2 = \left(3.704 + \frac{34.253}{10^3}\right)\frac{kg}{s}$$

$$m_2 = 3.738\frac{kg}{s}$$

6. Specific enthalpy of water sprayed into the room air is read from tables as saturated water at 12°C h_f 50.38 kJ/kg. The absolute pressure of water does not affect h_f. In this case, h_f replaces h_g.

$$h_2 = \left(\frac{m_1h_1 + m_3h_g}{m_2}\right)\frac{kJ}{kg}$$

$$46.32 = \left(\frac{3.704 \times h_1 + \frac{34.253}{10^3} \times 50.38}{3.738}\right)\frac{kJ}{kg}$$

Find h_1 as this is the preheated air condition prior to humidification:

$$h_1 = \frac{(46.32 \times 3.738 - 1.726)}{3.704}\frac{kJ}{kg}$$

$$h_1 = 46.28\frac{kJ}{kg}$$

This is within calculation accuracy of the final air condition h_2. Thus the factory air is humidified adiabatically, meaning at constant specific enthalpy.

7. Plot the outdoor and factory air conditions on a psychrometric chart as in Figure 4.12.
8. Draw a line across the chart at the room specific enthalpy 46.3 kJ/kg.
9. Draw a line horizontally from the outdoor air condition to intersect with the 46.3 kJ/kg specific enthalpy line.
10. The intersection occurs at 42.8°C d.b. This is the preheated factory air temperature prior to humidification.
11. Calculate the preheater coil power:

$$Q = m_1(h_2 - h_1)\text{kW}$$

$$Q = 3.704 \times (46.32 - (-0.076))\text{kW}$$

$$Q = 171.851\text{ kW}$$

12. The alternative humidification system utilising a steam humidifier in the air handling plant would require the preheater to raise outdoor air to 20°C where heater coil's power:

$$Q = m_1(h_2 - h_1)\text{kW}$$

$$Q = 3.704 \times (23.11 - (-0.076))\text{kW}$$

$$Q = 85.881\text{kW}$$

13. Steam boiler input power:

$$Q = m_3h_g$$

$$Q = \frac{34.253}{10^3}\frac{\text{kg}}{\text{s}} \times 2500.8\frac{\text{kJ}}{\text{kg}} \times \frac{\text{kWs}}{\text{kJ}}$$

$$Q = 85.66 \text{ kW}$$

14. Total heat input from the preheater coil and steam boiler:

$$Q = (85.881 + 85.66) \text{ kW}$$

$$Q = 171.541 \text{ kW}$$

This is the same power demand as for the preheater only design. The advantage of the steam injection system is that it would be impractical to preheat the factory air to 42.8°C to enable it to be adiabatically humidified.

Questions

These questions require the use of CIBSE psychrometric charts, data tables or a psychrometric App. Sketch simplified psychrometric charts for rough work and for recording the data for each question.

1. Air in a room is at 22°C d.b. and 15°C w.b. Find these conditions from a chart and verify it from the data tables: percentage saturation, specific volume, dew point, specific enthalpy and moisture content.
2. Outdoor air at 0°C d.b. and 100% saturation is heated with a low pressure hot water coil to 30°C d.b. Sketch the psychrometric cycle on a chart and identify all the condition data for both entry and exit states.
3. Air at 12°C d.b. and 20% saturation is heated through an increase of 25 K and then adiabatically humidified to 90% saturation. Sketch the cycle and identify all the condition data for the end states.
4. Summer outdoor air at 30°C d.b., 21°C w.b. is cooled to 12°C d.b. and 90% saturation. Sketch the psychrometric cycle and identify all the condition data for the end points.
5. Outdoor air at 28°C d.b., 22°C w.b. passes through a direct expansion cooling coil where the refrigerant evaporates at 5°C. Assume that the air side dew point of the coil is 5°C and that 100% contact factor is maintained. Sketch the psychrometric cycle and identify all the condition data for the coil air on and air off states. Calculate the reductions in specific enthalpy and moisture content per kg of air produced.
6. Recirculated room air at 21°C d.b., 50% saturation is mixed in equal amounts with summer outdoor air at 31°C d.b., 22°C w.b. Sketch the mixing process and state the condition of the mixed air.
7. An air handling unit receives recirculated room air at 23°C d.b., 50% saturation and a flow rate of 2 m^3/s, and fresh air at -5°C d.b., 100% saturation with a flow rate of 0.5 m^3/s. The mixed air is heated with a LPHW heater coil to 35°C d.b. and then adiabatically humidified to 24°C d.b. Sketch the psychrometric cycle, identify each condition point and calculate the heater coil load in kW.
8. Calculate the heater coil duty when air is heated from 10°C d.b., 8°C w.b. to 40°C d.b. when the inlet air volume flow rate is 3 m^3/s. Use the inlet specific volume to calculate the air mass flow rate.
9. A cooling coil has an air inlet condition of 29°C d.b., 21.8°C w.b. and an air outlet state of 13°C d.b., 90% saturation. Sketch the cycle and mark on the sketch all the condition data. Calculate the cooling duty of the coil when the inlet air flow to the coil is 4 m^3/s.
10. A single duct air conditioning system takes 1.5 m^3/s of external air at −3°C d.b., 80% saturation and mixes it with 5 m^3/s of recirculated room air at 20°C d.b., 50% saturation. The mixed air is heated to 32°C d.b. prior to being supplied to the rooms. Sketch the cycle and calculate the heater coil duty.
11. An air handling unit mixes 0.8 m^3/s of fresh air at 32°C d.b., 23°C w.b. with 4 m^3/s of recirculated room air at 22°C d.b., 55% saturation. The mixed air passes through a chilled water cooling coil whose

dew point is 6°C d.b. Incomplete contact between the air and dew point surfaces causes 10% of the mixed air to bypass the cooling effect. This 10% air flow mixes with the 90% that contacts the wet surfaces and is cooled to the coil dew point. Sketch the mixing and cooling process and identify all the data. Calculate the refrigeration capacity of the cooling coil in kW and ton refrigeration, given that 1 ton refrigeration is 3.517 kW, and the rate of moisture removal from the air in kg/h.

12. Outside air at −5°C d.b., 80% saturation enters a preheater coil and leaves at 24°C d.b. The fresh air inlet volume flow rate is 2 m^3/s. Find the outdoor air wet bulb temperature and specific volume, the heated air moisture content and percentage saturation. Calculate the heater coil duty.
13. A cooling coil has chilled water passing through it at a mean temperature of 10°C. An air flow of 1.5 m^3/s at 28°C d.b., 23°C w.b. enters the coil and leaves at 15°C d.b. Find the leaving air wet bulb temperature and percentage saturation. Calculate the refrigeration capacity of the coil.
14. 2 m^3/s of air that has been recirculated from an air conditioned room, is at 22°C d.b., 50% saturation. It is mixed with 0.5 m^3/s of fresh air that is at 10°C d.b., 6°C w.b. Calculate the dry bulb air temperature and moisture content of the mixed air. Plot the process on a psychrometric chart and read the specific enthalpy, specific volume and wet bulb temperature of the mixed air.
15. The cooling coil of a packaged air conditioner in a hotel bedroom has refrigerant in it at a temperature of 16°C. Room air at 31°C d.b., 40% saturation enters the coil and leaves at 20°C d.b. at a flow rate of 0.5 m^3/s. Is the air dehumidified by the conditioner? Find the room air wet bulb temperature and specific volume. Calculate the total cooling load in the room.
16. Air in an occupied room is measured to be 24°C d.b. and 16°C w.b. with a sling psychrometer. Calculate the following physical properties and verify them from CIBSE tables and the psychrometric chart, commenting upon any differences found: saturation vapour pressure, saturation moisture content, vapour pressure, moisture content, percentage saturation, specific volume, density, specific enthalpy, dew point.
17. Outdoor air is at 1°C d.b. and 0.5°C w.b. Calculate the physical properties of the air and verify them from CIBSE tables, the psychrometric chart and an App.
18. Outdoor air is at 32°C d.b. and 22°C w.b. Calculate the physical properties of the air and verify them from all the reference sources.

5 System design

Learning objectives

Study of this chapter will enable the reader to:

1. know the reasons for ventilation;
2. identify suitable quantities for ventilation air;
3. calculate the air volume and mass flow rate needed to control sensible heat gains;
4. calculate air density;
5. calculate room air change rate;
6. calculate winter supply air temperature;
7. decide suitable combinations of quantity and temperature for supply air systems;
8. manipulate the design formulae;
9. validate the accuracy of design calculations;
10. calculate the moisture content of supply air used to control latent heat gains;
11. use psychrometric data;
12. use the sensible to total heat ratio line for a room;
13. find the air flows for the distribution ductwork;
14. know how to create zones of different static air pressure;
15. use schematic logic diagrams for air flows through air handling duct systems;
16. calculate air flows between rooms;
17. carry out air quantity and condition calculations for complete systems;
18. plot psychrometric processes;
19. calculate and use psychrometric data;
20. construct psychrometric chart processes from incomplete data;
21. calculate heater and cooler coil loads;
22. undertake design calculations for different countries;
23. identify where condensation occurs;
24. understand mass cooling to reduce energy consumption with fabric energy storage floors and labyrinths.

Key terms and concepts

Air density 144; air mass flow rate 148; air volume flow rate 144; condensation 167; cooling coil load 159; dual duct system 176; ductwork schematic 153; evaporation 150; exhaust air 142; fabric energy storage 169; fan coil units 165; fresh air 162; heat balance 143; heating coil load 161; labyrinth 171; latent heat 150; lower explosive limit 141; mass cooling 169; mixed air 154; moisture 150; occupational exposure limit 141; outdoor air 141; plant air flow 153; psychrometric processes 160; recirculated air 162; room air change rate 148; sensible heat 143; sensible to total heat ratio 152; single duct system 148; static air pressure 153; supply air moisture content 151; supply air temperature 143; system logic 153; transfer between rooms 153; upper explosive limit 141; ventilation 141; warm air heating system 147.

Introduction

System design brings together the constituent parts of an air conditioning system covered in the other chapters. The fresh air inlet quantity is explained in terms of its dilution property for safety and human comfort. The air flow quantity supplied to a conditioned space is calculated from the heating and cooling loads on the building. This forms the basis for air conditioning design.

Sensible and latent heat gains and losses are used. Equations are derived from basic principles. An unknown supply air temperature can be calculated from knowledge of the other factors. Calculation examples are constructed from simple principles into full design applications. Data that are appropriate to different parts of the world are used.

Psychrometric charts are used for complete system designs for both winter and summer states. Schematic duct layouts are introduced to account for all the air flows through a building. Practical problem solving is used for realistic cases. Designers are faced with a fresh challenge by each new application and there are often several correct solutions. The answers shown are illustrative and will encourage the user to apply their principles.

The advantages gained by fabric energy storage walls, floor and labyrinths are explained with examples.

Ventilation requirements

Ventilation air is needed to sustain safe and comfortable conditions for the occupants. The quantity of outdoor air is related to human activity level and the need to remove or dilute atmospheric pollutants. Occupational exposure limits *OEL* are based on an average for an eight hour working day, or a peak 10 minute period. The carbon dioxide CO_2 limit is 5000 parts per million (ppm) or 0.5%.

$$OEL = \frac{5000 \text{ parts } CO_2}{10^6 \text{ parts air}} \times \%$$

$$OEL = 0.5\%$$

Other pollutants have much lower eight hour *OEL*. The ammonia (NH_3) limit is 25 ppm, carbon monoxide (CO) 50 ppm, hydrogen sulphide (H_2S) 10 ppm, nitrogen dioxide (NO_2) 2 ppm and sulphur dioxide (SO_2) 2 ppm. Their threshold of noticeable smell is often even lower and NH_3 is 5 ppm, H_2S 0.1 is ppm and chlorine (Cl) is 0.02 ppm.

Gas and vapour in air can form explosive mixtures. The lower explosive limit *LEL* for methane (CH_4) is 5%, meaning that below 5% methane in air, the mixture is not combustible. The upper explosive limit *UEL* for CH_4 is 15%, so an explosive mixture is created between 5% and 15%. To create the explosion, the mixture has to be exposed to a surface, spark or flame that is at its ignition temperature of 538°C. Gas detectors and alarms can be installed in the building or in service ducts. However, upon a gas or pollutant

Table 5.1 Recommended outdoor air supply rates for air conditioned spaces

Type of space	$\frac{l}{s}$ *per person*	$\frac{l}{m^2 s}$ *floor area*
Factory	8	0.8
Open plan office	8	1.3
Shop	8	3
Theatre	8	—
Private office	12	1.3
Conference room	18–25	—
Residence	12	—
Bar	18	—
Dining room	18	—
Heavy smoking area	25	6
Corridor	—	1.3
House kitchen	—	10
Restaurant kitchen	—	20
Toilets	—	10

Data in this table are approximate and only to be used within this book

leak, fresh air ventilation is relied upon to disperse the fuel and lower the local temperature. Recommended outdoor air supply rates for air conditioned spaces are given in Table 5.1.

Where the polluted air in a room cannot be recirculated to other areas, such as from kitchens and toilets, it is exhausted directly to the outdoor atmosphere. Heat exchangers can be used for energy recovery. Such areas are 100% fresh air ventilated. Most comfort applications will utilise the recirculation of air from conditioned spaces to retain the already correct temperature and humidity. The fresh air will be typically 5% to 25% of the total air in circulation.

Air handling equations

The air handling system is normally designed to remove the sensible and latent heat gains from the conditioned space. These are calculated from the Chapter 3 equations or by dedicated computer software. Sensible and latent heat gain extraction should be at the same rate as their occurrence. When this happens, the conditioned space remains at constant temperature and humidity. The designer calculates the maximum heating, cooling and moisture plant loads. These determine the plant size. The control engineer arranges for the plant to react to maintain the desired conditions within an acceptable band of values.

SH = sensible heat gain or loss to the space kW

LH = latent heat gain or loss to the space kW

t_r = room air temperature °C d.b.

t_s = supply air temperature °C d.b.

SHC = specific heat capacity of air 1.0048 $\frac{kJ}{kgK}$ at 20°C.

ρ = density of air 1.1906 $\frac{kg}{m^3}$ at 20°C d.b., 50% saturation

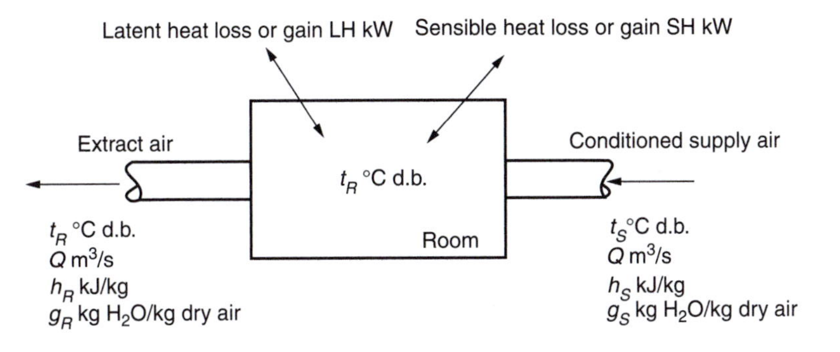

5.1 Basic data for air flow design.

Figure 5.1 shows the arrangement of conditioned supply air entering a room at t_S, absorbing the sensible and latent heat gains and leaving through the extract duct at the room air temperature t_r. While either the supply air temperature or quantity is appropriate to the gains, the room condition remains stable. Any change in the weather, solar or internal gains or occupancy will result in a changed room air temperature. The room air temperature detector may be located in the extract air duct or in the conditioned space. This instigates a change in the fluid flow rate through the heating or cooling coil when a variable temperature system is used, or a variation to the supply air flow rate in a VAV system.

A heat balance can be made for summer cooling:

Heat gain to room = heat gained by air flow through the room

$$SH = \text{air mass flow rate} \times \text{increase in specific enthalpy}$$

$$SH = m\frac{\text{kg}}{\text{s}} \times (h_r - h_S)\frac{\text{kJ}}{\text{kg}} \times \frac{\text{kWs}}{\text{kJ}}$$

The sensible heat gain SH kW and the room air specific enthalpy h_r are known but both the mass flow m and condition h_S of the supply air are to be found. It is the volume flow rate Q m^3/s and temperature t_S°C d.b. of the supply air that is more useful to the designer. Q and t_S are interrelated and have implications for the thermal comfort of the occupants and for the type of air terminal device to be used. There is no easy answer to the formulation of a suitable design and an iterative, or trial and error, routine may be necessary.

$$SH = \text{air mass flow rate} \times SHC \times \Delta t$$

$$SH = m\frac{\text{kg}}{\text{s}} \times SHC\frac{\text{kJ}}{\text{kgK}} \times (t_r - t_S)\text{K}$$

$$m = Q\frac{\text{m}^3}{\text{s}} \times \rho\frac{\text{kg}}{\text{m}^3}$$

$$SH = Q\frac{\text{m}^3}{\text{s}} \times \rho\frac{\text{kg}}{\text{m}^3} \times SHC\frac{\text{kJ}}{\text{kgK}} \times (t_r - t_S)\text{K}$$

The quantity of supply air is to be calculated but density depends upon temperature. From the general gas equation:

$$PV = mRT$$

$$P = \text{absolute pressure } \frac{\text{N}}{\text{m}^2}$$

$$V = \text{volume m}^3$$

$$m = \text{mass of gas kg}$$

$$R = \text{gas constant } 287.1 \frac{\text{J}}{\text{kgK}}$$

$$T = \text{absolute temperature K}$$

$$\rho = \frac{m \text{ kg}}{V \text{ m}^3}$$

$$\rho = \frac{P}{RT}$$

The gas constant R, humidity and pressure should be constants along the supply air duct:

$$\rho \propto \frac{1}{T}$$

$$\text{Standard air density at } T_1, \rho_1 \propto \frac{1}{T_1}$$

$$\text{Density at } T_2, \rho_2 \propto \frac{1}{T_2}$$

By ratio:

$$\rho_2 = \rho_1 \frac{T_1}{T_2}$$

Density of air can be corrected for other supply air temperatures:

$$\rho_2 = 1.1906 \times \frac{(273 + 20)}{(273 + t_s)} \frac{\text{kg}}{\text{m}^3}$$

Substitute this into the heat balance formula:

$$SH = Q \frac{\text{m}^3}{\text{s}} \times \rho_2 \frac{\text{kg}}{\text{m}^3} \times SHC \frac{\text{kJ}}{\text{kgK}} \times (t_r - t_s)\text{K}$$

$$SH = Q \times 1.1906 \times \frac{(273 + 20)}{(273 + t_s)} \frac{\text{kg}}{\text{m}^3} \times 1.0048 \frac{\text{kJ}}{\text{kgK}} \times (t_r - t_s)\text{K} \times \frac{\text{kWs}}{\text{kJ}}$$

$$SH = 351Q \frac{(t_r - t_s)}{(273 + t_s)} \text{ kW}$$

Rearrange to find Q:

$$Q = \frac{SH \text{ kW}}{351} \times \frac{(273 + t_s)}{(t_r - t_s)} \frac{\text{m}^3}{\text{s}} \quad \text{this is for summer cooling}$$

$$Q = \frac{SH \text{ kW}}{351} \times \frac{(273 + t_s)}{(t_s - t_r)} \frac{\text{m}^3}{\text{s}} \quad \text{this is for winter heating}$$

This is the general formula that will be used to find the combination of supply air temperature t_s°C and volume flow rate Q m^3/s that will satisfy the room cooling and heating loads.

EXAMPLE 5.1

An office has a sensible heat gain of 10 kW when the room air temperature is 20°C d.b. Calculate the necessary volume flow rate of supply air to maintain the room at the design temperature when the supply air temperature can be 10°C d.b.

$$Q = \frac{SH\ \text{kW}}{351} \times \frac{(273 + t_S)}{(t_r - t_S)} \frac{\text{m}^3}{\text{s}}$$

$$Q = \frac{10\ \text{kW}}{351} \times \frac{(273 + 10)}{(20 - 10)} \frac{\text{m}^3}{\text{s}}$$

$$Q = 0.806 \frac{\text{m}^3}{\text{s}}$$

EXAMPLE 5.2

A lecture theatre is 12 m × 12 m × 6 m. It is maintained at 20°C d.b. and has four air changes per hour of cooled outdoor air supplied at 15°C d.b. Calculate the maximum cooling loads that the equipment can meet.

$$Q = \frac{N\ \text{air changes}}{\text{h}} \times V\ \text{m}^3 \times \frac{1\ \text{h}}{3600\ \text{s}}$$

$$Q = \frac{4 \times 12 \times 12 \times 6}{3600} \frac{\text{m}^3}{\text{s}}$$

$$Q = 0.96 \frac{\text{m}^3}{\text{s}}$$

$$Q = \frac{SH\ \text{kW}}{351} \times \frac{(273 + t_S)}{(t_r - t_S)} \frac{\text{m}^3}{\text{s}}$$

$$SH = \frac{Q \times 351 \times (t_r - t_S)}{(273 + t_S)}$$

$$SH = \frac{0.96 \times 351 \times (20 - 15)}{(273 + 15)} \text{kW}$$

Maximum cooling capacity $SH = 5.85$ kW

The summer supply air temperature needed for fixed values of sensible heat gain and volume flow rate can be found by rearrangement:

$$Q = \frac{SH\text{kW}}{351} \times \frac{(273 + t_S)}{(t_r - t_S)} \frac{\text{m}^3}{\text{s}}$$

$$Q \times 351 \times (t_r - t_S) = SH \times (273 + t_S)$$

$$Q \times 351 \times t_r - Q \times 351 \times t_s = SH \times 273 + SH \times t_s$$

$$Q \times 351 \times t_r - SH \times 273 = Q \times 351 \times t_s + SH \times t_s$$

$$Q \times 351 \times t_r - SH \times 273 = t_s(Q \times 351 + SH)$$

$$t_s = \frac{351Qt_r - 273SH}{351Q + SH}$$

EXAMPLE 5.3

An open plan office is 20 m × 15 m × 3 m. It is maintained at 22°C d.b. and has eight air changes per hour of cooled supply air. The cooling plant load is 26 kW. Calculate the supply air temperature.

$$Q = \frac{N \text{ air changes}}{\text{h}} \times V \text{ m}^3 \times \frac{1 \text{ h}}{3600 \text{ s}}$$

$$Q = \frac{8 \times 20 \times 15 \times 3}{3600} \frac{\text{m}^3}{\text{s}}$$

$$Q = 2 \frac{\text{m}^3}{\text{s}}$$

$$t_s = \frac{351Qt_r - 273SH}{351Q + SH}$$

$$t_s = \frac{351 \times 2 \times 22 - 273 \times 26}{351 \times 2 + 26}$$

$$t_s = 11.5°\text{C}$$

Check the accuracy of the new formula by substituting for t_s:

$$Q = \frac{SH \text{ kW}}{351} \times \frac{(273 + t_s)}{(t_r - t_s)} \frac{\text{m}^3}{\text{s}}$$

$$Q = \frac{26}{351} \times \frac{(273 + 11.5)}{(22 - 11.5)} \frac{\text{m}^3}{\text{s}}$$

$$Q = 2.007 \frac{\text{m}^3}{\text{s}} \text{ this is correct to within 0.4\% due to rounding}$$

The same procedure is adopted for winter heating. The difference being that the supply air temperature is warmer than that of the room air ($t_s - t_r$).

EXAMPLE 5.4

A room has a winter sensible heat loss of 12 kW when the room air temperature is 21°C d.b. Calculate the necessary volume flow rate of supply air to maintain the room at the design temperature when the supply air temperature is 35°C.

$$Q = \frac{SH \text{ kW}}{351} \times \frac{(273 + t_s)}{(t_s - t_r)} \frac{\text{m}^3}{\text{s}}$$

$$Q = \frac{12}{351} \times \frac{(273+35)}{(35-21)} \frac{m^3}{s}$$

$$Q = 0.752 \frac{m^3}{s}$$

EXAMPLE 5.5

A ducted warm air heating system serves a public hall of 18 m × 11 m × 4 m. It is maintained at 20°C d.b. and has 1.5 air changes per hour of recirculated air supplied at 30°C d.b. Calculate the maximum heat loss from the building that the system can meet.

$$Q = \frac{N \text{ air changes}}{h} \times V\ m^3 \times \frac{1\ h}{3600\ s}$$

$$Q = \frac{1.5 \times 18 \times 11 \times 4}{3600} \frac{m^3}{s}$$

$$Q = 0.33 \frac{m^3}{s}$$

$$Q = \frac{SH\ kW}{351} \times \frac{(273+t_s)}{(t_s - t_r)} \frac{m^3}{s}$$

$$SH = \frac{Q \times 351 \times (t_s - t_r)}{(273+t_s)}$$

$$SH = \frac{0.33 \times 351 \times (30-20)}{(273+30)} kW$$

Maximum heating capacity $SH = 3.823$ kW

It may be assumed that the winter Q m^3/s is the same as the summer Q m^3/s in a constant volume system. The same air conditioning system is providing both conditions and operating continuously throughout the year. Some commercial buildings with very high interior heat loads from people, lights and computer systems in moderate climates can require cooling all year; and also in hot or humid climates. An error in this assumption is that the summer and winter supply air temperatures and densities are different. The higher winter supply temperature increases the air volume flow rate. The winter supply air temperature that is needed for fixed values of sensible heat loss and volume flow rate can be found by rearrangement.

$$Q = \frac{SH\ kW}{351} \times \frac{(273+t_s)}{(t_s - t_r)} \frac{m^3}{s}$$

$$Q \times 351 \times (t_s - t_r) = SH \times (273 + t_s)$$

$$Q \times 351 \times t_s - Q \times 351 \times t_r = SH \times 273 + SH \times t_s$$

$$Q \times 351 \times t_s - SH \times t_s = Q \times 351 \times t_r + SH \times 273$$

$$t_s(351Q - SH) = 351Qt_r + 273SH$$

$$t_s = \frac{351Qt_r + 273SH}{351Q - SH}$$

EXAMPLE 5.6

An open plan office is 20 m × 15 m × 3 m. It is maintained at 18°C d.b. and has eight air changes per hour of supply air in both summer and winter. The room heat loss is 38 kW. Calculate the supply air temperature.

$$Q = 2\frac{m^3}{s} \text{ from example 5.3}$$

$$t_s = \frac{351Qt_r + 273SH}{351Q - SH}$$

$$t_s = \frac{351 \times 2 \times 18 + 273 \times 38}{351 \times 2 - 38}$$

$$t_s = 34.7°C$$

Check the accuracy of the new formula by substituting for t_s.

$$Q = \frac{SH\text{kW}}{351} \times \frac{(273 + t_s)}{(t_s - t_r)} \frac{m^3}{s}$$

$$Q = \frac{38}{351} \times \frac{(273 + 34.7)}{(34.7 - 18)} \frac{m^3}{s}$$

$$Q = 1.995\frac{m^3}{s}, \text{ which is correct to within 0.3\% due to rounding}$$

The air mass flow rate will be needed for heater and cooler coil load calculation. Either the air density or specific volume is used.

EXAMPLE 5.7

A single duct air conditioning system serves a public hall of 960 m^3 that has a sensible heat gain of 20 kW and a winter heat loss of 18 kW. It is maintained at 21°C d.b. during winter and summer. The summer supply air temperature is 13°C d.b. The vapour pressure of the supply air is 1023 Pa in summer and winter. The supply air fan is fitted upstream of the heater and cooler batteries and operates in an almost constant air temperature that is close to that of the room air. Calculate the mass flow rate of the supply air in summer, room air change rate, the winter supply air temperature and comment upon the volume flow rate under winter heating conditions.

In summer,

$$Q = \frac{SH \text{ kW}}{351} \times \frac{(273 + t_s)}{(t_r - t_s)} \frac{m^3}{s}$$

$$Q = \frac{20 \text{ kW}}{351} \times \frac{(273 + 13)}{(21 - 13)} \frac{m^3}{s}$$

$$Q = 2.037 \frac{\text{m}^3}{\text{s}}$$

$$N = Q\frac{\text{m}^3}{\text{s}} \times \frac{3600\text{ s}}{1\text{ h}} \times \frac{1\text{ air change}}{V\text{ m}^3}$$

$$N = 2.037\frac{\text{m}^3}{\text{s}} \times \frac{3600\text{ s}}{1\text{ h}} \times \frac{1\text{ air change}}{960\text{ m}^3}$$

$$N = 7.6\frac{\text{air changes}}{\text{h}}$$

Specific volume of humid air at standard atmospheric pressure is:

$$v = \frac{287.1 \times (273 + 13)}{101325 - 1023}\frac{\text{m}^3}{\text{kg}}$$

$$v = 0.8186\frac{\text{m}^3}{\text{kg}}$$

Summer supply air mass flow rate:

$$m = \frac{Q}{v}$$

$$m = 2.037\frac{\text{m}^3}{\text{s}} \times \frac{\text{kg}}{0.8186\text{ m}^3}$$

$$m = 2.489\frac{\text{kg}}{\text{s}}$$

$2.037\frac{\text{m}^3}{\text{s}}$ is assumed to be the supply air volume flow rate in winter. This may not be strictly accurate due to changes in air density with the supply air temperature. The winter supply air temperature is found from:

$$t_s = \frac{351Qt_r + 273SH}{351Q - SH}$$

$$t_s = \frac{351 \times 2.037 \times 21 + 273 \times 18}{351 \times 2.037 - 18}$$

$$t_s = 28.6°\text{C}$$

Winter supply air specific volume:

$$v = \frac{287.1 \times (273 + 28.6)}{101325 - 1023}\frac{\text{m}^3}{\text{kg}}$$

$$v = 0.863\frac{\text{m}^3}{\text{kg}}$$

Winter supply air mass flow rate will be the same as the summer value as the fan is handling constant temperature air. The winter supply air volume flow rate will be:

$$Q = m\frac{\text{kg}}{\text{s}} \times v\frac{\text{m}^3}{\text{kg}}$$

$$Q = 2.489 \frac{kg}{s} \times 0.863 \frac{m^3}{kg}$$

$$Q = 2.148 \frac{m^3}{s}$$

Substitute $2.148 \frac{m^3}{s}$ for $2.037 \frac{m^3}{s}$ and recalculate the winter supply air temperature to be 28.2°C rather than 28.6°C.

The quantity of air that is supplied to offset the sensible heat transfer is also utilised to balance the latent heat exchange. There is normally a latent heat gain to the conditioned space from the occupants due to their respiration and moisture evaporation through the skin. The supply air absorbs the moisture produced by the occupants and leaves the room at the design room moisture content and percentage saturation. Some industrial applications may require the supply air to add moisture to the conditioned space, as paper or textiles may be absorbing moisture and removing water from the air.

The latent heat equivalent of the evaporated moisture is used for convenience but the psychrometric cycle requires the use of air moisture content kg H_2O/kg dry air.

$$g_r = \text{room air moisture content } \frac{\text{kg } H_2O}{\text{kg dry air}}$$

$$g_s = \text{supply air moisture content } \frac{\text{kg } H_2O}{\text{kg dry air}}$$

$$Q = \text{supply air volume flow rate from thermal load } \frac{m^3}{s}$$

$$h_{fg} = \text{latent heat of vapourisation of water at 0°C } 2453.61 \frac{kJ}{kg}$$

A mass balance can be made between the moisture evaporated in the room and the moisture absorbed by the supply air flow:

moisture evaporated in room = moisture added to supply air

LH = latent heat gain to room from evaporated moisture

$$LH = m \frac{kg}{s} \times (g_r - g_s) \frac{\text{kg } H_2O}{\text{kg dry air}} \times h_{fg} \frac{kJ}{\text{kg } H_2O} \times \frac{kWs}{kJ}$$

$$LH = Q \frac{m^3}{s} \times 1.1906 \frac{kg}{m^3} \times \frac{(273+20)}{(273+t_s)} \times (g_r - g_s) \frac{\text{kg } H_2O}{\text{kg dry air}} \times 2453.61 \frac{kJ}{\text{kg } H_2O} \times \frac{kWs}{kJ}$$

$$LH = Q \times 1.1906 \times \frac{(273+20)}{(273+t_s)} \times (g_r - g_s) \times 2453.61 \text{ kW}$$

$$LH = Q \times \frac{855932}{(273+t_s)} \times (g_r - g_s) \text{ kW}$$

$$Q = \frac{LH\text{kW}}{(g_r - g_s)} \times \frac{(273+t_s)}{860000} \frac{m^3}{s}$$

The 855932 can be written as 860000 with an insignificant 0.5% error. In cases such as a single person office the difference between g_r and g_s will be too small to calculate, measure it with a sensing instrument or control it with a humidifier. A 5% change in room percentage saturation is likely to be unimportant to the occupants, so calculation accuracy can be relaxed.

EXAMPLE 5.8

An office has a sensible heat gain of 12 kW and four occupants each having a latent heat output of 40 W when the room air condition is 20°C d.b., 50% saturation. Calculate the necessary volume flow rate of supply air and its moisture content to maintain the room at the design state when the supply air temperature can be 12°C d.b. The room air moisture content is 0.007376 kg H_2O/kg dry air.

Calculate the supply air quantity from the sensible heat load only:

$$Q = \frac{SH\ \text{kW}}{351} \times \frac{(273 + t_s)}{(t_r - t_s)} \frac{\text{m}^3}{\text{s}}$$

$$Q = \frac{12}{351} \times \frac{(273 + 12)}{(20 - 12)} \frac{\text{m}^3}{\text{s}}$$

$$Q = 1.218 \frac{\text{m}^3}{\text{s}}$$

$$LH = 4\ \text{people} \times 40 \frac{W}{\text{person}} \times \frac{1\ \text{kW}}{10^3\ \text{W}}$$

$$LH = 0.16\ \text{kW}$$

Use the latent heat equation to find the supply air moisture content that is needed:

$$Q = \frac{LH\text{kW}}{(g_r - g_s)} \times \frac{(273 + t_s)}{860000} \frac{\text{m}^3}{\text{s}}$$

$$(g_r - g_s) = \frac{LH}{Q} \times \frac{(273 + t_s)}{860000} \frac{\text{m}^3}{\text{s}}$$

$$g_s = g_r - \frac{LH}{Q} \times \frac{(273 + t_s)}{860000}$$

$$g_s = 0.007376 - \frac{0.16}{1.218} \times \frac{(273 + 12)}{860000}$$

$$g_s = 0.007332 \frac{\text{kg H}_2\text{O}}{\text{kg dry air}}$$

The sensible to total heat ratio S/T can be used on the psychrometric chart during the design of the heating, cooling and humidity control processes.

$$\frac{S}{T} = \frac{\text{sensible heat gain or loss}}{\text{total heat gain or loss}}$$

$$\frac{S}{T} = \frac{SH}{SH + LH}$$

This ratio is used for heating and cooling plant operations. Total heat transfer is the sum of $(SH + LH)$ kW, either may be positive or negative. The ratio is always less than unity.

$$\frac{S}{T} \leq 1$$

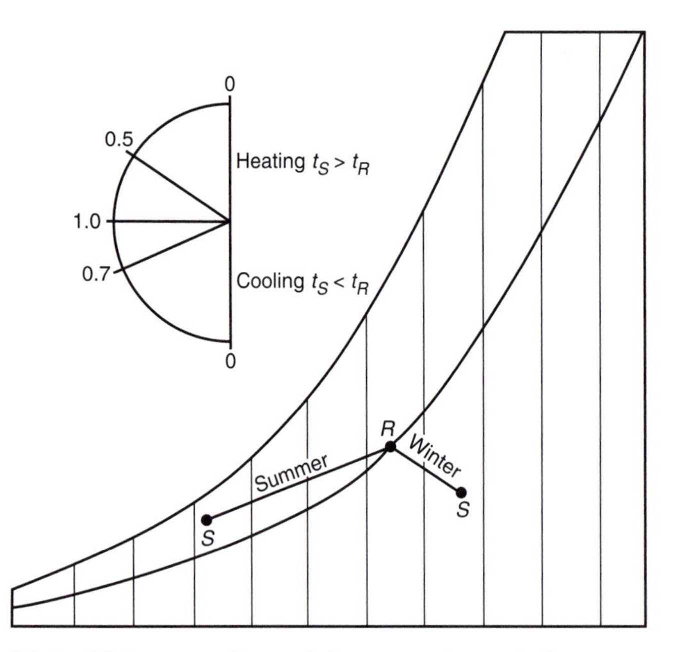

5.2 Sensible heat to total heat ratio line on a psychrometric chart.

The S/T ratio is measured on a quadrant on a psychrometric chart. Draw a straight line from the centre of the quadrant radius to the perimeter. This produces a sloping line that connects the room and supply air conditions at the peak heating or cooling load. The lower quadrant is used when cooling the room and the upper quadrant when the supply air is heating the conditioned space. Transfer the sloping line onto the chart with drawing instruments. Figure 5.2 demonstrates the use of S/T ratio lines. Plotting points on the chart can often be aided with the S/T line when calculation of the air condition cannot be easily made.

EXAMPLE 5.9

A room has a sensible heat gain of 38 kW and a latent heat gain of 2 kW. The room air condition is 20°C d.b. and 50% saturation. The supply air temperature is 12°C d.b. Plot the room and supply conditions on a psychrometric chart.

$$\frac{S}{T} = \frac{38}{38+2}$$

$$\frac{S}{T} = 0.95$$

The intersection of the S/T ratio line and 12°C d.b. occurs at the supply air condition and is shown in Figure 5.3.

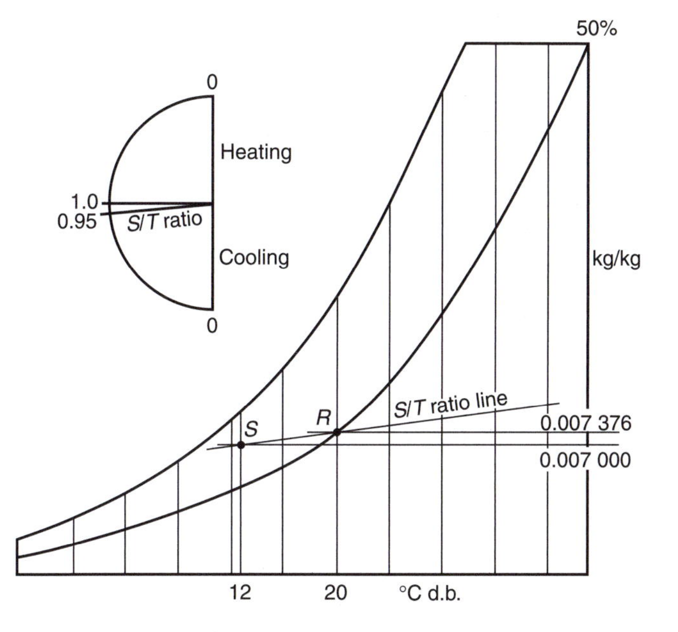

5.3 Solution to example 5.9.

Plant air flow design

A schematic drawing of the air ductwork logic aids comprehension. The quantity of air flow is specified for each section of duct for sizing purposes and these data are an integral part of the overall system design. All air flows need to be specified, particularly where transfers take place between rooms. Accuracy down to a single litre per second (l/s) is required. The designer's accuracy may not be matched by reality when the commissioning stage tolerance of 5% is allowed.

The effect of air transfers between rooms creates draughts around doors, whistling noise through gaps and suction forces on doors. Air transfer grilles in walls or doors may be acceptable but these also pass speech and other sounds. Air ductwork that connects rooms can transfer speech and noise unless they have acoustic attenuation. The commissioning engineer uses the design air flow data, and a current record of all the information is essential. Design changes to air flows are used to update the records.

Static air pressure that is to be maintained in a space is a function of the supply and extract air duct pressures and air flows into and out from the space. Design total air flow into a space must balance the quantity extracted. Temperature differences in the various parts of the air circulation network cause density changes, and volume corrections are applied.

Figure 5.4 is the schematic logic diagram of the air flow quantities for a single duct system serving an office area. Ignore the figures in brackets. They are used later. Some conditioned air is exhausted to atmosphere through the toilet accommodation. Notice that all the flows of air are accounted for and that movement is deliberately in the direction of the toilets by restricting the extract grille quantities in the office and corridor. Fresh air inlet quantity is based upon the occupancy requirement, say, 50 people at 15 l/s per person, or 750 l/s. The building is assumed to be air sealed from the outdoor environment. There is no continuous air flow through the external perimeter, doors or windows. This is unlikely in practice as found by the PROBE reports where a tight standard of building leaks at around 8 m^3/h per m^2 floor area, when pressurised to 50 Pa; however, this is a higher pressure than would normally be maintained in most buildings.

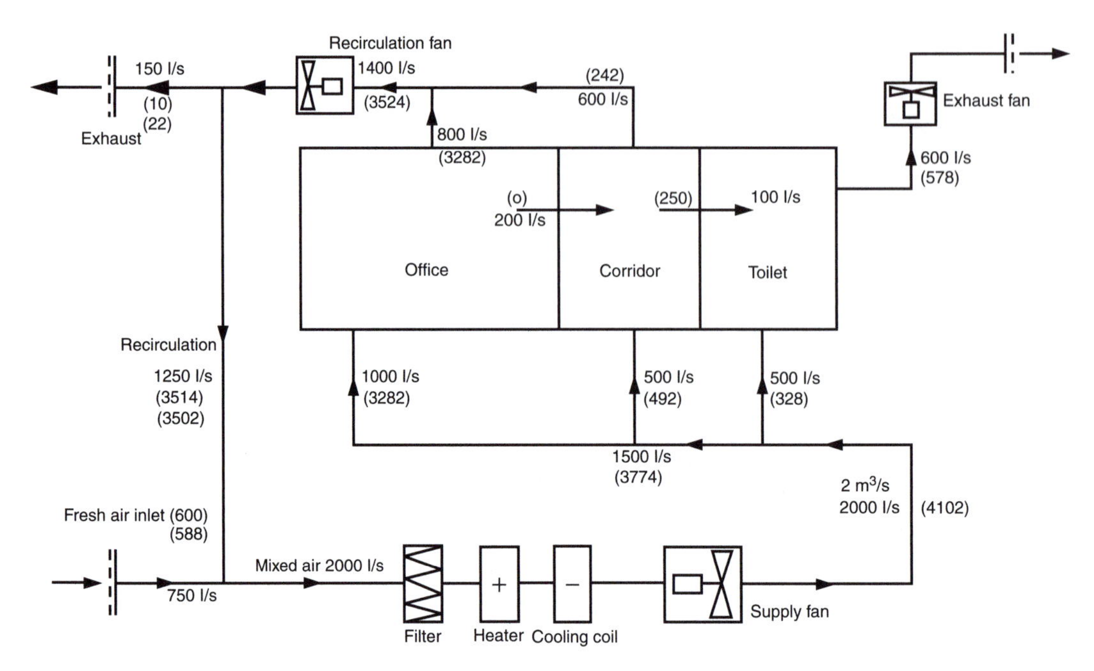

5.4 Schematic logic diagram of air flows to rooms. (Solution to example 5.10 is in brackets.).

EXAMPLE 5.10

Calculate the summer air flows for all the ducts shown in Figure 5.4 using the following data:

Room	*Volume, m^3*	*SH, kW*	*Occupants*
Office	2000	20	40
Corridor	270	3	5
Toilet	150	2	4

Supply air temperature is 15°C d.b. Room air temperatures are 20°C d.b. Toilet is to have a minimum of three air changes per hour. Fresh air requirement is 12 l/s per person. Office and corridor air static pressures are equal. The toilet door has a grille of free area of 0.1 m^2 into the corridor. Maximum air velocity through the grille is to be 2.5 m/s.

Office:

$$Q = \frac{20}{351} \times \frac{(273+15)}{(20-15)} \frac{\text{m}^3}{\text{s}}$$

$$Q = 3.282 \frac{\text{m}^3}{\text{s}},\ 3282 \frac{\text{l}}{\text{s}}$$

Corridor:

$$Q = 0.492\frac{\text{m}^3}{\text{s}},\ 492\frac{\text{l}}{\text{s}}$$

Toilet:

$$Q = 0.328\frac{\text{m}^3}{\text{s}},\ 328\frac{\text{l}}{\text{s}}$$

$$N = Q\frac{\text{m}^3}{\text{s}} \times \frac{3600\ \text{s}}{1\ \text{h}} \times \frac{1\ \text{air change}}{V\ \text{m}^3}$$

$$N = 0.328\frac{\text{m}^3}{\text{s}} \times \frac{3600\ \text{s}}{1\ \text{h}} \times \frac{1\ \text{air change}}{150\ \text{m}^3}$$

$$N = 7.9\frac{\text{air change}}{\text{h}}$$

Total supply air:

$$Q = (3.282 + 0.492 + 0.328)\frac{\text{m}^3}{\text{s}}$$

$$Q = 4.102\frac{\text{m}^3}{\text{s}},\ 4102\frac{\text{l}}{\text{s}}$$

Fresh air quantity for each room:

$$\text{Office } Q_f = 40 \text{ occupants} \times 12\frac{\text{l}}{\text{person s}}$$

$$\text{Office } Q_f = 480\frac{\text{l}}{\text{s}}$$

$$\text{Corridor } Q_f = 5 \text{ occupants} \times 12\frac{\text{l}}{\text{person s}}$$

$$\text{Corridor } Q_f = 60\frac{\text{l}}{\text{s}}$$

$$\text{Toilet } Q_f = 4 \text{ occupants} \times 12\frac{\text{l}}{\text{person s}}$$

$$\text{Toilet } Q_f = 48\frac{\text{l}}{\text{s}}$$

$$\text{Total outside air intake} = (480 + 60 + 48)\frac{\text{l}}{\text{s}}$$

$$\text{Total outside air intake} = 588\frac{\text{l}}{\text{s}}$$

$$\text{Airflow through toilet door grille} = 0.1\ \text{m}^2 \times 2.5\frac{\text{m}}{\text{s}} \times \frac{10^3\text{l}}{\text{m}^3}$$

$$\text{Airflow through toilet door grille} = 250\frac{\text{l}}{\text{s}}$$

Air flows are shown in brackets in Figure 5.4. Check that the correct fresh air flow rate is supplied into each room. The office is the most important room for fresh air as it requires:

$$\text{Office outside air} = \frac{480}{3282} \times 100\%$$

$$\text{Office outside air} = 14.6\% \text{ fresh air}$$

This is higher than the average for all the rooms:

$$\text{Average outside air} = \frac{588}{4102} \times 100\%$$

$$\text{Average outside air} = 14.3\% \text{ fresh air}$$

$$\text{Fresh air proportion of office supply air} = 14.3\% \times 3282 \frac{l}{s}$$

$$\text{Fresh air proportion of office supply air} = 470 \frac{l}{s}$$

But it was designed to be 480l/s so it will be deficient in the supply of outdoor air. Increase the fresh air intake to the plant to:

$$\text{Outside air intake} = 14.6\% \times 4102 \frac{l}{s}$$

$$\text{Outside air intake} = 600 \frac{l}{s}$$

This ensures the designer's intentions are met. Updated air flows are shown in Figure 5.4.

Coordinated system design

The essential principles of system design can now be applied to practical problems. The difference between exercises that are structured for learning purposes and realistic designs is that there may be several possible correct solutions in the real cases. Selection of data and assumptions is open to interpretation. Solutions provided here are appropriate but the reader may arrive at unique answers that can also be correct.

A range of applications is analysed by examples and questions. Use the CIBSE psychrometric charts, data tables and App to find any missing data. There is no need to calculate data from first principles if it is published. Make any necessary assumptions to find a correct solution.

EXAMPLE 5.11

The office module shown in Figure 5.5 is to have a single duct air conditioning system. There is to be a roof mounted air handling and refrigeration plant. The building has five floors of offices and each floor has eight similar layouts. Use a psychrometric chart and the data provided to find the peak summer and winter design loads and air conditions.

Summer room air 22°C d.b., 50% saturation.
Summer outdoor air 30°C d.b., 23°C w.b.
Winter room air 18°C d.b., 50% saturation.

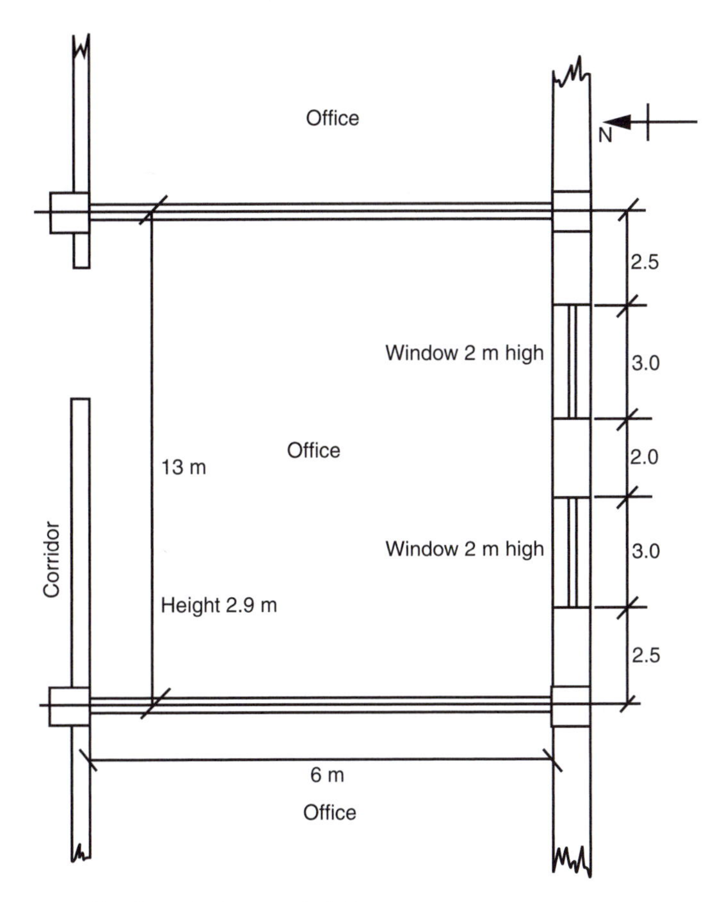

5.5 Office floor plan in example 5.11.

Winter outdoor air −2°C d.b., −3°C w.b.
The office has six occupants, 90 W sensible, 50 W latent each.
The fresh air provision is 12 l/s per person.
There are four computers of 150 W each.
There is no infiltration of outdoor air into the office.
The peak cooling load through the south facing glazing is 333 W/m^2 at noon on 22 September.
A solar gain correction factor of 0.77 applies when the light coloured slatted blinds are used.
Ignore heat gains through the structure.
Surrounding rooms are at the same conditions.
Glazing U value is 5.7 W/m^2 K.
External wall U value is 0.4 W/m^2 K.
Use the simplest forms of heat gain and loss calculation to obtain approximate plant loads.
Peak summer cooling load:

$$\text{Glazing load} = (2 \times 3 \times 2)\ \text{m}^2 \times 0.77 \times 333 \frac{\text{W}}{\text{m}^2}$$

$$\text{Glazing load} = 3077\ \text{W}$$

$$\text{Internal gain} = 6 \text{ occupants} \times 90\text{ W} + 4 \text{ PCs} \times 150\text{ W}$$

$$\text{Internal gain} = 1140\text{ W}$$

$$\text{Peak summer gain} = (3077 + 1140)\text{ W}$$

$$\text{Peak summer gain} = 4217\text{ W}$$

$$\text{Office } V = 13\text{ m} \times 6\text{ m} \times 2.8\text{ m}$$

$$\text{Office } V = 226.2\text{ m}^3$$

$$\text{Heat gain} = \frac{4217\text{ W}}{226.2\text{ m}^3}$$

$$\text{Heat gain} = 18.6\frac{\text{W}}{\text{m}^3}$$

This will be typical for similar offices in the UK facing south. Heat gains and losses vary from 5 W/m^3 to 30 W/m^3 depending upon orientation, thermal insulation, glazing area and internal heat generation from people, lights and equipment.

Winter heat loss:

$$\text{Glazing loss} = 2 \times 3\text{ m} \times 2\text{ m} \times 5.7\frac{\text{W}}{\text{m}^2\text{ K}} \times (18 - (-2))\text{K}$$

$$\text{Glazing loss} = 1368\text{ W}$$

$$\text{Wall loss} = (13\text{ m} \times 2.9\text{ m} - 2 \times 3\text{ m} \times 2\text{ m}) \times 0.4\frac{\text{W}}{\text{m}^2} \times 20\text{ K}$$

$$\text{Wall loss} = 206\text{ W}$$

$$\text{Design heat loss} = (1368 + 206)\text{W}$$

$$\text{Design heat loss} = 1574\text{ W}$$

In summer, try t_s 13°C d.b.

$$Q = \frac{SH\text{kW}}{351} \times \frac{(273 + t_s)}{(t_r - t_s)}\frac{\text{m}^3}{\text{s}}$$

$$Q = \frac{4.217}{351} \times \frac{(273 + 13)}{(22 - 13)}\frac{\text{m}^3}{\text{s}}$$

$$Q = 0.382\frac{\text{m}^3}{\text{s}}, 382\ \frac{\text{l}}{\text{s}}$$

$$N = Q\frac{\text{m}^3}{\text{s}} \times \frac{3600\text{ s}}{1\text{ h}} \times \frac{1\text{ air change}}{V\text{ m}^3}$$

$$N = 0.382\frac{\text{m}^3}{\text{s}} \times \frac{3600\text{ s}}{1\text{ h}} \times \frac{1\text{ air change}}{226.2\text{ m}^3}$$

$$N = 6.1\frac{\text{air changes}}{\text{h}}$$

This is within the acceptable range of 4–20 air changes per hour for most rooms. It would be possible to use a higher supply air temperature, 15°C d.b. at Q of 0.494 m^3/s and N of 7.9 if the lower supply temperature would cause draughts.

$$LH = 6 \times 50 \frac{W}{person} \times \frac{1\ kW}{10^3\ W}$$

$$LH = 0.3 kW$$

$$Q = \frac{LH kW}{(g_r - g_s)} \times \frac{(273 + t_s)}{860000} \frac{m^3}{s}$$

$$g_r = 0.008366 \frac{kg}{kg}$$

$$(g_r - g_s) = \frac{LH}{Q} \times \frac{(273 + t_s)}{860000} \frac{m^3}{s}$$

$$g_s = g_r - \frac{LH}{Q} \times \frac{(273 + t_s)}{860000}$$

$$g_s = 0.008366 - \frac{0.3}{0.382} \times \frac{(273 + 13)}{860000}$$

$$g_s = 0.008105 \frac{kg}{kg}$$

$$\text{Fresh air supply} = 12 \frac{l}{s} \times 6 \text{ people}$$

$$\text{Fresh air supply} = 72 \frac{l}{s}$$

$$\text{Fresh air proportion} = \frac{72}{382} \times 100\%$$

$$\text{Fresh air proportion of supply air} = 19\%$$

To find the mixed air condition entering the cooling coil, proportion the outdoor air and recirculated room air flows:

$$t_m = 22 + 19\% \times (30 - 22)°C \text{ d.b.}$$

$$t_m = 23.5°C \text{ d.b.}$$

Plot the mixed air conditions on the psychrometric chart, a sketch of which is shown in Figure 5.6, t_m 17.2°C w.b., specific enthalpy h_m 48 kJ/kg, v 0.853 m^3/kg.

Plot the supply air condition on the chart at 13°C d.b., 0.0081kg/kg, 11.8°C w.b., h_s is 33.5 kJ/kg.

$$\text{Cooling coil load} = 0.382 \frac{m^3}{s} \times (48 - 33.5) \frac{kJ}{kg} \times \frac{kg}{0.835\ m^3} \times \frac{kWs}{kJ}$$

$$\text{Cooling coil load} = 6.494 \text{ kW}$$

Find the winter supply air temperature for Q 0.382 m^3/s:

$$t_s = \frac{351 Q t_r + 273 SH}{351 Q - SH}$$

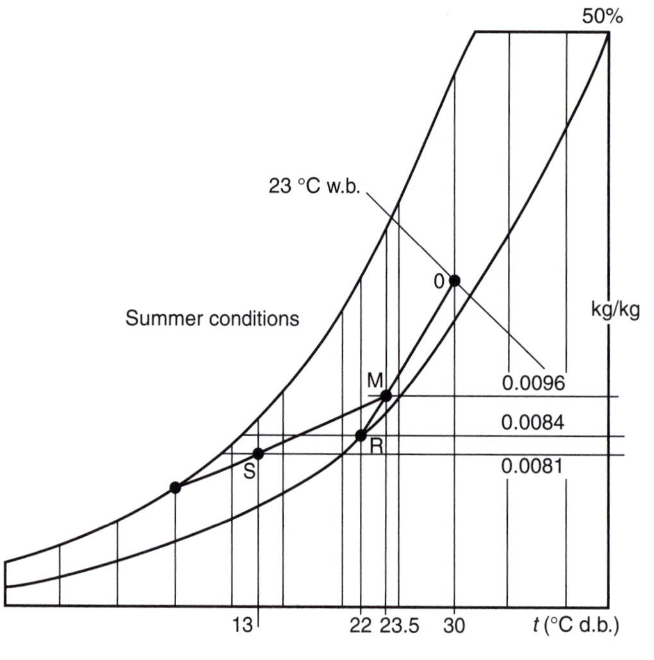

5.6 Summer cycle for example 5.11.

$$t_s = \frac{351 \times 0.382 \times 18 + 273 \times 1.574}{351 \times 0.382 - 1.574}$$

$$t_s = 21.5°\text{C d.b.}$$

Plot the winter room and outdoor air conditions on the psychrometric chart. Join them with a straight line and calculate the mixed air state.

$$t_m = 18 - 0.19 \times (18 - (-2))°\text{C d.b.}$$

$$t_m = 14.2°\text{C d.b.}$$

From the earlier calculation, ignoring the change in t_s:

$$(g_r - g_s) = \frac{0.261}{10^3} \frac{\text{kg}}{\text{kg}}$$

From psychrometric data, winter $g_r = 0.006492$ kg/kg

$$g_s = \left(0.006492 - \frac{0.261}{10^3}\right) \frac{\text{kg}}{\text{kg}}$$

$$g_s = 0.006231 \frac{\text{kg}}{\text{kg}}$$

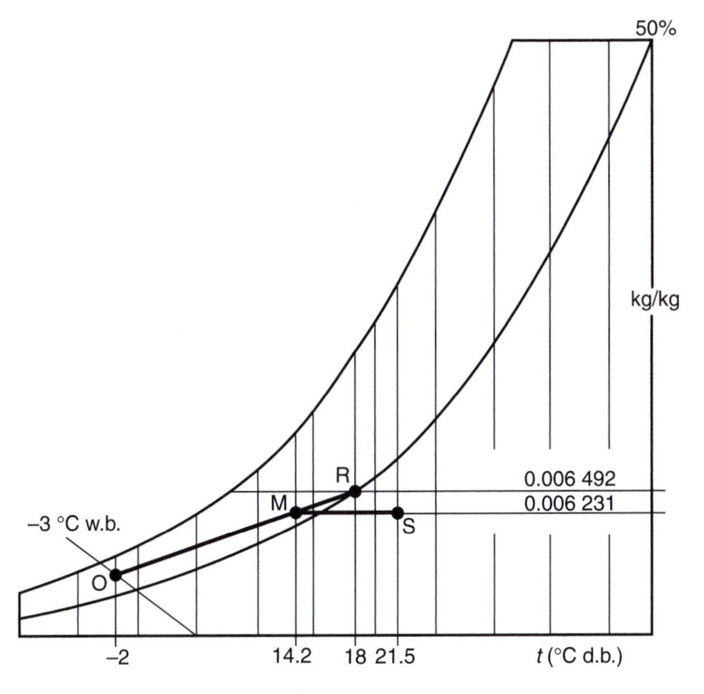

5.7 Winter cycle for example 5.11.

A sketch of the winter heating cycle is shown in Figure 5.7, t_m 14.2°C w.b., specific enthalpy h_m 30 kJ/kg, v_m 0.823 m^3/kg. Plot the supply air condition on the chart at 21.5°C d.b., 0.006231 kg/kg, 13.5°C w.b., h_s 37.5kJ/kg.

$$\text{Heating coil load} = 0.382 \frac{m^3}{s} \times (37.5 - 30) \frac{kJ}{kg} \times \frac{kg}{0.823 m^3} \times \frac{kW\,s}{kJ}$$

$$\text{Heating coil load} = 3.481\text{ kW}$$

Figure 5.8 shows the duct air flows for one room of this type. Nearly all air conditioning systems are in the single duct air handling category. The principal variation is the addition of terminal heating and cooling coils in fan coil, variable air volume or induction systems. The use of air volume flow rate regulation, VAV systems, does not affect the psychrometric peak loads, only the part load operation. Where terminal heating or cooling coils are used, the central air handling plant may only be supplying conditioned fresh air. Recirculated air remains within the room.

Outdoor air may provide some sensible cooling to the room or it may be at the desired room temperature. The fresh air can be designed to provide all the humidity control. It is common to do sensible cooling only with recirculation units in computer halls. The computer hall outdoor air supply plant controls room humidity to ±5% of the desired value. In all applications, the supply air condition will be determined from the room or zone having the greatest demand upon it. This will be the space requiring the most extreme central plant supply temperature and moisture content. Other rooms that are connected to the same duct will use their terminal coils to provide locally correct conditions.

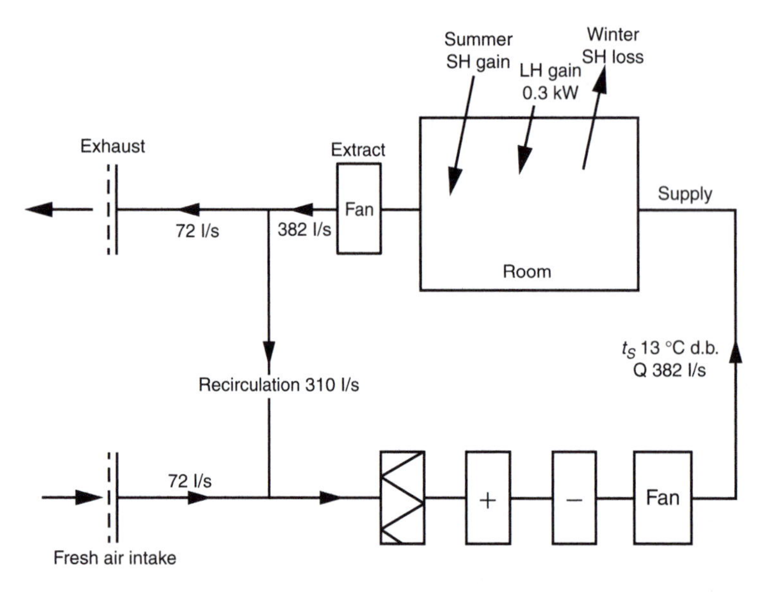

5.8 Air handling schematic for example 5.11.

EXAMPLE 5.12

A Sidney building has two north facing office floors served from a single duct air conditioning system with fan coil units. Use a psychrometric chart and the data provided to find the peak summer and winter design loads, air flows and conditions.

Summer room air 21°C d.b., 50% saturation.
Summer outdoor air 35°C d.b., 24°C w.b.
Winter room air 21°C d.b., 50% saturation.
Winter outdoor air 6°C d.b., 3°C w.b.

Office 1 has 60 m of perimeter length, a south facing glazing area of 100 m^2, is 7 m wide and 3 m high. It has 40 occupants and 25 computers of 120 W each.

Office 2 has 30 m of perimeter length, glazing of 60 m^2, is 5 m wide and 4 m high. It has 15 occupants and 10 computers of 150 W each.

Occupancy heat gains are 90 W sensible and 50 W latent each.
Outdoor air provision is 12 l/s per person.
There is no infiltration of outdoor air into the offices.
Peak cooling load through the south facing glazing is 450 W/m^2 at noon on 21 February.
A solar gain correction factor of 0.55 applies to the glass type and shading blinds.
Ignore heat gains through the structure.
Surrounding rooms are at the same conditions.
Glazing U value is 2.8 W/m^2K.
External wall U value is 0.6 W/m^2K.

Use the simplest forms of heat gain and loss calculation to obtain approximate plant loads and recommend a suitable design for the fan coil units.

Office 1

Peak summer cooling load:

$$\text{Glazing load} = 100\ \text{m}^2 \times 0.55 \times 450 \frac{\text{W}}{\text{m}^2}$$

$$\text{Glazing load} = 24.75\ \text{kW}$$

$$\text{Internal gain} = 40\ \text{occupants} \times 90\ \text{W} + 25\ \text{PCs} \times 120\ \text{W}$$

$$\text{Internal gain} = 6.6\ \text{kW}$$

$$\text{Peak summer gain} = (3077 + 1140)\ \text{W}$$

$$\text{Peak summer gain} = 31.35\ \text{kW}$$

$$\text{Office}\ V = 60\ \text{m} \times 7\ \text{m} \times 3\ \text{m}$$

$$\text{Office}\ V = 1260\ \text{m}^3$$

$$\text{Heat gain} = \frac{31350\ \text{W}}{1260\ \text{m}^3}$$

$$\text{Heat gain} = 24.9 \frac{\text{W}}{\text{m}^3}$$

Winter heat loss:

$$\text{Glazing loss} = 100\ \text{m}^2 \times 2.8 \frac{\text{W}}{\text{m}^2\ \text{K}} \times (21 - 6)\text{K}$$

$$\text{Glazing loss} = 4.2\ \text{kW}$$

$$\text{Wall loss} = \left(60\ \text{m} \times 3\ \text{m} - 100\ \text{m}^2\right) \times 0.6 \frac{\text{W}}{\text{m}^2} \times 15\ \text{K}$$

$$\text{Wall loss} = 0.72\ \text{kW}$$

$$\text{Design heat loss} = (4.2 + 0.72)\text{kW}$$

$$\text{Design heat loss} = 4.92\ \text{kW}$$

Office 2

Peak summer cooling load:

$$\text{Glazing load} = 60\ \text{m}^2 \times 0.55 \times 450 \frac{\text{W}}{\text{m}^2}$$

$$\text{Glazing load} = 14.85\ \text{kW}$$

$$\text{Internal gain} = 15\ \text{occupants} \times 90\ \text{W} + 10\ \text{PCs} \times 150\ \text{W}$$

$$\text{Internal gain} = 2.85\ \text{kW}$$

$$\text{Peak summer gain} = 17.7 \text{ kW}$$

$$\text{Office } V = 30 \text{ m} \times 5 \text{ m} \times 4 \text{ m}$$

$$\text{Office } V = 600 \text{ m}^3$$

$$\text{Heat gain} = \frac{17700 \text{ W}}{600 \text{ m}^3}$$

$$\text{Heat gain} = 29.5 \frac{\text{W}}{\text{m}^3}$$

Winter heat loss:

$$\text{Glazing loss} = 60 \text{ m}^2 \times 2.8 \frac{\text{W}}{\text{m}^2\text{K}} \times (21 - 6)\text{K}$$

$$\text{Glazing loss} = 2.52 \text{ kW}$$

$$\text{Wall loss} = \left(30 \text{ m} \times 4 \text{ m} - 60 \text{ m}^2\right) \times 0.6 \frac{\text{W}}{\text{m}^2} \times 15 \text{ K}$$

$$\text{Wall loss} = 0.54 \text{ kW}$$

$$\text{Design heat loss} = (2.52 + 0.54)\text{kW}$$

$$\text{Design heat loss} = 3.06 \text{ kW}$$

In summer, try t_s 12°C d.b. for Office 1:

$$Q = \frac{SH\text{kW}}{351} \times \frac{(273 + t_s)}{(t_r - t_s)} \frac{\text{m}^3}{\text{s}}$$

$$Q = \frac{31.35}{351} \times \frac{(273 + 12)}{(21 - 12)} \frac{\text{m}^3}{\text{s}}$$

$$Q = 2.828 \frac{\text{m}^3}{\text{s}}, 2828 \frac{\text{l}}{\text{s}}$$

$$N = Q \frac{\text{m}^3}{\text{s}} \times \frac{3600 \text{ s}}{1 \text{ h}} \times \frac{1 \text{ air change}}{V \text{ m}^3}$$

$$\text{N} = 2.828 \frac{\text{m}^3}{\text{s}} \times \frac{3600 \text{ s}}{1 \text{ h}} \times \frac{1 \text{ air change}}{1260 \text{ m}^3}$$

$$\text{N} = 8 \frac{\text{air changes}}{\text{h}}$$

$$\text{LH} = 40 \times 50 \frac{\text{W}}{\text{person}} \times \frac{1 \text{ kW}}{10^3 \text{ W}}$$

$$\text{LH} = 2 \text{ kW}$$

From data, $g_r = 0.007857 \frac{\text{kg}}{\text{kg}}$

$$(g_r - g_s) = \frac{LH}{Q} \times \frac{(273 + t_s)}{860000} \frac{\text{m}^3}{\text{s}}$$

$$g_s = 0.007857 - \frac{2}{2.828} \times \frac{(273 + 12)}{860000}$$

$$g_s = 0.007857 - 0.000234$$

$$g_s = 0.007622 \frac{kg}{kg}$$

Now calculate the summer supply air conditions needed for Office 2 using $t_s = 12°$ C d.b.

$$Q = 1.6 \frac{m^3}{s}$$

$$N = 9.6 \frac{\text{air changes}}{h}$$

$$LH = 0.75 \text{ kW}$$

$$g_r = 0.007857 \frac{kg}{kg}, \text{ the same as Office 1}$$

$$g_s = 0.007857 - \frac{0.75}{1.6} \times \frac{(273 + 12)}{860000}$$

$$g_s = 0.007702 \frac{kg}{kg}$$

Supply air moisture content required for Office 1 is lower than that for Office 2, so g_s of 0.007622 kg/kg will be used for the air that leaves the central air handling plant. This means that the humidity control of Office 2 will not be correct, however, the error will not be noticeable and probably not measurable.

Figure 5.9 shows the air handling schematic for the two offices. It is likely that several fan coil units would be installed in each office to provide local control of each part of the room. Manufacturers' literature is used to find units that deliver suitable air flow rates. If each unit can handle the supply air for 6 m of perimeter, than 10 are needed for Office 1 and five for Office 2. The cooling coil in the central air handling plant will remove the excess heat from the fresh air intake and the fan coil units will take out the local heat gains. The mixing of conditioned fresh air and recirculated room air takes place within each fan coil unit.

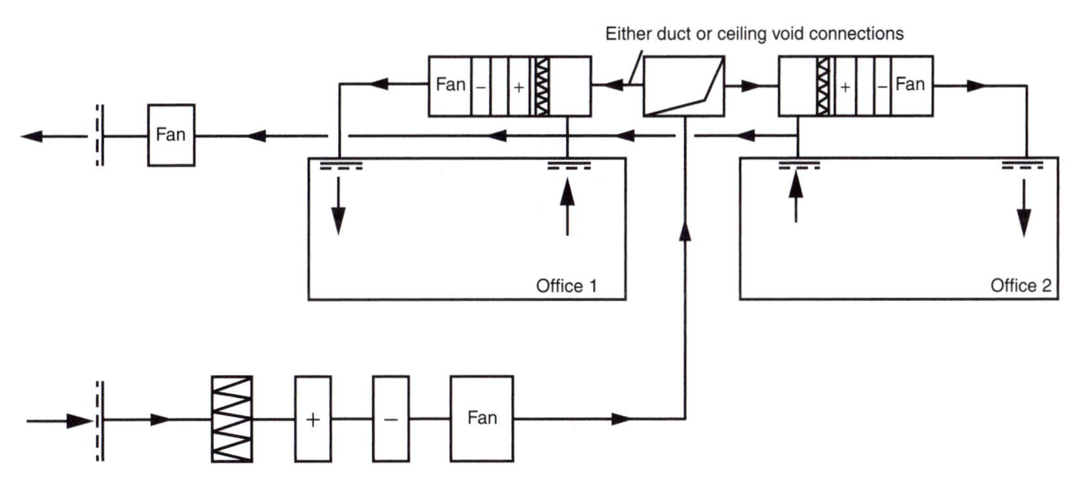

5.9 Air handling plant schematic for the fan coil unit system in example 5.12.

Find the mixed air condition entering an Office 1 unit:

$$OA = 12\frac{l}{s} \times 40 \text{ occupants}$$

$$OA = 480\frac{l}{s}$$

$$OA = 0.48\frac{m^3}{s}$$

$$OA \text{ proportion of } SA = \frac{0.48}{2.828} \times 100\%$$

$$OA \text{ proportion of } SA = 17\%$$

$$\text{Each fan coil unit will pass } Q = \frac{2.828}{10}\frac{m^3}{s}$$

$$\text{Each fan coil unit will pass } Q = 0.283\frac{m^3}{s}, \text{ of which 17\% is } OA, 48\frac{l}{s}.$$

The summer cycle can be drawn on the psychrometric chart, Figure 5.10, and unknown points or lines are found by experimental construction. Plot the known outside *O*, room *R* and supply air *S* conditions. Construct a horizontal line from *O* to the room air temperature 21°C d.b., but 19.9°C w.b. This is condition *C*. It is the fresh air state that leaves the central air handling plant. Maintaining *C* at the room air temperature means that the ducted air system is at room temperature and will have negligible heat gains or losses through the building.

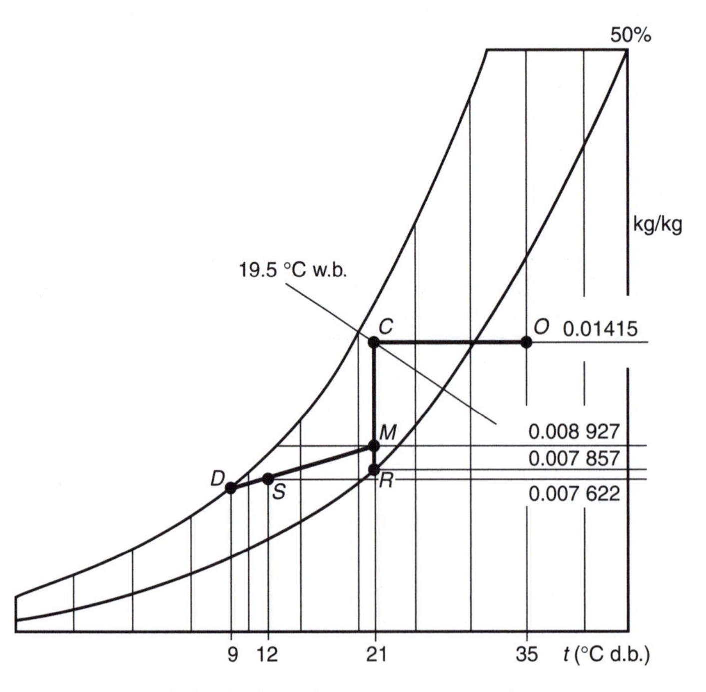

5.10 Summer cycle for the fan coil unit system in example 5.12.

Construct a vertical line from C to R. Recirculated room air R and fresh air C mix in the fan coil unit along this line. The mixture M must contain 17% fresh air and 83% recirculated room air. This mixing process takes place along line CR such that CM will be 83% of CR. The fresh air proportion MR will be 17% of CR. The ratio can be calculated along the moisture content axis as it has equal increments.

$$g_m = 0.007857 + 0.17 \times (0.01415 - 0.007857)\frac{kg}{kg}$$

$$g_m = 0.008927\frac{kg}{kg}$$

Condensation will not take place in the mixing chamber of the fan coil unit as the outside air dew point is not reached. This mixed air M passes to the terminal unit cooling coil where it is further cooled and dehumidified to the correct supply condition S.

Construct a straight line from M, through S, to the saturation curve at 9°C. This is the room coil dew point D. The supply air leaves the room cooling coil at 12°C d.b., 10.8°C w.b., S. The data for the two cooling coils are:

(a) At O, 35°C d.b., 24°C w.b., $71.2\frac{kJ}{kg}$, $0.8923\frac{m^3}{kg}$, $0.01415\frac{kg}{kg}$;

(b) At C, 21°C d.b., 19.9°C w.b., $57.2\frac{kJ}{kg}$, $0.01415\frac{kg}{kg}$;

(c) At M, 21°C d.b., 15.8°C w.b., $43.74\frac{kJ}{kg}$, $0.845\frac{m^3}{kg}$, $0.008927\frac{kg}{kg}$;

(d) At S, 12°C d.b., 10.8°C w.b., $31.23\frac{kJ}{kg}$, $0.007622\frac{kg}{kg}$

Calculate the coil loads:

$$SH = m\frac{kg}{s} \times (h_r - h_s)\frac{kJ}{kg} \times \frac{kWs}{kJ}$$

The room unit cooling coil load M to S is:

$$SH = 0.283\frac{m^3}{s} \times (43.74 - 31.23)\frac{kJ}{kg} \times \frac{kg}{0.845\ m^3} \times \frac{kW\ s}{kJ}$$

$$SH = 4.19\ kW$$

There are 10 room units in Office 1.

The OA plant cooling coil load O to C is:

$$SH = 2.828\frac{m^3}{s} \times (71.2 - 57.2)\frac{kJ}{kg} \times \frac{kg}{0.8923\ m^3} \times \frac{kWs}{kJ}$$

$$SH = 44.371\ kW$$

Find the office winter supply air temperature for Q $0.283\frac{m^3}{s}$:

$$t_s = \frac{351Qt_r + 273SH}{351Q - SH}$$

$$t_s = \frac{351 \times 0.283 \times 21 + 273 \times 3.06}{351 \times 0.283 - 3.06}$$

$$t_s = 30.3°\text{C d.b.}$$

Supply air moisture content in winter is the same as during summer:

$$g_s = 0.007702 \frac{\text{kg}}{\text{kg}}$$

Plot the winter outdoor air condition *O* on the psychrometric chart. Raise the outdoor air supply to the room air temperature of 21°C d.b. but at 10.5°C w.b. condition *H*. Draw a vertical line from *H* to *R*. This represents the heated fresh air mixing with the recirculated room air that enters the fan coil unit. *M* is the mixed air and 17% of it comes from *H* and 83% from *R*. This mixing process takes place along line *HR* such that *HM* will be 83% of *HR*. The ratio can be calculated along the moisture content axis as it has equal increments.

$$g_m = 0.007857 - 0.17 \times (0.007857 - 0.003432) \frac{\text{kg}}{\text{kg}}$$

$$g_m = 0.007105 \frac{\text{kg}}{\text{kg}}$$

Draw a horizontal line from *M* to the supply air temperature *S* at 30.3°C d.b., but 0.007105 kg/kg. *M* to *S* is the heating process through the coil in the room terminal unit. The supply air leaves the room heating coil at 30.3°C d.b., 17.5°C w.b. and 0.007105 kg/kg. This is below the required supply air moisture content of 0.007622 kg/kg. Estimate the room air percentage saturation that will be produced, g_r, by adding the difference $(g_r - g_s)$ to g_s.

$$g_r = (0.007105 + 0.000234) \frac{\text{kg}}{\text{kg}}$$

$$g_r = 0.007339 \frac{\text{kg}}{\text{kg}}$$

Plot the expected room air condition 21°C d.b., 0.007339 kg/kg and find that it will be at around 47% saturation. This is sufficiently close to the desired 50% for humidification to be avoided. Any value above 40% is likely to be acceptable for comfort purposes. The data for the two heating coils are:

(a) At *O*, 6°C d.b., 3°C w.b., $14.66 \frac{\text{kJ}}{\text{kg}}, 0.7947 \frac{\text{m}^3}{\text{kg}}, 0.003432 \frac{\text{kg}}{\text{kg}};$

(b) At *H*, 21°C d.b., 10.5°C w.b., $29.91 \frac{\text{kJ}}{\text{kg}}, 0.003432 \frac{\text{kg}}{\text{kg}};$

(c) At *M*, 21°C d.b., 14°C w.b., $38.95 \frac{\text{kJ}}{\text{kg}}, 0.8423 \frac{\text{m}^3}{\text{kg}}, 0.007105 \frac{\text{kg}}{\text{kg}};$

(d) At *S*, 30.3°C d.b., 17.5°C w.b., $48.5 \frac{\text{kJ}}{\text{kg}}, 0.007105 \frac{\text{kg}}{\text{kg}}$

A sketch of the winter heating cycle is shown in Figure 5.11. The heating load on each fan coil unit is:

$$\text{Heating coil load} = 0.283 \frac{\text{m}^3}{\text{s}} \times (48.5 - 38.95) \frac{\text{kJ}}{\text{kg}} \times \frac{\text{kg}}{0.8423\ \text{m}^3} \times \frac{\text{kWs}}{\text{kJ}}$$

$$\text{Heating coil load} = 3.209\ \text{kW}$$

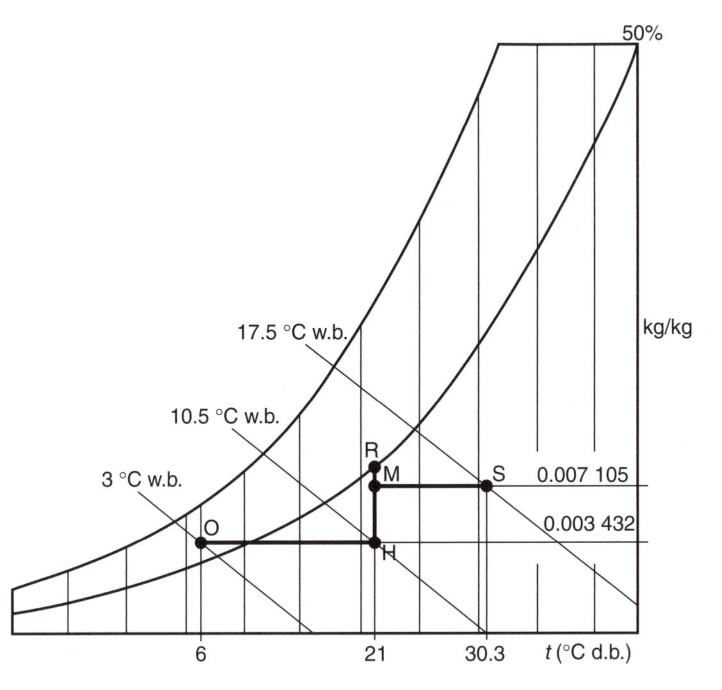

5.11 Winter cycle for the fan coil unit system in example 5.12.

There are 10 room units in Office 1. *OA* plant heating coil load *O* to *H*:

$$\text{Heating coil load} = 2.828\frac{\text{m}^3}{\text{s}} \times (29.91 - 14.66)\frac{\text{kJ}}{\text{kg}} \times \frac{\text{kg}}{0.7947\ \text{m}^3} \times \frac{\text{kWs}}{\text{kJ}}$$

$$\text{Heating coil load} = 54.268\ \text{kW}$$

Mass cooling

This is the oldest building technique for keeping warm in winter and cool in summer by surrounding the living area with mass rock, brick and concrete. Revisit the White Tower analysis in Chapter 1. Architecture since the 1940s forgot or avoided this principle in the haste to erect lightweight houses, schools and multi-storey apartment and office towers using thin skinned timber, steel, lightweight concrete and large areas of glass in the pursuit of modernism. Revisit The Shard analysis in Chapter 1. Now that CO_2 emissions are a minimisation target, designers return to the natural ventilation and high thermal mass solutions from history.

Advantage might be actively taken from the ability of mass elements of a building to absorb, store and release heat energy during cyclic weather and internal occupancy variations to reduce plant energy use (Low Energy Cooling, n.d.). Heavyweight and hollow concrete floors, ceilings, the central concrete core of the lift shafts plus exposed internal brick walls are all used when architecture, engineering and interior design permit (CIBSE Guide B, 2005, page 2-76). Figure 5.12 shows some of the principles.

Mass cooling, fabric energy storage (FES), requires the heavyweight concrete structure to be exposed to the room air. This has thermal advantages but some disadvantages. There is nowhere to conceal the ugly structure, lighting fittings, cables, air conditioning ducts, water pipes, control sensors, and smoke,

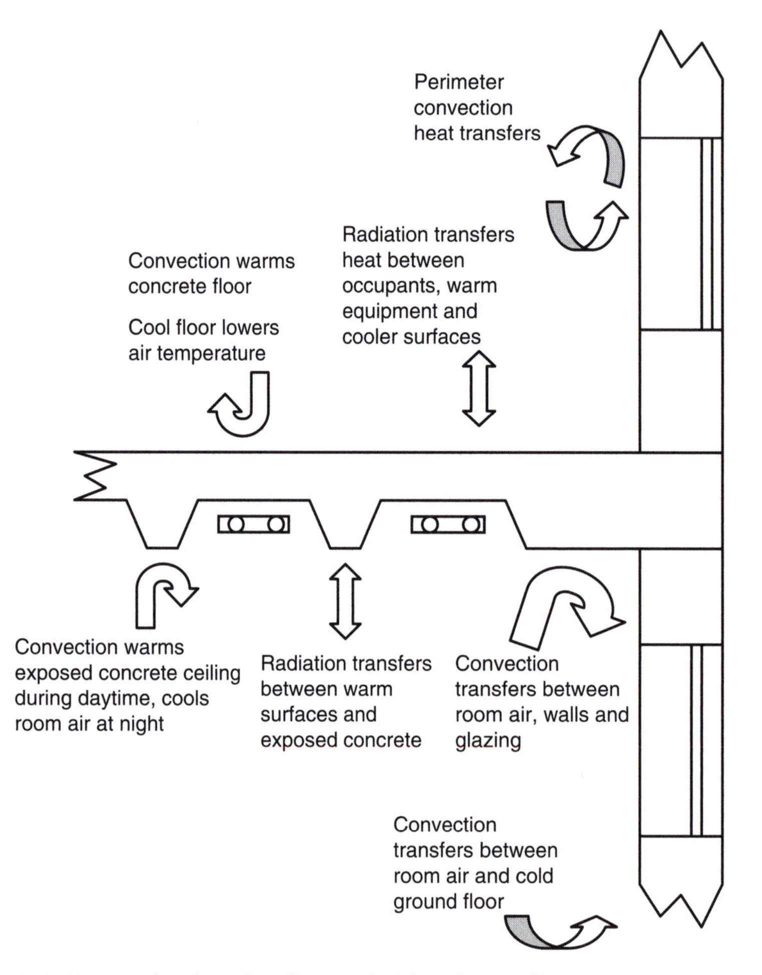

5.12 Cross-section through multi-storey building showing fabric energy storage routes, FES.

heat and security detectors. Hard surfaces cause multiple reflections of all sound waves hitting them from voices, foot traffic, phone conversations, mechanical and electrical equipment plus any ingress of traffic noise. With no acoustic absorption by hard surfaces, the aural environment can be harsh. There is reduced privacy from conversations. The room may have noticeable echoes. While we can accept high background noise levels for an hour or so, they can be annoying in the long term. A noisy room creates a continuous white noise level that has no discernible frequency; it is just a blanket of noise. The interior may appear cheap, unfinished, dark and forbidding, industrial, prison or cave-like. Concrete surfaces can be painted to minimise loose dust and seal indentations against accumulating dirt, but are not easy to clean, and may become unhygienic making them unsuitable for health care buildings, needing regular repainting. Passive FES, exposed concrete, stone and brick surfaces may reduce peak indoor air temperature by around 5°C when compared with plastered, carpeted and false ceiling designs. They delay the effect of solar radiation and convection heat gains until, for example, daytime office workers have gone home.

Heat flows into the structure relate to admittance, Y values W/m^2K , surface areas m^2 and difference between the internal air and surface temperatures, K, as well as radiation heat exchanges from the Stefan

Boltzmann equation. The best results from convection heat transfers occur with turbulent air flow across rough concrete surfaces and maximise the convection heat transfer coefficient,

$$h_{si} = \frac{1}{R_{si}} \frac{W}{m^2K}.$$

Passive radiation heat transfer takes place between, for example, solar warmed exterior walls and the cooler interior walls, floors and ceilings. Solar radiation through glazing rapidly warms interior surfaces. Low temperature long wave radiation from interior surfaces passes outwards through glazing depending upon reflectivity and transmissibility of the glass. High thermal mass may take 24 hours to respond to weather variations and changes in internal heat loads from people and computers. Overcrowding during ad-hoc conferences can overcome the natural ability of the FES to maintain conditions within an acceptable band of comfort.

Radiation heat transfers are beyond control by a BMS or other room air temperature control system, as these do not assess mean radiant temperature of the room with a globe sensor. Consequently, radiation heat gains and losses within the occupied space are fortuitous and have to be controlled or modified by the room user operating exterior and interior sun blinds, curtains and deciding where to sit in the room.

Hollow core concrete floors

Supply air ducted through hollow core concrete floors is used to moderate room temperature using fabric energy storage (TermoDeck International Ltd, n.d.). Figure 5.13 shows the working principle.

Stable concrete temperature supplements other passive control methods, and may be used with natural or mechanical ventilation strategies. Passing outside air through the hollow floors overnight lowers their stored temperature during hot weather. It may be employed where close temperature, humidity and airflow control is not critical, such as for comfort in a less than fully air conditioned building. Structural columns and beams support the lightweight hollow concrete floor sections. These may all be exposed to maximise natural temperature moderation.

Labyrinth cooling

Outside air supplied through an underground concrete labyrinth can be used to condition air prior to AHU heating and cooling, or at other suitable times, to avoid the need for plant energy use, other than for fans of course (Cement and Concrete Association of Australia, 2002).

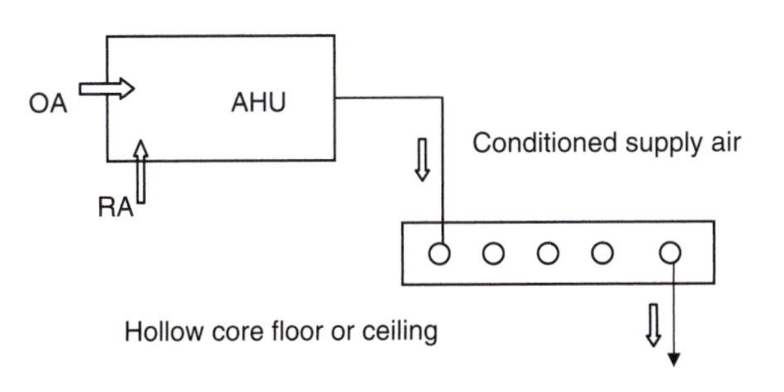

5.13 Hollow core concrete floor or ceiling.

Passing air through concrete labyrinths, hollow floors, highly obstructed raised floors and suspended ceiling voids, maybe all night to cool an air conditioned building, is very expensive of fan energy. Audited performance of the Bourke Street building (Chapter 3) showed that energy consumption by the air handling unit fans greatly exceeded energy used by the chillers, pumps and cooling towers in providing refrigerated cooling. Less energy might be used by turning off AHU fans and using refrigeration to keep a building cool. Which method is more useful for saving energy depends upon all factors involved. Plant running during off peak hours uses lower cost power and that may be a deciding factor.

EXAMPLE 5.13

A low energy theatre for 300 people in a tropical climate has a below ground concrete labyrinth. Incoming outside air passes through the labyrinth to precool the theatre prior to occupation, reduce refrigeration plant energy and cool the building overnight after occupation. Theatre dimensions are 30 m × 30 m × 8 m. Average day temperature is 35°C and 20°C at night. Indoor air is to be maintained at 24°C. Daytime design cooling load is 250 W/m^2, occupancy heat emission is 90 W each. Average concrete temperature in the labyrinth is 20°C. Comfort air conditions are to be maintained from 0900 h until 23.00 h. Supply air temperature at peak cooling load is to be 15°C. Fan total pressure is 500 Pa, fan motor and belt drive efficiency is 60%. *COP* of the refrigeration system is 2.5 not including the AHU fan. Analyse the energy implications for the design over a 24 hour period.

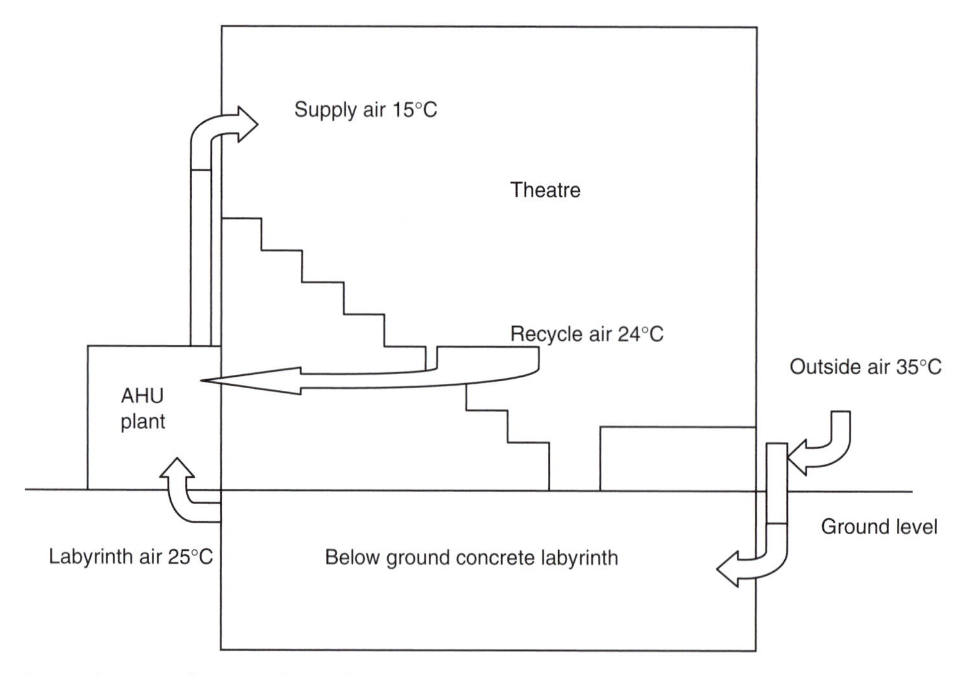

5.14 Below ground concrete labyrinth pre-cooling.

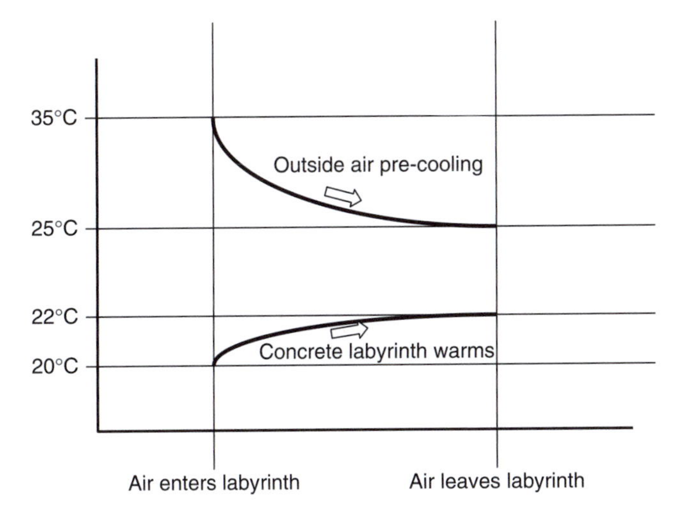

5.15 Labyrinth outside air pre-cooling.

Figure 5.14 shows a schematic of the ducted labyrinth outside air pre-cooling system while Figure 5.15 plots the heat transfer changes to the outside air and the slight warming of the concrete.

$$\text{Air conditioning hours} = 09.00\text{ h to }23.00\text{ h}$$

$$\text{Air conditioning hours} = 14\text{ h}$$

$$\text{Night cooling hours} = 10\text{ h}$$

$$\text{Theatre volume } V = (30 \times 30 \times 8)\text{ m}^3$$

$$\text{Theatre volume } V = 7200\text{ m}^3$$

$$SH\text{ gain} = \frac{\text{W}}{\text{m}^2\text{ floor area}} + \text{lights} + \text{people}$$

$$SH\text{ gain} = \left(30\text{ m} \times 30\text{ m} \times 250\frac{\text{W}}{\text{m}^2}\right) \times \frac{1\text{ kW}}{10^3\text{ W}} + 5\text{ kW} + 30 \times \frac{90\text{ W}}{\text{person}} \times \frac{1\text{ kW}}{10^3\text{ W}}$$

$$SH\text{ gain} = 257\text{ kW}$$

Calculate the required supply air flow:

$$Q = \frac{257\text{ kW}}{(24-15)} \times \frac{(273+15)}{355}\frac{\text{m}^3}{\text{s}}$$

$$Q = 23\frac{\text{m}^3}{\text{s}}$$

$$N = \frac{Q}{V} \times 3600\frac{\text{air changes}}{\text{h}}$$

$$N = \frac{23}{7200} \times 3600\frac{\text{air changes}}{\text{h}}$$

$$N = 11.5\frac{\text{air changes}}{\text{h}}$$

$$\text{AHU fan motor power} = 23\frac{m^3}{s} \times 500\ \text{Pa} \times \frac{1\ \text{N}}{1\ \text{Pa}\ m^2} \times \frac{1}{0.6} \times \frac{1\ \text{Ws}}{1\ \text{Nm}} \times \frac{1\ \text{kW}}{10^3\ \text{W}}$$

$$\text{AHU fan motor power} = 19.2\ \text{kW}$$

Assess the peak day-time energy saving if the labyrinth heat exchanger reduces incoming outdoor air by 10 K.

$$\text{Cooling energy saved} = 23\frac{m^3}{s} \times 1.012\frac{\text{kJ}}{\text{kg K}} \times 1.2\frac{\text{kg}}{m^3} \times 10\ \text{K} \times \frac{1}{2.5} \times \frac{1\ \text{kWs}}{1\ \text{kJ}}$$

$$\text{Cooling energy saved} = 112\ \text{kWe}$$

That is, 112 kW electrical input power to the refrigeration system at peak cooling performance.

$$\text{Electrical energy saved} = 112\ \text{kWe} \times \frac{14\ \text{h}}{\text{day}}$$

$$\text{Electrical energy saved} = 1568\ \frac{\text{kWh}}{\text{day}}$$

The labyrinth should be cooled overnight back to around 20°C. Assuming the AHU fan runs all night,

$$\text{Night fan energy} = 19\ \text{kW} \times \frac{10\ \text{h}}{\text{night}}$$

$$\text{Night fan energy} = 190\ \frac{\text{kWh}}{\text{night}}$$

$$\text{Electrical energy saved} = (1568 - 190)\ \frac{\text{kWh}}{24\ \text{h}}$$

$$\text{Electrical energy saved} = 1378\ \frac{\text{kWh}}{24\ \text{h}}$$

Running the AHU fan at night at reduced speed from a variable frequency controller until the desired labyrinth temperature is achieved and then switching it off would save more energy. It may be possible to allow natural ventilation to cool the labyrinth, or use a different, smaller fan and avoid passing air through the AHU and Theatre. The 1600 m^2 floor area Federation Square labyrinth is said to reduce incoming 36°C outside air by 12 K; and the labyrinth is above ground as the first level of the building is over railway tracks and so is subject to outside air variations from the shaded railway tunnels.

Other considerations for the building designer and user include rain water drainage for the below ground structure, whether below ground car parking is needed, where basement services plant can be located, access into the labyrinth for regular inspection and cleaning, whether damp and mould conditions will develop, part load performance and whether mild outdoor air should bypass the labyrinth.

In conclusion, this theatre benefits significantly from labyrinth outside air pre-cooling; energy demand is reduced and night fan energy might be at an off-peak tariff.

Annual energy cost savings need to be compared with the additional capital cost of the labyrinth in the foundation structure to calculate return on investment. Federation Square labyrinth reportedly cost AU$3M, UK£1.8M, which seems a lot for a heat exchanger.

Questions

1. An office has a sensible heat gain of 22 kW when the room air temperature is 23°C d.b. Calculate the necessary volume flow rate of supply air to maintain the room at the design temperature when the supply air temperature can be 14°C d.b.
2. A lecture theatre is 18 m × 10 m × 4 m. It is maintained at 21°C d.b. and has six air changes per hour of cooled outdoor air supplied at 16°C d.b. Calculate the maximum cooling loads that the equipment can meet.
3. A hotel lounge is 15 m × 15 m × 4 m. It is maintained at 22°C d.b. and has nine air changes per hour of cooled supply air. The cooling plant load is 19 kW. Calculate the required supply air temperature.
4. An exhibition hall has a winter sensible heat loss of 96 kW when the room air temperature is 18°C d.b. Calculate the necessary volume flow rate of supply air to maintain the room at the design temperature when the supply air temperature is 40°C.
5. A ducted warm air heating system serves a shop of 30 m × 20 m × 4 m. It is maintained at 18°C d.b. and has three air changes per hour of recirculated air supplied at 25°C d.b. Calculate the maximum heat loss from the building that the system can meet.
6. An open plan office is 30 m × 12 m × 2.8 m. It is maintained at 19°C d.b. and has six air changes per hour of supply air for both summer and winter. The room heat loss is 27 kW. Calculate the required supply air temperature.
7. A single duct air conditioning system serves a theatre of 1500 m^3 that has a sensible heat gain of 52 kW and a winter heat loss of 39 kW. It is maintained at 22°C d.b. during winter and summer. The summer supply air temperature is 14°C d.b. The vapour pressure of the supply air is 1044 Pa in summer and winter. The supply air fan is fitted upstream of the heater and cooler coils and operates in an almost constant air temperature that is close to that of the room air. Calculate the mass flow rate of the supply air in summer, room air change rate, the winter supply air temperature and comment upon the volume flow rate under winter heating conditions.
8. An office has a sensible heat gain of 16 kW and seven occupants each having a latent heat output of 50 W when the room air condition is 23°C d.b., 50% saturation. Calculate the necessary volume flow rate of supply air and its moisture content to maintain the room at the design state when the supply air temperature can be 14°C d.b. The room air moisture content is 0.008905 kg/kg.
9. A room has a sensible heat gain of 66 kW and a latent heat gain of 3 kW. The room air condition is 21°C d.b. and 50% saturation. The supply air temperature is 13°C d.b. Plot the room and supply conditions and sensible to total heat ratio line on a psychrometric chart.
10. Calculate the summer air flows for all the ducts shown in Figure 5.4 using the following data;

Room	*Volume, m^3*	*SH, kW*	*Occupants*
Office	3500	65	70
Corridor	400	12	6
Toilet	550	3	4

$t_s = 14°C$ d.b.

$t_r = 19°C$ d.b.

Toilet to have a minimum of six air changes per hour.
OA requirement $= 12 \frac{l}{\text{person s}}$

The office and toilet doors each have a grille of free area of 0.25 m^2 into the corridor. The maximum air velocity through the grille is to be 2.75 m/s.

11. A space has sensible heat gains of 60 kW and 3 kW latent. The space condition is 20°C d.b., 50% saturation. The supply air temperature is 15°C d.b. Calculate the supply air quantity and moisture content needed.
12. A room 5 m × 12 m × 3 m is to be air conditioned. The maximum sensible heat gain is 7 kW when the latent heat gain is 1 kW. The room condition is to be maintained at 21°C d.b., 50% saturation by a ventilation rate of eight air changes per hour. Calculate the required supply air condition.
13. A shop has solar heat gains of 8 kW and an internal air condition of 20°C d.b., 50% saturation. There will be 25 occupants emitting 110 W of sensible heat and 50 W of latent heat each. There are 2 kW of heat gains from lighting and 1 kW of heat output from refrigerated display cabinets. If the supply air can be at 16°C d.b., calculate its quantity and moisture content.
14. An office floor is to have a single duct air conditioning system. Use a psychrometric chart and the data provided to find the peak summer and winter design loads and air conditions.

    Summer room air 23°C d.b., 50% saturation.
    Summer outdoor air 29°C d.b., 22°C w.b.
    Winter room air 19°C d.b., 50% saturation.
    Winter outdoor air −1°C d.b., −2°C w.b.
    Summer supply air temperature 15°C d.b.
    Office volume 900 m^3.
    Glazing area 45 m^2.
    External wall area 100 m^2.
    Office has 25 occupants, 90 W sensible, 50 W latent each.
    Outside air provision is $12 \frac{l}{\text{person s}}$.
    There are 12 computers of 180 W.
    There is one air change per hour of infiltration by the outdoor air into the office.
    The peak cooling load through the west facing glazing is 314 W/m^2 at 1600 h on 21 June.
    A solar gain correction factor of 0.48 applies to the reflective glazing.
    Ignore heat gains through the structure.
    Surrounding rooms are at the same conditions.
    Glazing U value is $2.7 \frac{W}{m^2\ K}$.
    External wall U value is $0.6 \frac{W}{m^2\ K}$.
    Use the simplest forms of heat gain and loss calculation to obtain approximate plant loads.

15. A Melbourne, Australia, hotel has a dual duct air conditioning system. Use a psychrometric chart and the data provided to find the peak summer and winter design loads, air conditions and supply air flows from the hot and cold ducts for one room.

    Summer room air 22°C d.b., 50% saturation.
    Summer outdoor air 39°C d.b., 23°C w.b.
    Winter room air 20°C d.b., 50% saturation.
    Winter outdoor air 3°C d.b., 2°C w.b.
    Summer supply air temperature 16°C d.b.
    Air leaves the plant cooling coil in summer at 13°C d.b.
    Air leaves the plant heating coil in winter at 25°C d.b.
    Glazing area is 12 m^2 and external wall area is 30 m^2.
    Room has two occupants, 90 W sensible, 50 W latent each.
    OA provision is $12 \frac{l}{\text{person s}}$.

There is no infiltration of outdoor air.
The peak cooling load through the north facing glazing is 347 W/m^2 at noon on 21 January.
A solar gain correction factor of 0.66 applies to the shading.
Ignore heat gains through the structure.
Surrounding rooms are at the same conditions.
Glazing U value is $5.7 \frac{W}{m^2 K}$.
External wall U value is $0.8 \frac{W}{m^2 K}$.
Room volume is 150 m^3.
Use the simplest forms of heat gain and loss calculation to obtain approximate plant loads.

16. Here's a challenge. On a hot sunny day, preferably during a series of them, visit a variety of buildings which you have permission to enter and write a report on how successful they are at maintaining indoor air comfort conditions. A suitable range of building types include a holiday tent, caravan, building site cabin, portable office or beach hut. Compare brick cavity traditional houses, with timber frame, with weatherboard homes. Find how uninsulated metal clad industrial buildings, that is, large tin sheds, cope with hot weather. Are naturally ventilated brick and concrete built commercial and academic buildings comfortable? Do stately homes, castles and religious buildings that have stood for hundreds of years maintain indoor comfort or do they remain cold indoors? Would you be willing to work in a multi-storey commercial building that does not have air conditioning? Feel free to develop other locations for assessment and in any country.

6 Ductwork design

Learning objectives

Study of this chapter will enable the user to:

1. understand the meaning of static, velocity and total air pressure terms and units;
2. apply Bernoulli's theorem to air flow in ducts;
3. know how to measure air pressure in ducts;
4. calculate the air pressures used in ductwork design;
5. analyse the changes in pressure when air flows through changes in duct size;
6. calculate and plot graphs of the changes of air pressure that occur at fans;
7. define the fan pressure terms;
8. design air duct systems;
9. use a spreadsheet for duct sizing calculations;
10. understand the methods used for the measurement of air flows in duct systems.

Key terms and concepts

Air density 180; air duct sizing 205; duct fittings 181; fan static pressure 197; fan total pressure 196; fan velocity pressure 196; flow measurement devices 199; index route 207; limiting air velocity 206; manometer 199; pressure gradient 182; pitot-static tube 179; static pressure 179; static regain 189; total pressure 179; velocity pressure loss factors 183.

Introduction

The design procedure for air duct systems is described, from the use of Bernoulli's equation through to a workbook that can be used for duct routes. Air pressure terms and methods of measurement are explained. Detailed pressure changes at duct fittings and fans are calculated. The equations for the flow of air in ducts that were developed in Chapter 10 are used for duct sizing and have been used for the production of a duct sizing chart for a limited range of flows.

Air pressure in a duct

Air flowing through a duct exerts two types of pressure on its surroundings. There is dynamic pressure due to its motion or kinetic energy; this is velocity pressure p_V. The second is the bursting pressure of the air trying to escape from the enclosure or of the surrounding air trying to enter the enclosing duct; this is static pressure p_S and it acts in all directions. The sum of these two pressures is the total pressure p_t.

Total pressure = static pressure + velocity pressure

$$p_t = p_s + p_v$$

Losses of pressure due to the frictional resistance of the duct, its bends, branches, changes in cross-section, dampers, filters and air heating or cooling coils, cause loss of total pressure. Bernoulli's theorem states that for a fluid flowing from position 1 to position 2, the total pressure energy at 1 equals the total pressure energy at 2 plus the loss of energy due to friction. All energy lost in friction is dissipated as heat and in some cases can cause a noticeable rise in the air temperature.

The balance of air pressures between 1 and 2:

$p_{t1} = p_{t2} + \Delta p$

p_{t1} = total pressure at 1

p_{t2} = total pressure at 2

Δp = pressure drop due to friction

Figure 6.1 shows how the total, velocity and static pressures are measured. A water U tube manometer or electronic micro manometer is connected to a probe in the airway. Static pressure is measured from

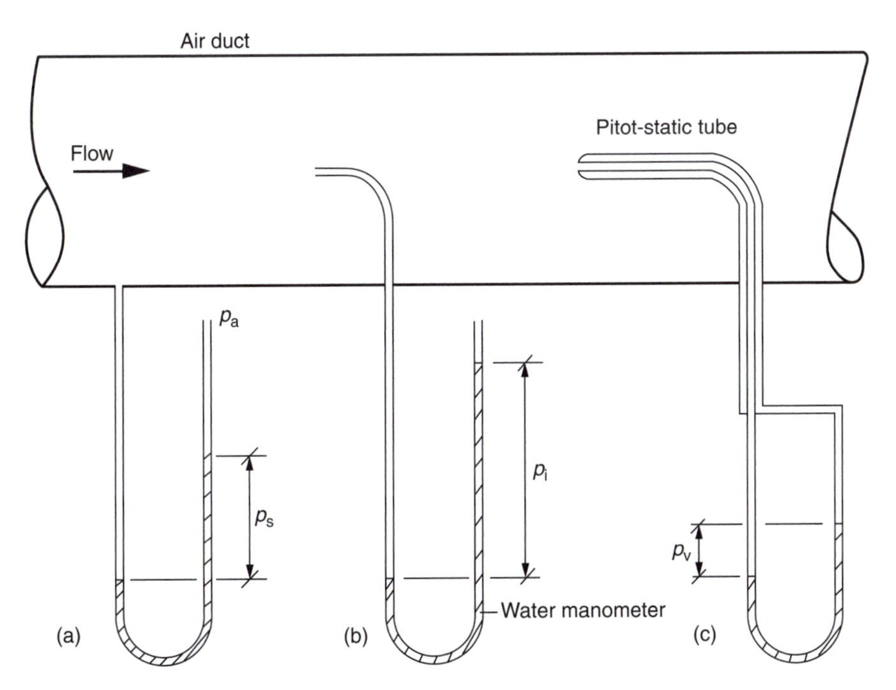

6.1 Methods of pressure measurement in an air duct.

a tapping in the side of the duct. In this case, the duct static air pressure is above the atmospheric air pressure p_a. When this pressure is measured on the inlet side of a fan, static pressure may be below atmospheric.

The total pressure is found from the probe facing the air stream as this records both the dynamic and static pressures. The difference between the total and static pressures is the velocity pressure.

$$p_t = p_s + p_v$$

$$p_v = p_t - p_s$$

$$p_v = 0.5\ \rho v^2$$

Air density ρ is 1.2 kg/m^3 at 20°C d.b., 43% saturation and 101325 Pa, the standard atmospheric pressure at sea level. The densities at 101325 Pa but other temperatures or percentage saturations, are found by calculation or by reading the specific volume from the psychrometric chart and inverting it. Density at any other condition is found from the atmospheric pressure p_a Pa and air temperature t°C d.b.

$$\rho = 1.2 \times \frac{p_a}{101325} \times \frac{(273+20)}{(273+t)} \frac{\text{kg}}{\text{m}^3}$$

The pressure p_a includes any deviation from the surrounding atmospheric pressure due to the static air pressure within an enclosure or duct p_s.

$$p_a = (101325 + p_s)\ \text{Pa}$$

Note that p_s can be positive or negative. Pressures measured with water manometers, H mm H_2O, are converted into pascals by:

$$p_s = \rho g H$$

$$\rho\ \text{water} = 10^3 \frac{\text{kg}}{\text{m}^3}$$

$$p_s = 10^3 \frac{\text{kg}}{\text{m}^3} \times 9.807 \frac{\text{m}}{\text{s}^2} \times H\ \text{mmH}_2\text{O} \times \frac{1\ \text{m}}{10^3\ \text{mm}} \times \frac{1\ \text{N s}^2}{1\ \text{kg m}} \times \frac{1\ \text{Pa m}^2}{1\ \text{N}}$$

$$p_s = (9.807\ H\ \text{mm H}_2\text{O})\ \text{Pa}$$

The air velocity in a circular duct is found from the volume flow rate Q m^3/s and duct internal diameter d m.

$$v = \frac{4Q}{\pi d^2} \frac{\text{m}}{\text{s}}$$

EXAMPLE 6.1

Calculate the total pressure of air flowing at 0.2 m^3/s in a 250 mm internal diameter duct if the air temperature is 22°C d.b., the static pressure of the air in the duct is 25 mm water gauge above the atmospheric pressure of 101450 Pa.

Atmospheric pressure within the duct:

$$p_a = (101450 + p_s)\,\text{Pa}$$

$$p_s = (9.807 \times 25\ \text{mm}\ H_2O)\,\text{Pa}$$

$$p_s = 245.2\ \text{Pa}$$

$$p_a = (101450 + 245.2)\,\text{Pa}$$

$$p_a = 101695.2\ \text{Pa}$$

Ignore decimal parts of pascal pressures:

$$p_s = 245\ \text{Pa}$$

$$p_a = 101695\ \text{Pa}$$

$$\rho = 1.2\frac{p_a}{101325} \times \frac{(273+20)}{(273+t)}\frac{\text{kg}}{\text{m}^3}$$

$$\rho = 1.2\frac{101695}{101325} \times \frac{293}{(273+22)}\frac{\text{kg}}{\text{m}^3}$$

$$\rho = 1.196\frac{\text{kg}}{\text{m}^3}$$

$$v = \frac{4Q}{\pi d^2}\ \frac{\text{m}}{\text{s}}$$

$$v = \frac{4 \times 0.2}{\pi \times 0.25^2}\ \frac{\text{m}}{\text{s}}$$

$$v = 4.074\ \frac{\text{m}}{\text{s}}$$

$$p_v = 0.5\rho v^2$$

$$p_v = 0.5 \times 1.196\frac{\text{kg}}{\text{m}^3} \times 4.074^2\frac{\text{m}^2}{\text{s}^2} \times \frac{1\ \text{N s}^2}{1\ \text{kg m}} \times \frac{1\ \text{Pa m}^2}{1\ \text{N}}$$

$$p_v = 9.925\ \text{Pa}$$

Ignore the decimal part:

$$p_v = 10\ \text{Pa}$$

$$p_t = p_s + p_v$$

$$p_t = (245 + 10)\,\text{Pa}$$

$$p_t = 255\ \text{Pa}$$

Variation of pressure along a duct

Pressure gradients are produced along the length of a duct as shown in Figure 6.2.

The gradient caused along the constant diameter duct is calculated from the pressure drop rate due to friction calculated from the methods used in Chapter 10. The gradients of the total and static pressure lines

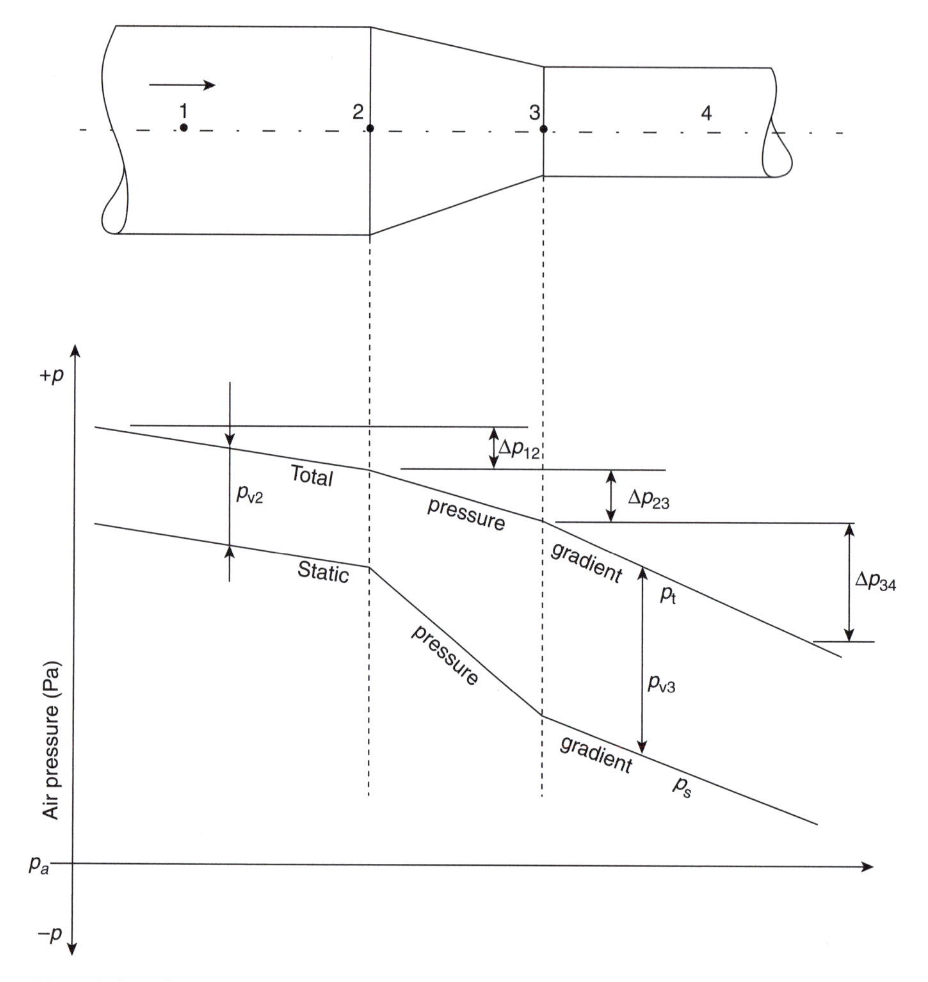

6.2 Variation of air pressures through a reducer.

are equal. Friction losses can be calculated as total or static pressure drops. It is convenient to calculate all pressure drops as reductions in the available total pressure.

Duct fittings, such as the reducer shown, cause loss of pressure through surface friction and additional turbulence because of the shape of the enclosure. Such losses are found from experiment and require the use of unique factors that are multiplied by the velocity pressure. Data for commonly used duct fittings are given in Figures 6.3 and 6.4 from CIBSE Guide C (2007).

When air enters a duct system, there is an entry contraction friction loss factor $0.5p_{v1}$, that is, half the velocity pressure in the duct, and this is in addition to loss through the intake louvres and wire mesh screen behind the grille.

When air discharges from a duct system, there is a friction loss factor $1p_{v2}$, that is, equal to the velocity pressure in the prior duct and this is in addition to loss through the discharge louvres. All the total pressure remaining in the discharge duct, or leaving the grille, is converted into velocity energy in the throw of air into the conditioned space or to outdoors. This value of velocity pressure accounts for the fan energy used creating velocity energy in addition to overcoming other frictional resistances and is expressed in terms of a loss of total pressure. All friction losses are losses of total pressure.

To produce a pressure gradient diagram similar to Figure 6.2, draw the ductwork to be analysed directly above a graph as shown. Clearly mark each point where a change of section takes place and identify each with a number. The sequence is as follows.

1. Find the total pressure at node 1 p_{t1} Pa and plot it to an appropriate scale above, or below, atmospheric pressure p_a Pa.
2. Calculate the pressure drop between nodes 1 and 2 in the straight duct Δp_{12}. Note that Δp_{12} means the pressure drop due to friction between nodes 1 and 2.
3. Subtract Δp_{12} from p_{t1} to find the total pressure at point 2 p_{t2} Pa.
4. Draw the total pressure straight line from p_{t1} to p_{t2}.
5. Find the velocity pressure loss factor for the duct fitting K from Figures 6.3 or 6.4 or the reference (CIBSE Guide C, 2007).
6. Find the velocity pressure to be used for the frictional resistance calculation for the fitting. In this case, it is the velocity pressure in the smaller duct p_{v3}.

Contractions

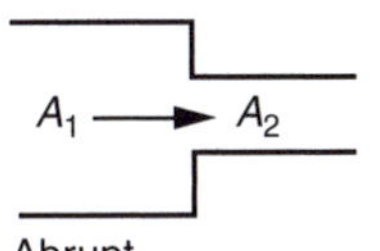

A_2/A_1	0.25	0.40	0.55	0.65
k	0.37	0.28	0.19	0.12

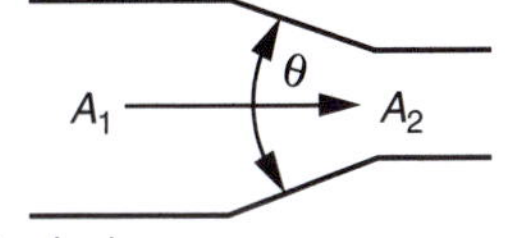

θ	30°	45°	60°
k	0.02	0.04	0.07

Enlargements

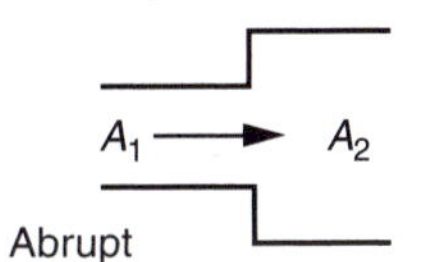

A_1/A_2	0.25	0.40	0.55	0.70
k_1	0.56	0.36	0.20	0.09

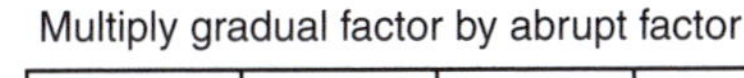

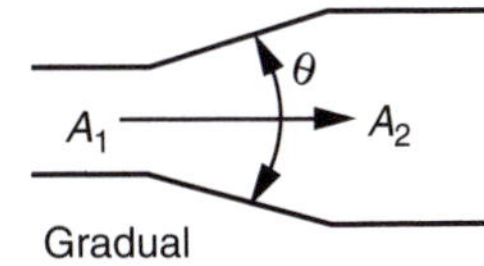

θ	20°	30°	40°
k_2	0.6	0.8	1.0

6.3 Velocity pressure loss factors for air duct fittings (CIBSE Guide C, 2007). Factors are multiplied by the velocity pressure in the smaller area, $\Delta p_{12} = k_1 k_2 p_{v1}$.

45° Fresh air inlet louvres

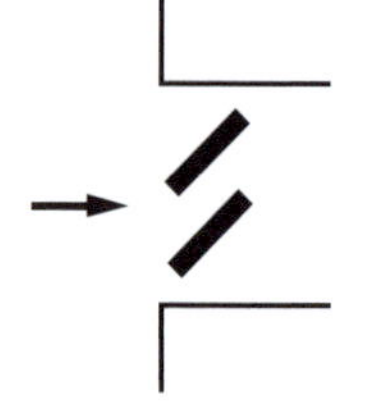

Free air ratio	0.5	0.6	0.7	0.8
k	4.5	3.0	2.1	1.4

Mitre bends

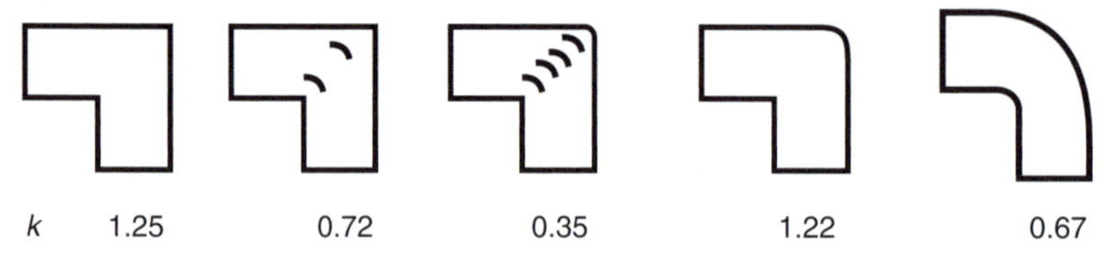

Rectangular branch

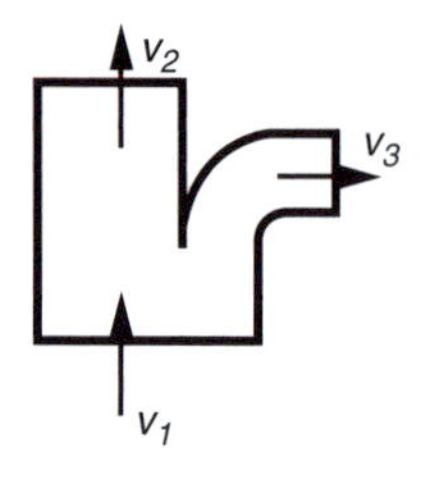

$\frac{v_2}{v_1}$	k_2	$\frac{v_3}{v_1}$	K_3 x bend *k*
0.6	0.34	0.6	3.5 "
0.8	0.18	0.8	1.8 "
1.0	0.09	1.0	1.0 "
1.2	0.05	1.2	1.0 "

6.4 Dynamic pressure loss factors for air duct fittings (CIBSE Guide C, 2007).

7. Multiply the fitting *K* factor by the velocity pressure p_{v3} to calculate the pressure drop through the fitting Δp_{23}.
8. Subtract pressure drop Δp_{23} from the total pressure p_{t2} to find the total pressure at point 3 p_{t3}.
9. Plot p_{t3} and connect the total pressure straight line from p_{t2}.
10. Calculate the pressure drop between points 3 and 4 in the straight duct Δp_{34}.
11. Subtract Δp_{34} from p_{t3} to find the total pressure at point 4 p_{t4} Pa.
12. Draw the total pressure straight line from p_{t3} to p_{t4}.
13. Calculate the velocity pressure at point 1, p_{v1} Pa. Note that $p_{v3} = p_{v2}$ as the velocity remains constant in an equal diameter duct along its length.
14. Subtract p_{v1} from both p_{t1} and p_{t2}. The differences are the static pressures at points 1 and 2, p_{s1} and p_{s2}.
15. Plot p_{s1} and p_{s2} and connect them with a straight line.

16. Subtract p_{v3} from p_{t3} and find static pressure at point 3 p_{s3}.
17. Draw a straight line from p_{s2} to p_{s3}.
18. Subtract p_{v3} from p_{t4} and find p_{s3}. Note that $p_{v3} = p_{v4}$.
19. Draw a straight line from p_{s3} to p_{s4}.
20. Mark the upper gradient as the total pressure line.
21. Mark the lower line as the gradient of static pressure.
22. Mark the differences between the two gradients as velocity pressures.
23. Identify each of the corresponding points of the duct diagram and the graph.
24. Mark on the graph all the calculated pressures.

This procedure applies to other air duct cases and to the changes of pressure across a fan. Notice that for this reduction in diameter of the duct, the static pressure reduces. This is because the velocity pressure has increased. The available total pressure has also fallen due to friction losses. If the duct diameter increases between points 2 and 3, the static pressure may rise due to a fall in the velocity pressure. Such a regain of static pressure depends upon how much frictional pressure drop occurs. Static regain can also take place at branches due to a reduction in the air flow quantity and velocity. Such regains of static pressure may be usefully employed in overcoming the resistance of downstream ducts or terminal equipment such as grilles, dampers or heater coils.

EXAMPLE 6.2

Calculate the pressures that occur when 4 m^3/s flows through a 5 m long, 1 m diameter duct that then reduces to 600 mm diameter and remains at 600 mm for 5 m. The air total pressure at the commencement of the 1 m duct is 300 Pa above atmospheric. The reducer is the 45° concentric type shown in Figure 6.3. Air density is 1.2 kg/m^3. Frictional pressure loss rates are 0.22 Pa/m in the 1 m diameter and 3 Pa/m in the 600 mm diameter ducts.

Using node and sequence numbers listed earlier, the solution is shown in Figure 6.5:

1. $p_{t1} = 300$ Pa, plot graph axes.
2. For the 1 m duct that is 5 m long:

$$\Delta p_{12} = 0.22 \frac{\text{Pa}}{\text{m}} \times 5 \text{ m}$$

$$\Delta p_{12} = 1 \text{ Pa}$$

3. Find p_{t2}, p_{s2} and p_{v3}:

$$p_{t2} = p_{t1} - \Delta p_{12}$$

$$p_{t2} = (300 - 1) \text{ Pa}$$

$$p_{t2} = 299 \text{ Pa}$$

4. Draw the total pressure line 1 to 2.
5. The reducer has a velocity pressure loss factor of 0.04 and this is multiplied by the smaller duct velocity pressure p_{v3}.

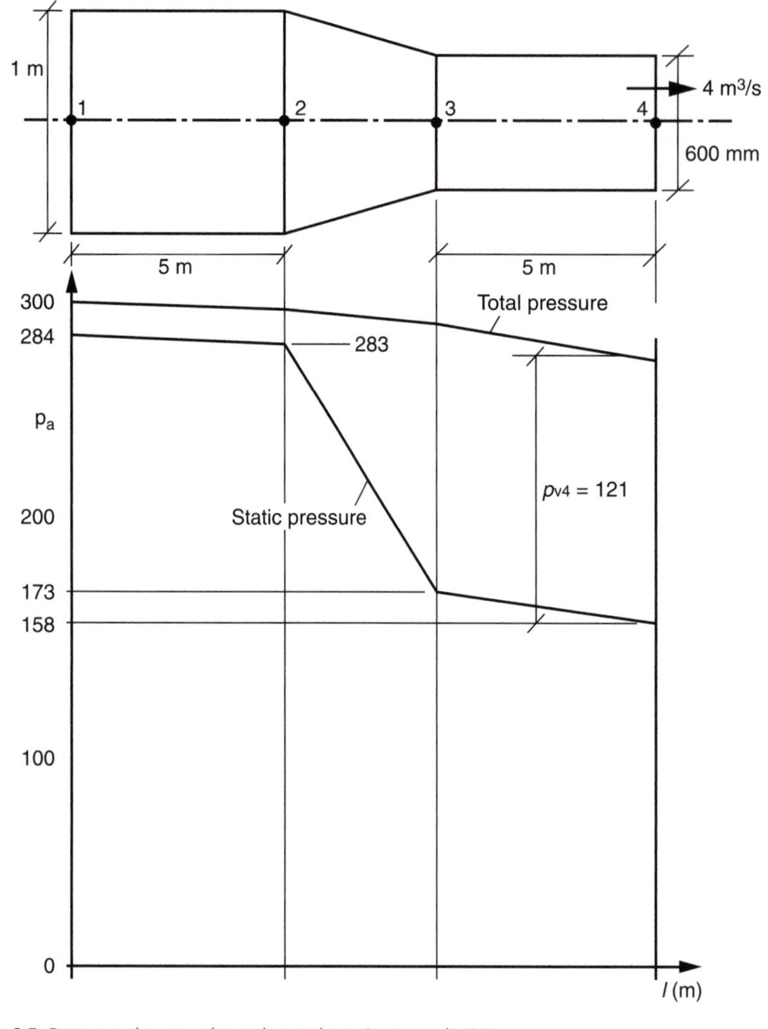

6.5 Pressure changes through a reducer in example 6.2.

6. $v_3 = \frac{4Q}{\pi d^2} \frac{\text{m}}{\text{s}}$

$$v_3 = \frac{4 \times 4}{\pi \times 0.6^2} \frac{\text{m}}{\text{s}}$$

$$v_3 = 14.2 \frac{\text{m}}{\text{s}}$$

$$p_{v3} = 0.5 \rho v^2$$

$$p_{v3} = \left(0.5 \times 1.2 \times 14.2^2\right) \text{Pa}$$

$$p_{v3} = 121 \text{ Pa}$$

7. Pressure drop through reducer:

$$\Delta p_{23} = \text{factor} \times p_{v3}$$

$$\Delta p_{23} = 0.04 \times 121 \text{ Pa}$$

$$\Delta p_{23} = 5 \text{ Pa}$$

8. $p_{t3} = p_{t2} - \Delta p_{23}$

$$p_{t3} = (299 - 5)\text{ Pa}$$

$$p_{t3} = 294 \text{ Pa}$$

9. Plot total pressure line 2 to 3.
10. Pressure drop through 600 mm duct:

$$\Delta p_{34} = 3\frac{\text{Pa}}{\text{m}} \times 5 \text{ m}$$

$$\Delta p_{34} = 15\text{Pa}$$

11. $p_{t4} = p_{t3} - \Delta p_{34}$

$$p_{t4} = (294 - 15) \text{ Pa}$$

$$p_{t4} = 279 \text{ Pa}$$

12. Draw total pressure line 3 to 4.
13. $v_1 = \dfrac{4 \times 4}{\pi \times 1^2} \dfrac{\text{m}}{\text{s}}$

$$v_1 = 5.1\frac{\text{m}}{\text{s}}$$

$$p_{v1} = \left(0.5 \times 1.2 \times 5.1^2\right) \text{Pa}$$

$$p_{v1} = 16 \text{ Pa}$$

$$p_{v2} = 16 \text{ Pa}$$

14. $p_{t1} = p_{s1} + p_{v1}$

$$p_{s1} = p_{t1} - p_{v1}$$

$$p_{s1} = (300 - 16)\text{ Pa}$$

$$p_{s1} = 284 \text{ Pa}$$

$$p_{s2} = p_{t2} - p_{v2}$$

$$p_{s2} = (299 - 16) \text{ Pa}$$

$$p_{s2} = 283 \text{ Pa}$$

15. Plot static pressures 1 and 2 and draw a line between them.
16. $p_{s3} = p_{t3} - p_{v3}$

$$p_{s3} = (294 - 121) \text{ Pa}$$

$$p_{s3} = 173 \text{ Pa}$$

17. Draw a static pressure line from 2 to 3.

18. $p_{v3} = p_{v4}$

$$p_{s4} = p_{t4} - p_{v4}$$

$$p_{s4} = (279 - 121)\,\text{Pa}$$

$$p_{s4} = 158\ \text{Pa}$$

19. Draw a static pressure line from 3 to 4.

20–24. Figure 6.5 shows these answers plotted to scale.

The following solutions are not listed in strict accordance with the sequence. The reader may follow the sequence and validate the answer with the results shown.

EXAMPLE 6.3

Calculate the static regain and pressure changes that occur when 2 m^3/s flow through a 12 m long, 500 mm diameter duct that then enlarges to 800 mm diameter and remains at 800 mm for 10 m. Total pressure at the commencement of the 500 mm duct is 200 Pa above atmospheric. The enlarger is the 20° concentric type shown in Figure 6.3. Air density is 1.2 kg/m^3. Frictional pressure loss rates are 2 Pa/m in the 500 mm diameter and 0.2Pa/m in the 800 mm diameter ducts.

Use the node numbers in Figure 6.6.

$$p_{t1} = 200\text{Pa}$$

$$v_1 = \frac{2 \times 4}{\pi \times 0.5^2}\,\frac{\text{m}}{\text{s}}$$

$$v_1 = 10.2\,\frac{\text{m}}{\text{s}}$$

$$p_{v1} = \left(0.5 \times 1.2 \times 10.2^2\right)\text{Pa}$$

$$p_{v1} = 62\ \text{Pa}$$

$$p_{s1} = p_{t1} - p_{v1}$$

$$p_{s1} = (200 - 62)\,\text{Pa}$$

$$p_{s1} = 138\ \text{Pa}$$

Along the 500 mm duct that is 12 m long:

$$\Delta p_{12} = 2\frac{\text{Pa}}{\text{m}} \times 12\ \text{m}$$

$$\Delta p_{12} = 24\ \text{Pa}$$

Find p_{t2}, p_{s2} and p_{v3}:

$$p_{t2} = p_{t1} - \Delta p_{12}$$

$$p_{t2} = (200 - 24)\,\text{Pa}$$

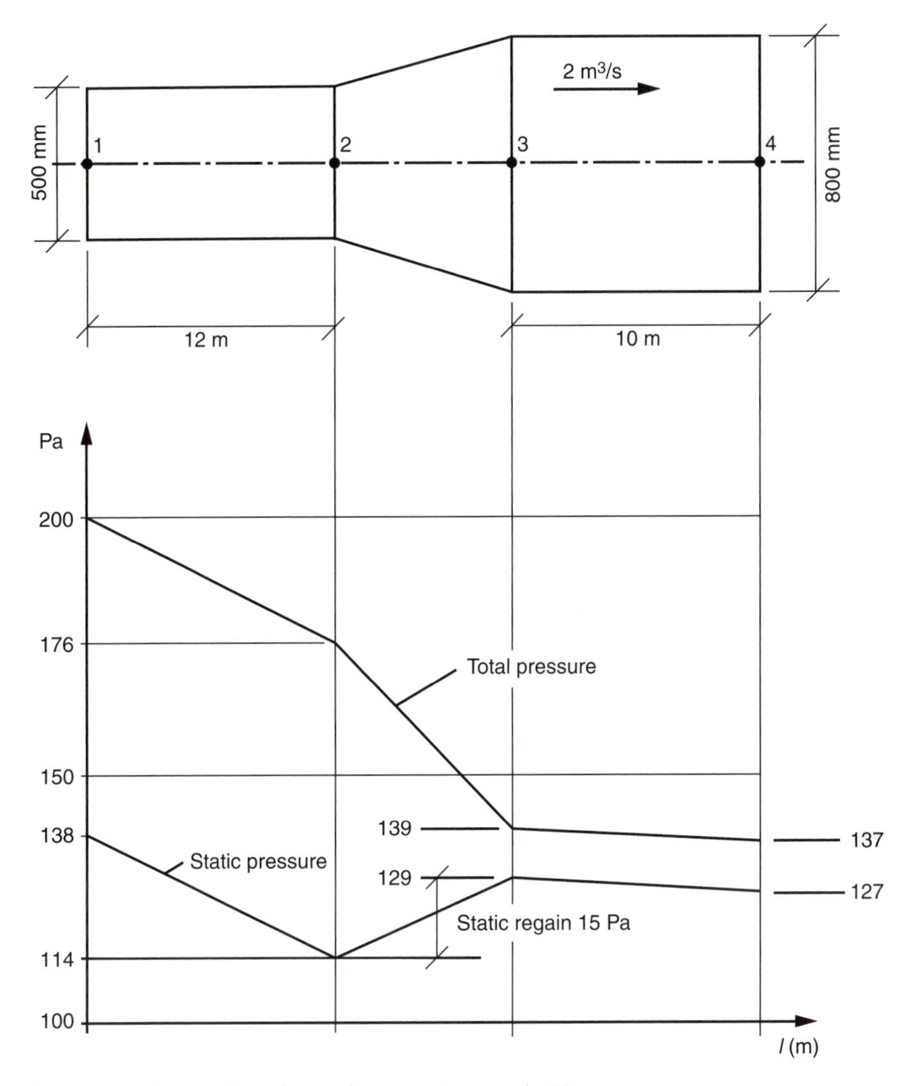

6.6 Pressure changes through an enlargement in example 6.3.

$$p_{t2} = 176 \text{ Pa}$$

$$p_{s2} = p_{t2} - p_{v2}$$

$$p_{v2} = p_{v1}$$

$$p_{s2} = (176 - 62)\,\text{Pa}$$

$$p_{s2} = 114 \text{ Pa}$$

$$v_3 = \frac{2 \times 4}{\pi \times 0.8^2} \frac{\text{m}}{\text{s}}$$

$$v_3 = 4 \frac{\text{m}}{\text{s}}$$

$$p_{v3} = \left(0.5 \times 1.2 \times 4^2\right) \text{Pa}$$

$$p_{v3} = 10 \text{ Pa}$$

The enlarger has a velocity pressure loss factor of 0.6 and this is multiplied by the smaller duct velocity pressure p_{v2}

$$\Delta p_{23} = 0.6 \times 62 \text{ Pa}$$

$$\Delta p_{23} = 37 \text{ Pa}$$

$$p_{t3} = p_{t2} - \Delta p_{23}$$

$$p_{t3} = (176 - 37) \text{Pa}$$

$$p_{t3} = 139 \text{ Pa}$$

$$p_{s3} = p_{t3} - p_{v3}$$

$$p_{s3} = (139 - 10) \text{Pa}$$

$$p_{s3} = 129 \text{ Pa}$$

Pressure drop through 800 mm duct:

$$\Delta p_{34} = 0.2 \frac{\text{Pa}}{\text{m}} \times 10 \text{ m}$$

$$\Delta p_{34} = 2 \text{ Pa}$$

$$p_{t4} = p_{t3} - \Delta p_{34}$$

$$p_{t4} = (139 - 2) \text{Pa}$$

$$p_{t4} = 137 \text{Pa}$$

$$p_{s4} = p_{t4} - p_{v4}$$

$$p_{v4} = p_{v3}$$

$$p_{s4} = (137 - 10) \text{Pa}$$

$$p_{s4} = 127 \text{ Pa}$$

Regain of static pressure:

$$SR = p_{s3} - p_{s2}$$

$$SR = (129 - 114) \text{Pa}$$

$$SR = 15 \text{ Pa}$$

Figure 6.6 shows these answers plotted to scale.

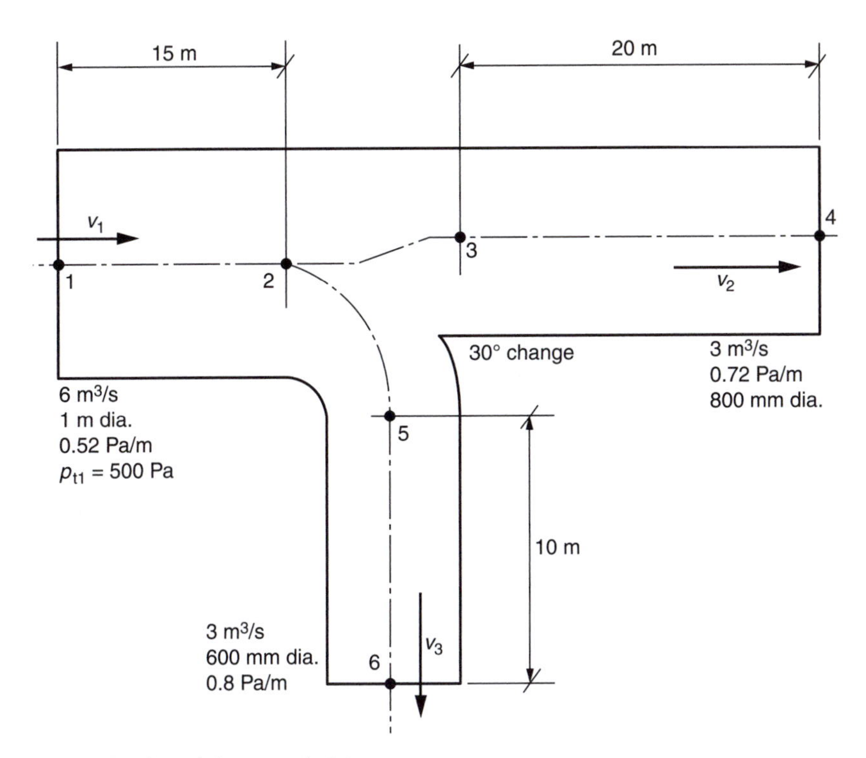

6.7 Air duct branch for example 6.4.

EXAMPLE 6.4

An air duct branch is shown in Figure 6.7. Use the data provided to calculate all the duct pressures at the nodes.

1 m diameter duct 1–2 is 15 m long and carries 6 m^3/s.
800 mm diameter duct 3–4 is 20 m long and carries 3 m^3/s.
600 mm diameter duct 5–6 is 10 m long and carries 3 m^3/s.
The branch is a short radius bend.
The straight through contraction has an angle of 30°.
Pressure drop rates are: duct 1–2, 0.52 Pa/m; duct 3–4, 0.72 Pa/m; duct 5–6, 0.8 Pa/m.
The total pressure at node 1 is 500 Pa.

$$p_{t1} = 500 \text{ Pa}$$

$$v_1 = \frac{6 \times 4}{\pi \times 1^2} \frac{\text{m}}{\text{s}}$$

$$v_1 = 7.6 \frac{\text{m}}{\text{s}}$$

$$p_{v1} = \left(0.5 \times 1.2 \times 7.6^2\right) \text{Pa}$$

$$p_{v1} = 35 \text{ Pa}$$

$$p_{s1} = p_{t1} - p_{v1}$$

$$p_{s1} = (500 - 35)\,\text{Pa}$$

$$p_{s1} = 465\ \text{Pa}$$

Along the 1 m duct that is 15 m long:

$$\Delta p_{12} = 0.52\frac{\text{Pa}}{\text{m}} \times 15\ \text{m}$$

$$\Delta p_{12} = 8\ \text{Pa}$$

$$p_{t2} = p_{t1} - \Delta p_{12}$$

$$p_{t2} = (500 - 8)\,\text{Pa}$$

$$p_{t2} = 492\ \text{Pa}$$

$$p_{s2} = p_{t2} - p_{v2}$$

$$p_{v2} = p_{v1}$$

$$p_{s2} = (492 - 35)\,\text{Pa}$$

$$p_{s2} = 457\ \text{Pa}$$

$$v_3 = \frac{3 \times 4}{\pi \times 0.8^2}\,\frac{\text{m}}{\text{s}}$$

$$v_3 = 6\ \frac{\text{m}}{\text{s}}$$

$$p_{v3} = \left(0.5 \times 1.2 \times 6^2\right)\text{Pa}$$

$$p_{v3} = 22\ \text{Pa}$$

The straight through part of the branch has a velocity ratio:

$$\frac{v_3}{v_2} = \frac{6}{7.6} = 0.8$$

Refer to Figure 6.4 to find the velocity pressure loss factor of 0.18 and this is multiplied by the smaller duct velocity pressure p_{v3}.

Pressure drop through the reducer:

$$\Delta p_{23} = \text{factor} \times p_{v3}$$

$$\Delta p_{23} = (0.18 \times 22)\,\text{Pa}$$

$$\Delta p_{23} = 4\ \text{Pa}$$

$$p_{t3} = p_{t2} - \Delta p_{23}$$

$$p_{t3} = (492 - 4)\,\text{Pa}$$

$$p_{t3} = 488\ \text{Pa}$$

$$p_{s3} = p_{t3} - p_{v3}$$

$$p_{s3} = (488 - 22)\,\text{Pa}$$

$$p_{s3} = 466\ \text{Pa}$$

Pressure drop through the 800 mm duct:

$$\Delta p_{34} = 20\ \text{m} \times 0.72 \frac{\text{Pa}}{\text{m}}$$

$$\Delta p_{34} = 14\ \text{Pa}$$

$$p_{t4} = p_{t3} - \Delta p_{34}$$

$$p_{t4} = (488 - 14)\,\text{Pa}$$

$$p_{t4} = 474\ \text{Pa}$$

$$p_{s4} = p_{t4} - p_{v4}$$

$$p_{s4} = (474 - 22)\,\text{Pa}$$

$$p_{s4} = 452\ \text{Pa}$$

Regain of static pressure:

$$SR = p_{s3} - p_{s2}$$

$$SR = (466 - 457)\,\text{Pa}$$

$$SR = 9\ \text{Pa}$$

Find the air velocity in the branch:

$$v_5 = \frac{3 \times 4}{\pi \times 0.6^2} \frac{\text{m}}{\text{s}}$$

$$v_5 = 10.6\ \frac{\text{m}}{\text{s}}$$

$$p_{v5} = \left(0.5 \times 1.2 \times 10.6^2\right)\text{Pa}$$

$$p_{v5} = 67\ \text{Pa}$$

The bend part of the branch has a velocity ratio:

$$\frac{v_5}{v_2} = \frac{10.6}{7.6} = 1.4$$

Refer to Figure 6.4 to find the velocity pressure loss factor of 1; this is multiplied by the bend factor 0.67 and the branch duct velocity pressure p_{v5}.

Pressure drop through the branch:

$$\Delta p_{25} = \text{factors} \times p_{v5}$$

$$\Delta p_{25} = (1 \times 0.67 \times 67)\,\text{Pa}$$

$$\Delta p_{25} = 45\ \text{Pa}$$

$$p_{t5} = p_{t2} - \Delta p_{25}$$

$$p_{t5} = (492 - 45)\,\text{Pa}$$

$$p_{t5} = 447\ \text{Pa}$$

$$p_{s5} = p_{t5} - p_{v5}$$

$$p_{s5} = (447 - 67)\,\text{Pa}$$

$$p_{s5} = 380\ \text{Pa}$$

Pressure drop through the 600 mm duct:

$$\Delta p_{56} = 10\ \text{m} \times 0.8\frac{\text{Pa}}{\text{m}}$$

$$\Delta p_{56} = 8\ \text{Pa}$$

$$p_{t6} = p_{t5} - \Delta p_{56}$$

$$p_{t6} = (447 - 8)\,\text{Pa}$$

$$p_{t6} = 439\ \text{Pa}$$

$$p_{s6} = p_{t6} - p_{v6}$$

$$p_{s6} = (439 - 67)\,\text{Pa}$$

$$p_{s6} = 372\ \text{Pa}$$

EXAMPLE 6.5

A centrifugal fan draws outdoor air through an inlet grille and 10 m duct, passing air through an air handling unit then a 10 m duct to discharge supply air into a conditioned room. Figure 6.8 shows the duct system and a scale drawing of the pressure variations through the system.

Data for the system are:

Section 1–2 10 Pa loss through grille plus $0.5p_{v2}$ entry loss.
Section 2–3, 10 m duct, $5\frac{\text{m}}{\text{s}}$, $\frac{\Delta p}{l}1\frac{\text{Pa}}{\text{m}}$.
Section 3–4, centrifugal fan.
Section 4–5, $v5\frac{\text{m}}{\text{s}}$, losses of 10 Pa.
Section 5–6, AHU loss 40 Pa.
Section 6–7, 10 m duct, $5\frac{\text{m}}{\text{s}}$, $\frac{\Delta p}{l}1\frac{\text{Pa}}{\text{m}}$.
Section 7–8, 10 Pa loss through grille plus $1p_{v7}$ discharge loss.

$$\rho = 1.2\frac{\text{kg}}{\text{m}^3}$$

$$p_{v2} = p_{v4} = p_{v7} = \left(0.5 \times 1.2 \times 5^2\right)\text{Pa}$$

$$p_{v2} = 15\ \text{Pa}$$

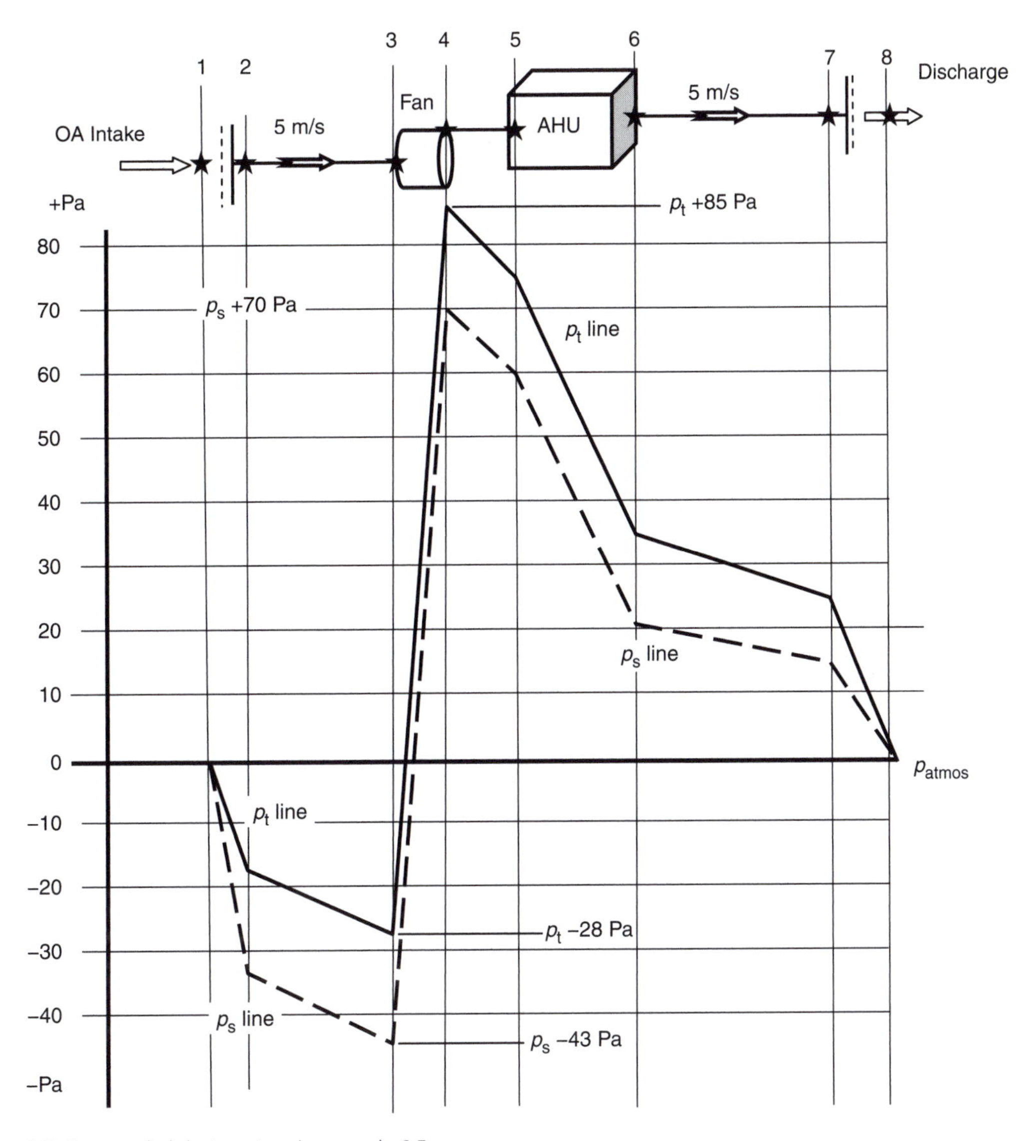

6.8 Open ended ducts system in example 6.5.

$$\text{Entry loss } \Delta p_{12} = (10 + 0.5 p_{v2})\,\text{Pa}$$

$$\text{Entry loss } \Delta p_{12} = (10 + 0.5 \times 15)\,\text{Pa}$$

$$\text{Entry loss } \Delta p_{12} = 18\ \text{Pa, rounding up decimal place}$$

$$\text{Discharge loss } \Delta p_{78} = (10 + 1 p_{v7})\,\text{Pa}$$

$$\text{Discharge loss } \Delta p_{78} = (10 + 1 \times 15)\,\text{Pa}$$

$$\text{Discharge loss } \Delta p_{78} = 25\ \text{Pa}$$

$$\text{Duct loss } \Delta p_{23} = 10\ \text{m} \times 1 \frac{\text{Pa}}{\text{m}}$$

Duct loss $\Delta p_{23} = 10$ Pa

Duct and transition losses $\Delta p_{45} = 10$ Pa

Duct loss $\Delta p_{67} = 10 \text{ m} \times 1 \frac{\text{Pa}}{\text{m}}$

Duct loss $\Delta p_{67} = 10$ Pa

Summary of pressure losses:

Δp_{12} 18 Pa, Δp_{23} 10 Pa, $\Delta p_{34} = FTP$, Δp_{45} 10 Pa, Δp_{56} 40 Pa, Δp_{67} 10 Pa, Δp_{78} 25 Pa

$$FTP = \sum \Delta p_{18}$$

$$FTP = (18 + 10 + 10 + 40 + 10 + 25)\,\text{Pa}$$

$$FTP = 113 \text{ Pa}$$

$$FVP = 15 \text{ Pa}$$

$$FSP = (113 - 15)\,\text{Pa}$$

$$FSP = 98 \text{ Pa}$$

Pressure changes at a fan

The fan produces a rise of total pressure in the whole system. This rise is the fan total pressure *FTP* Pa. *FTP* provides air pressure to overcome system resistance and air movement through and out of the system. Fan total pressure rise is equal to the drop of total pressure due to friction through the ductwork system, including entry and discharge loss factors. Fan total pressure is calculated from the total pressure in the fan outlet duct minus the total pressure in the fan inlet duct. Figure 6.9 shows the pressure changes across a centrifugal fan and duct system.

$$FTP = p_{t2} - p_{t1}$$

Fan velocity pressure *FVP* is defined as the velocity pressure in the discharge area from the fan.

$$FVP = p_{v2}$$

Fan static pressure *FSP* is defined as fan total pressure minus fan velocity pressure. Note that this is not necessarily the same as the change in static pressure across the fan connections.

$$FSP = FTP - FVP$$

The procedure is as follows:

1. Find air density.
2. Calculate air velocity and velocity pressure in each duct.
3. Convert all pressures to pascals.
4. Refer to Figure 6.9
5. Find inlet total pressure p_{t1} and static pressure p_{s1}.
6. Find outlet total pressure p_{t2} and static pressure p_{s2}.
7. Calculate *FTP*, *FVP* and *FSP*.

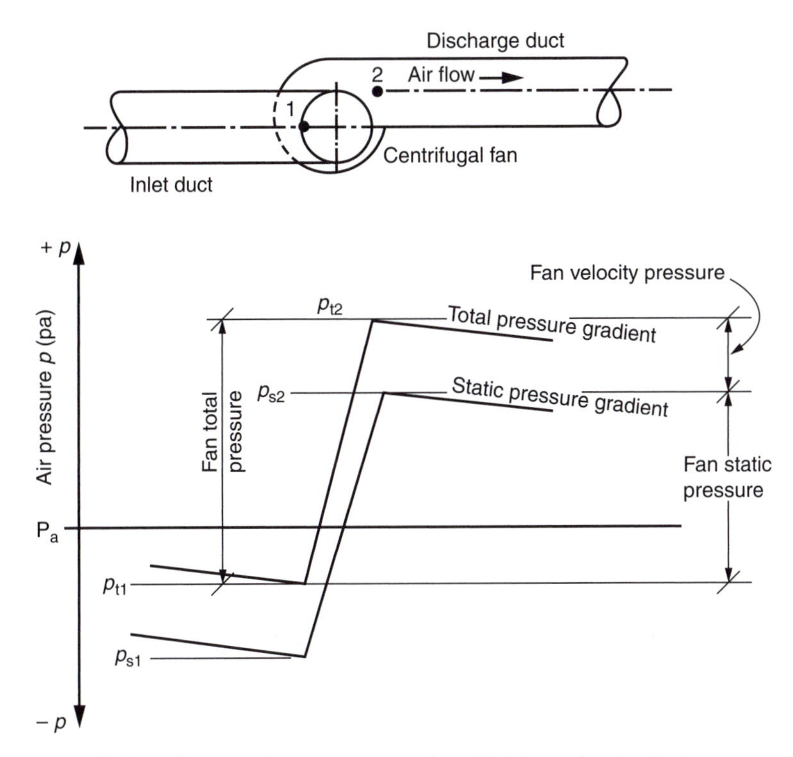

6.9 Definition of pressure increases across a fan with inlet and outlet ducts.

EXAMPLE 6.6

A centrifugal fan delivers 5 m^3/s into an outlet duct of 800 mm × 800 mm. The inlet duct to the fan is 700 mm diameter. The static pressure at the fan inlet was measured as −50 mm water gauge relative to standard atmospheric pressure. The ductwork system had a calculated resistance of 1500 Pa. Air density is 1.2 kg/m^3. Calculate the pressures either side of the fan.

Air density $1.2 \frac{kg}{m^3}$

$$v_1 = \frac{5 \times 4}{\pi \times 0.7^2} \frac{m}{s}$$

$$v_1 = 13 \frac{m}{s}$$

$$p_{v1} = \left(0.5 \times 1.2 \times 13^2\right) \text{Pa}$$

$$p_{v1} = 101 \text{ Pa}$$

$$v_2 = 5 \frac{m^3}{s} \times \frac{1}{0.8 \text{ m} \times 0.8 \text{ m}} \frac{m}{s}$$

$$v_2 = 7.8 \frac{m}{s}$$

$$p_{v2} = \left(0.5 \times 1.2 \times 7.8^2\right) \text{Pa}$$

$$p_{v2} = 37 \text{ Pa}$$

$$p_{s1} = -50 \text{ mm } H_2O$$

$$p_{s1} = (-50 \times 9.807) \text{ Pa}$$

$$p_{s1} = -490 \text{ Pa}$$

$$FTP = \text{duct system pressure drop}$$

$$FTP = 1500 \text{ Pa}$$

On the inlet side of the fan:

$$p_{t1} = p_{s1} + p_{v1}$$

$$p_{t1} = (-490 + 101) \text{Pa}$$

$$p_{t1} = -389 \text{ Pa}$$

$$FTP = p_{t2} - p_{t1}$$

$$p_{t2} = FTP + p_{t1}$$

$$p_{t2} = (1500 - 389) \text{Pa}$$

$$p_{t2} = 1111 \text{ Pa}$$

$$p_{s2} = p_{t2} - p_{v2}$$

$$p_{s2} = (1111 - 37) \text{Pa}$$

$$p_{s2} = 1074 \text{ Pa}$$

Summary:

$$FTP = 1500 \text{ Pa}$$

$$FVP = p_{v2} = 37 \text{ Pa}$$

$$FSP = (FTP - FVP) = (1500 - 37) = 1463 \text{ Pa}$$

Air ducts do leak (Chapter 8) and designers may wish to add a margin to the calculated *FTP* in selecting a fan and motor combination to ensure correct air delivery in air conditioned rooms; maybe 10%. Some designers calculate friction as losses of static pressure through the system and then add the discharge velocity pressure to find *FTP*. Whichever method is chosen, the commissioning engineer has the job of fine tuning the resulting installation with a variable frequency drive fan speed controller (VSD), volume control dampers (VCD) and a belt drive pulley on the motor shaft to achieve the design outcome to within a reasonable tolerance. Air duct system flow and pressures also vary considerably from their design specification due to the changing resistance of air filters (from clean to clogged), dirt accumulated in the duct system or at outside intake louvres (when do they get cleaned?), internal acoustic lining, changes from design drawings occurring during construction, intrusions into airways by sensors, rough duct edges at joints and possible irregularities from the design.

Flow measurement in a duct

An air duct installation may be fitted with a permanent flow meter such as a venturi nozzle (Figure 6.10), orifice plate (Figure 6.11), conical inlet (Figure 6.12), or flow grid (Figure 6.13). Each of these devices introduces a permanently installed frictional resistance. The air flow entering the meter must have minimum swirl due to upstream bends, branches or fans, as unstable flow produces pulsations in the measured air pressure and unsteady flow or velocity readings. Honeycomb flow straighteners can be used.

The air flow grid comprises one or more drilled pipes across the duct diameter. It measures the average velocity pressure. This single pressure is passed to a transducer that provides a 0 to 10 volt electrical signal to a computerised control or energy management system.

Portable air flow instruments such as the pitot-static tube, rotating vane anemometer, thermistor and hot wire anemometers are used in commissioning tests. A rotating vane anemometer combined with a venturi hood is used to measure the airflow from a supply grille.

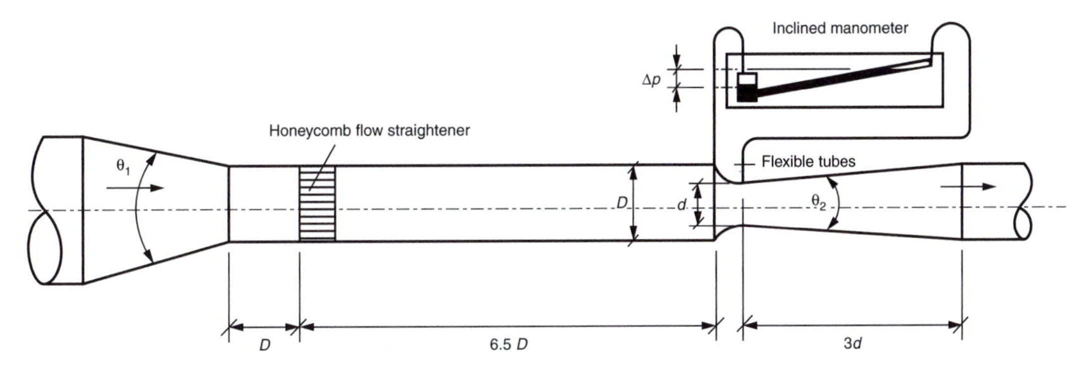

6.10 Venturi nozzle in-duct air flow meter, $D > 1.5d, \theta_1 = 7°, \theta_2 = 15°$.

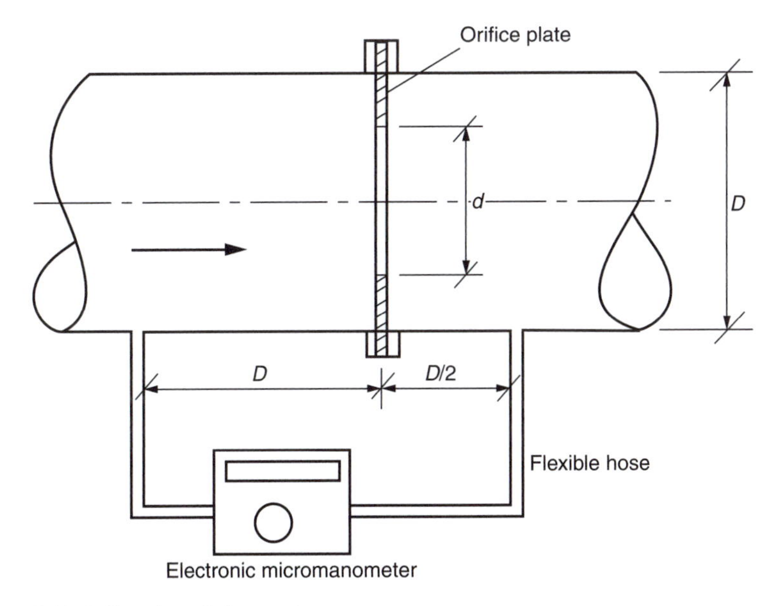

6.11 Orifice plate air flow meter.

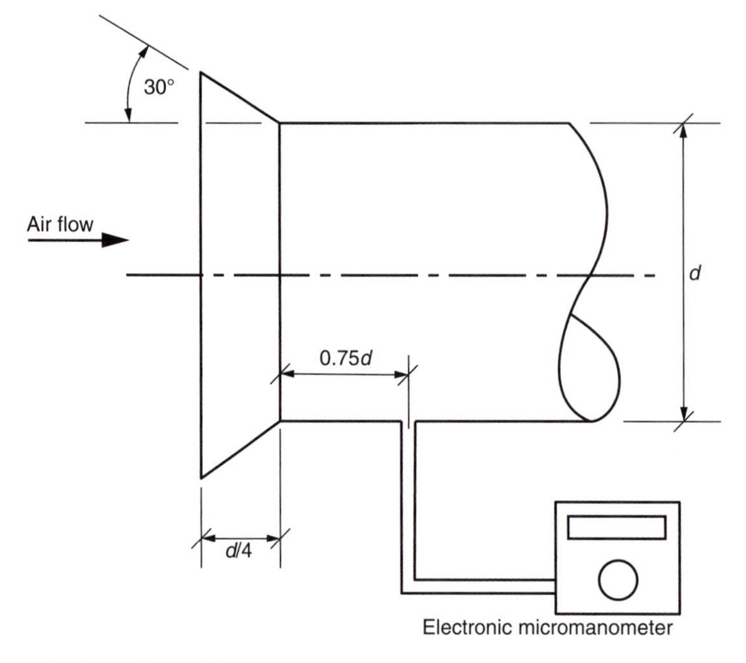

6.12 Conical inlet air flow meter.

Laboratory air flow meters for research or frequent tests by a manufacturer of air conditioning components have several diameters of upstream straight duct for flow straightening. The pitot-static tube is used for accurate flow measurements to enable calibration of a venturi or orifice plate meter.

The venturi meter causes relatively low frictional pressure loss but needs to have smooth internal surfaces and a shallow angle expansion duct to avoid flow disturbance. The orifice plate produces a higher pressure drop and the conical inlet is appropriate to the laboratory testing of fans.

The general expression for the volume flow rate of air through the venturi, orifice plate and conical inlet meters is:

$$Q = \text{coefficient} \times \frac{\pi d^2}{4} (2\rho \Delta p)^{0.5}$$

$Q = \text{air flow} \dfrac{\text{m}^3}{\text{s}}$

Coefficient = around 1 for a venturi

Coefficient = around 0.65 for an orifice plate

Coefficient = around 0.95 for a conical inlet

d = throat diameter, m

ρ = upstream air density, $\dfrac{\text{kg}}{\text{m}^3}$

Δp = pressure difference across meter, Pa

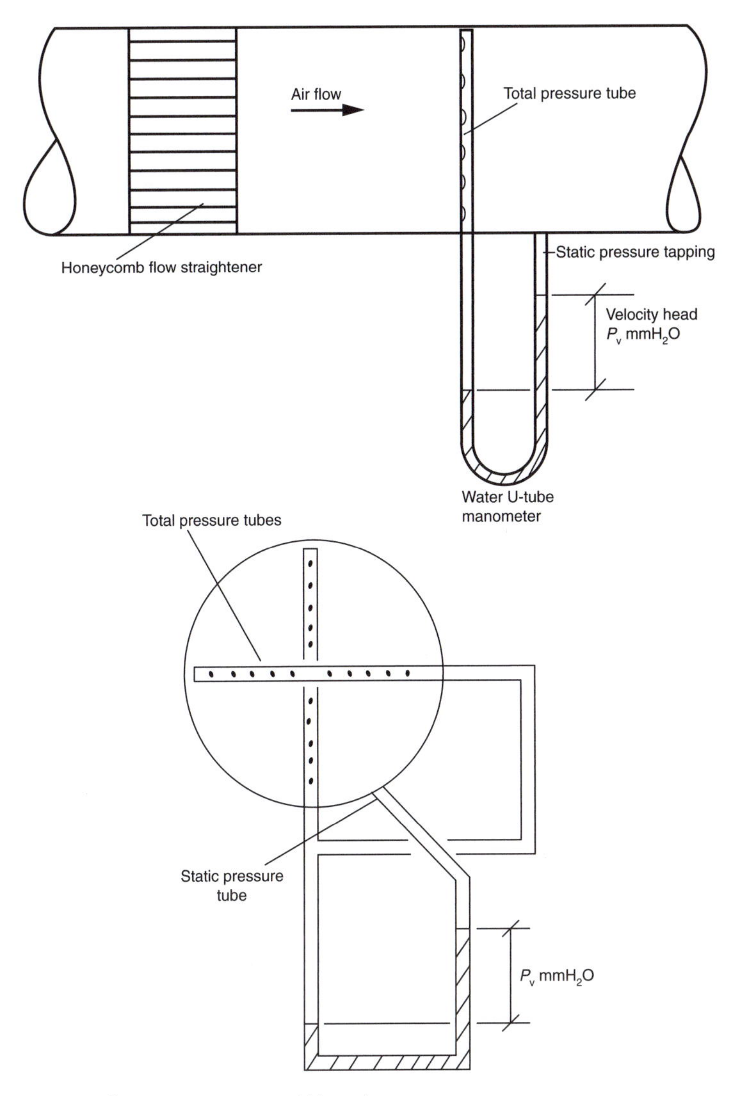

6.13 Air flow measurement grid in a duct.

EXAMPLE 6.7

A venturi meter in an air duct shows a pressure difference of 50 mm water gauge across its connections when passing air of density 1.2 kg/m^3. The throat of the venturi is 150 mm diameter, flow coefficient is 1, calculate the flow.

Venturi pressure drop $\Delta p = (9.807 \times 50)\,\text{Pa}$

Venturi pressure drop $\Delta p = 490\ \text{Pa}$

$$Q = \text{coefficient} \times \frac{\pi d^2}{4}(2\rho\Delta p)^{0.5}$$

$$Q = 1 \times \frac{\pi \times 0.15^2}{4}(2 \times 1.2 \times 490)^{0.5}\,\frac{m^3}{s}$$

$$Q = 0.606\,\frac{m^3}{s}$$

The pitot-static tube and inclined or electronic micro manometer are used as references for other methods of flow measurement. A National Physical Laboratory modified ellipsoidal-nosed pitot-static tube having an outside diameter not exceeding one forty-eighth of the airway diameter can be employed in air flows of up to 70 m/s and greater than 1 m/s. A traverse across three diameters in order to measure the velocity pressure at 24 locations is made for circular ducts. A total of 48 points can be used for flow measurement in rectangular ducts; Figures 6.14 and 6.15 show the measurement positions.

Average air velocity at the test section is:

$$v = \sqrt{\frac{2p_d}{\rho}}\,\frac{m}{s}$$

Where p_d is the SMR value of the dynamic pressures from the pitot-static traverse, that is, the square of the mean of the square roots of the j individual dynamic pressures:

$$p_d = \left(\frac{p_{v1}^{0.5} + p_{v2}^{0.5} + p_{vj}^{0.5}}{j}\right)^2$$

EXAMPLE 6.8

Calculate the air volume flow rate in a 350 mm internal diameter air conditioning duct when the pitot-static tube traverse revealed the following dynamic pressures in mm water gauge: 1.55, 1.58, 1.87, 1.96, 2.06, 1.23, 1.34, 0.85, 1.56, 1.54, 1.8, 1.9, 2, 1.2, 1.3, 0.8, 1.5, 1.52, 1.75, 1.85, 1.9, 1.18, 1.28, 0.75. The density of the air on the day of test was 1.2 kg/m^3.

Calculate the SMR value from the series of 24 dynamic pressures:

$$p_d = \left(\frac{p_{v1}^{0.5} + p_{v2}^{0.5} + p_{vj}^{0.5}}{j}\right)^2$$

$$p_d = \left(\frac{1.55^{0.5} + 1.58^{0.5} + \cdots 0.75^{0.5}}{24}\right)^2 \text{ mm } H_2O$$

$$p_d = 1.219 \text{ mm } H_2O$$

$$p_d = (9.807 \times 1.219)\,\text{Pa}$$

$$p_d = 12 \text{ Pa}$$

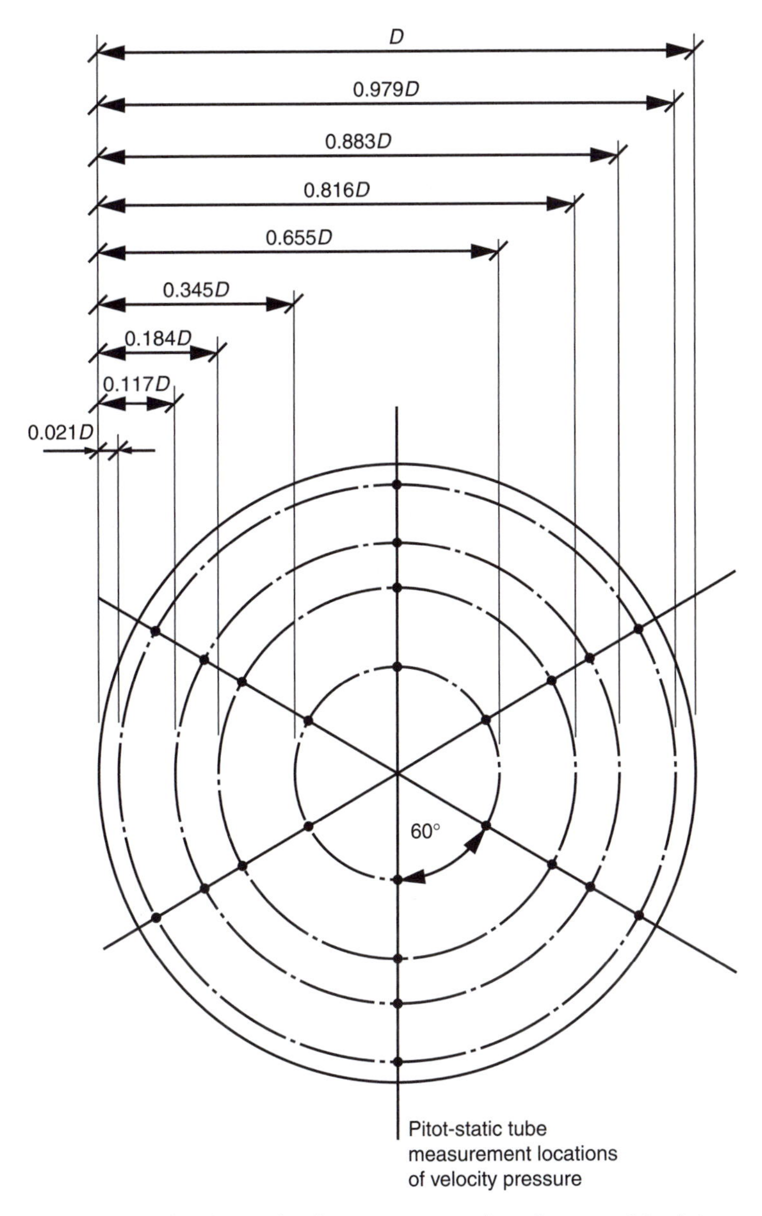

6.14 Locations for pitot-static tube traverse across three diameters, 24 points.

Average air velocity in the airway:

$$v = \sqrt{\frac{2p_d}{\rho}} \frac{m}{s}$$

$$v = \sqrt{\frac{2 \times 12}{1.2}} \frac{m}{s}$$

$$v = 4.47 \frac{m}{s}$$

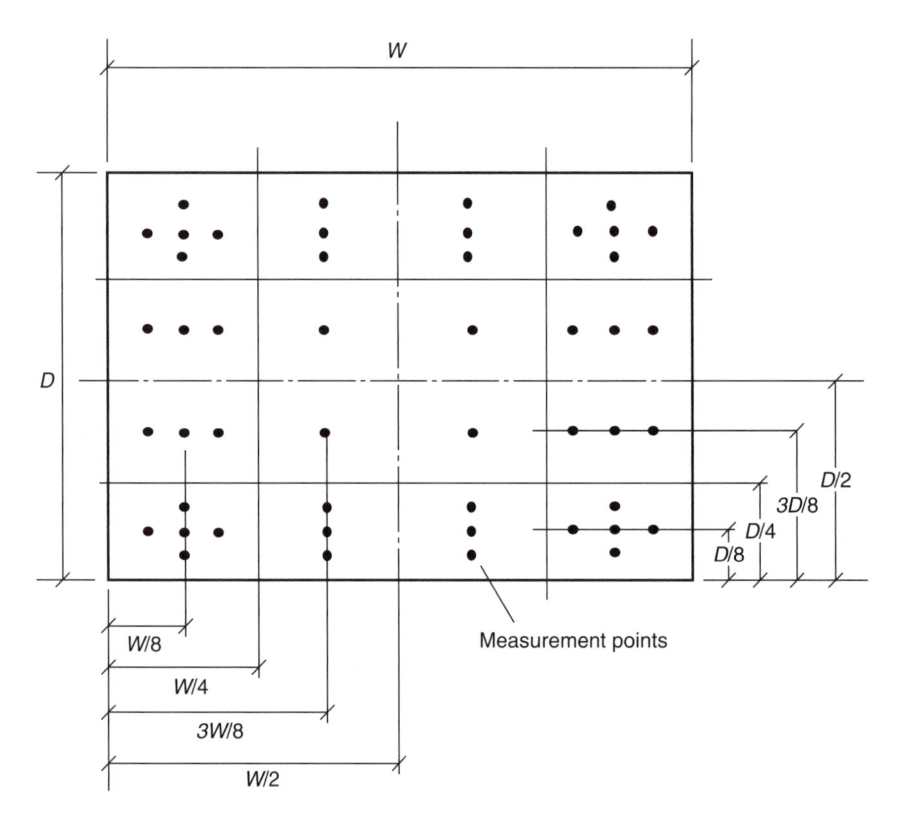

6.15 Locations for a 48 point pitot-static tube traverse in a rectangular airway.

Air volume flow rate:

$$Q = \frac{\pi d^2}{4} v$$

$$Q = \frac{\pi \times 0.35^2}{4} \times 4.47 \frac{m^3}{s}$$

$$Q = 0.43 \frac{m^3}{s}$$

EXAMPLE 6.9

If an air flow grid is installed in the air duct used in example 6.8, what average velocity pressure, velocity and volume flow rate would be indicated? Comment on the difference between the two calculation procedures.

The arithmetic average of the 24 dynamic pressures $\frac{p_{v1}^{0.5} + p_{v2}^{0.5} + p_{vj}^{0.5}}{j}$ is 1.51 mm H_2O.

$$p_d = 1.219 \text{mm } H_2O$$

$$p_d = (9.807 \times 1.219) \, \text{Pa}$$

$$p_d = 15 \text{ Pa}$$

$$\text{Average } v = \sqrt{\frac{2p_d}{\rho}} \frac{\text{m}}{\text{s}}$$

$$v = \sqrt{\frac{2 \times 15}{1.2}} \frac{\text{m}}{\text{s}}$$

$$v = 5 \frac{\text{m}}{\text{s}}$$

$$Q = \frac{\pi d^2}{4} v$$

$$Q = \frac{\pi \times 0.35^2}{4} \times 5 \frac{\text{m}^3}{\text{s}}$$

$$Q = 0.48 \frac{\text{m}^3}{\text{s}}$$

It would appear that the flow grid over estimates the flow rate by:

$$\text{Error} = \left(\frac{0.48 - 0.43}{0.43} \right) \times 100\%$$

$$\text{Error} = 11.6\%$$

The accuracy of the flow grid can be calibrated on site by checking it against a pitot-static traverse.

Duct system design

Design of a ventilation or air conditioning duct system:

1. Find the supply and extract air quantities for each room. The sensible heat formula is used to find the supply air flow rate. Extract air quantity may be slightly less than the supply quantity so that the room is pressurised and all air leakages become outwards. This avoids incoming draughts of unconditioned air. All the air flows into and out of the rooms must be accounted for as flow rates, litre/s. It is the flow rates between rooms that create draughts under doors and the differences between room static air pressures.
2. Draw a single line layout for both the supply and extract ductwork systems on the building plans.
3. Check that the duct system can be coordinated with the other services, the architectural features of the building and the structural design. Building Information Modelling (BIM) computer simulations are increasingly used in new complex construction projects. BIM facilitates data exchange with all designers to ensure a correctly working building and construction phase.
4. Ascertain that both horizontal and vertical service shafts are available for the intended route.
5. Decide the positions for supply air diffusers, terminal units and extract grilles. Refer to manufacturers' literature on discharge air velocity and throw of air from grilles. It is the position of the supply air grille that determines the air movement pattern generated within the room.
6. Draw a schematic layout of the ductwork system, Figure 6.16. This will be used as the working drawing for calculations and data.
7. Sketch the air handling plant to large size. This is to identify the plant items such as filters, heat exchangers, heating and cooling coils, and fans, showing their method of connection.

8. Sketch large details of any complicated items of ductwork, such as changes of section or tortuous routes within plant rooms.
9. Mark on the drawings the locations of air flow grids or other permanently installed flow meters, terminal air filters, fire dampers and air volume flow balancing dampers (VCD).
10. Locate the positions for air pressure control dampers.
11. Decide the possible positions for noise attenuators. These may be either side of the fans, at terminal units or grilles, or may be acoustic linings for some ducts.

Table 6.1 Limiting air velocities in ducts for low velocity system design

Application	*Main duct v,* $\frac{m}{s}$	*Branch duct v,* $\frac{m}{s}$
Hospital, concert hall, library, sound studio	5	3.5
Cinema, restaurant, hall	7.5	5
General office, dance hall, shop, exhibition hall	9	6
Factory, workshop, canteen	12.5	7.5

12. Decide the maximum air velocities for each part of the duct system from Table 6.1 depending upon the proximity of ducts to occupied rooms.
13. Summate all the air flows through the duct system. Where the air temperature remains constant, volume air flows can be added in litre/s or m^3/s. Air temperature changes due to zone or terminal heating, will produce changes in the air density and volume flow rate. In such cases, air mass flow rate in kg/s can be used.
14. Once all the duct air mass flow rates are fixed, the correct air density is used to calculate the volume flow rate in each duct.
15. The air mass flow rate supplied by the air handling plant will be equal to the sum of all the air mass flow rates leaving the ductwork.
16. Decide the limiting air pressure drop rate through the ducts. This may be chosen as 1 Pa/m for low velocity systems.
17. Find a suitable diameter for each duct from duct chart, data or calculated from:

$$Q = \frac{\pi d^2 v}{4}$$

$$d = \sqrt{\frac{4Q}{\pi v}} \text{ m}$$

$$Q = \text{duct air flow } \frac{m^3}{s}$$

$$v = \text{air velocity } \frac{m}{s}$$

$$d = \text{duct internal diameter, m}$$

18. Find approximate rectangular duct dimensions from:

$$\text{Circular duct area, } \frac{\pi d^2}{4} = \text{rectangular duct area, width} \times \text{depth}$$

19. Check that there is sufficient space within the false ceilings and service shafts for the ducts.
20. Assess the overall spaces needed for the service ducts to pass these air ducts and other services.
21. Decide whether the proposed duct system is practical.
22. If the ducts cannot be accommodated within the allocated spaces, it may be necessary to increase the air velocities. This will reduce the duct diameters but at the cost of increased frictional resistance, fan power and possible noise generation.
23. Record the design information for each section of duct on the schematic drawing.
24. Enter the data into a manual workbook or software.
25. Select the duct fittings to be used.
26. Record velocity pressure loss factors for each duct fitting.
27. Calculate the velocity pressure in each duct.
28. Multiply the duct length by the pressure loss rate for each section of duct and enter the duct Δp Pa on the workbook.
29. Multiply the duct fitting velocity pressure loss factor by the correct velocity pressure for each section and enter the fitting Δp Pa.
30. Enter all fixed pressure drops through items of plant such as filters, dampers, inlet and outlet grilles, diffusers, cooling and heating coils.
31. Identify the index circuit; this is the duct run offering the highest resistance to flow. The index route is the result of a series of resistances. It might not be the longest duct length. Other ducts are branches from this route and are in parallel with the index system.
32. Calculate the fan total pressure rise to offset the loss along the index route.
33. Draft the ductwork system onto the scale drawings of the building to validate the design and coordinate it with all the other design features, such as, structure, other services and architecture.

EXAMPLE 6.10

The duct system shown in Figure 6.16 is to be installed in a false ceiling over an office. An air handling plant, comprising fresh air inlet, filter, chilled water cooling coil and axial flow fan, is located in a plant room. Office A is supplied with 0.5 m^3/s and Office B has 1.25 m^3/s. Section 5–8 is in circular duct. Find suitable sizes for the ducts and state the performance specification for the fan.

Limiting air velocities are 2.5 m/s through the fresh air inlet grille and filter, 3.5 m/s through the cooling coil, 10 m/s through the fan and 6 m/s in ducts. The fresh air inlet is constructed from 45° louvres having a free area of 50%, wire mesh and a velocity pressure loss factor of 4.5. The filter, supply grille and cooling coil have air pressure drops of 75 Pa, 50 Pa and 20 Pa, respectively. Contractions in the duct are at an angle of 45° and the enlargement is at 30°. Ducts are to be sized for an air temperature of 20°C d.b. The data are to be entered onto the schematic drawing, Figure 6.16 and Table 6.2.

$$\rho = 1.2 \frac{\text{kg}}{\text{m}^3}$$

$$p_v = 0.5 \times 1.2 \times v^2 \text{ Pa}$$

$$p_{v2} = 0.5 \times 1.2 \times 2.5^2 \text{ Pa}$$

$$p_{v2} = 4 \text{ Pa}$$

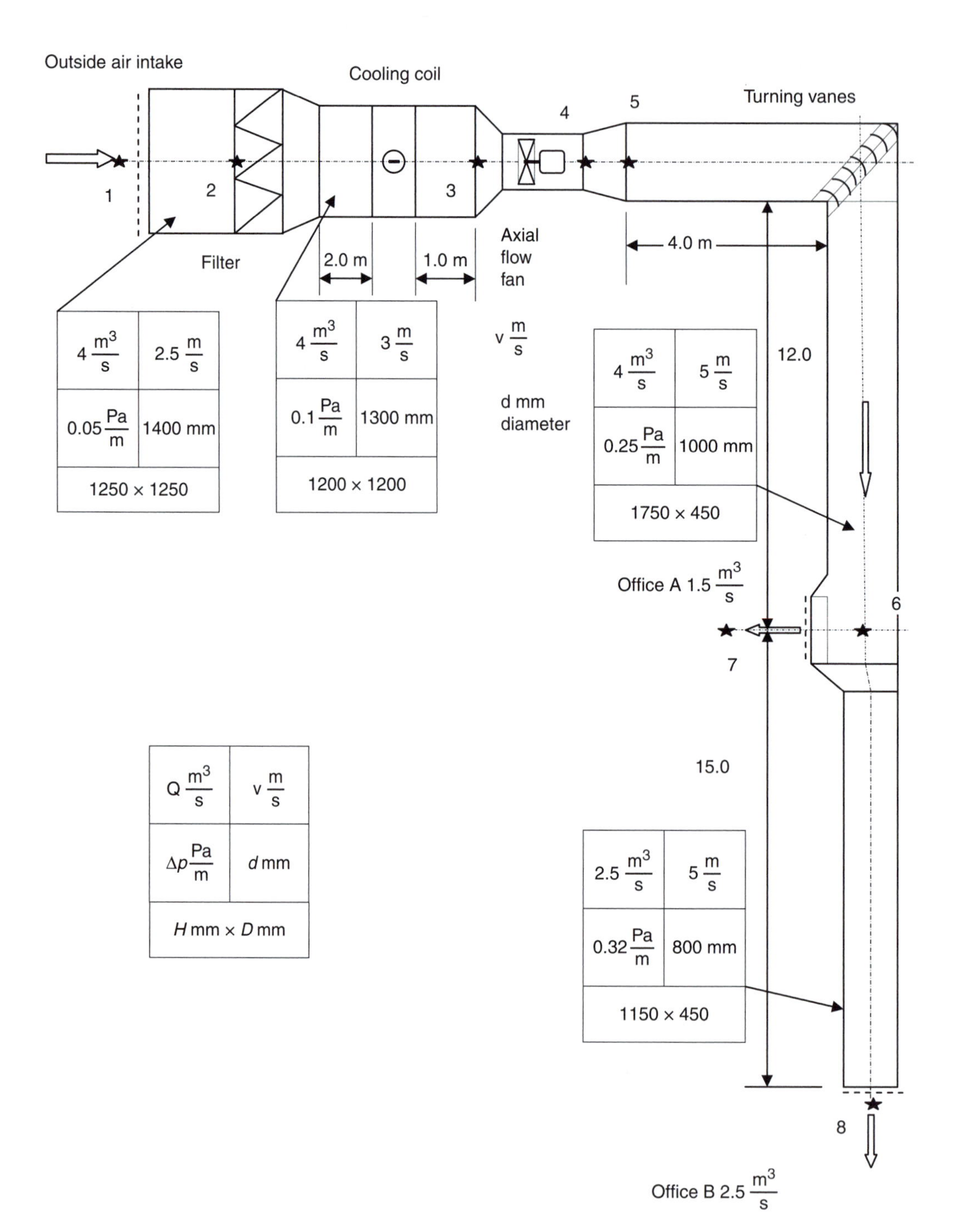

6.16 Schematic duct layouts in example 6.10.

Table 6.2 Duct sizing data for example 6.10

Section	*Length*	$\frac{\Delta p}{l}$	*v*	p_v	*k*	*Fitting*	*Duct*	*Total*	*Index*
	l m	$\frac{Pa}{m}$	$\frac{m}{s}$	*Pa*		*Pa*	*Pa*	*Pa*	
1–2	0	0	2.5	4	6	24	0	24	*
2–3	3	0.15	3.5	7	0.04	95	1	96	*
3–4	0	0	10	60	0.04	2	0	2	*
4–5	0	0	10	60	0.12	7	0	7	*
5–6	16	0.6	6	22	0.35	8	10	18	*
6–7	0	0	6	22	2.25	100	0	100	*
6–8	15	0.3	4	10	1	60	5	65	
				$\sum \Delta p_{17}$ Index route				247	

Fitting = pressure drop through the duct fittings Pa
Duct = pressure drop through the straight duct Pa

Pressure drop through the fresh air inlet grille:

$$\text{Pressure loss factor } k_{12} = \text{entry loss}(0.5) + \text{grille}(4.5) + \text{wire mesh}(1)$$

$$k_{12} = 0.5 + 4.5 + 1$$

$$k_{12} = 6$$

$$\Delta p_{12} = k \times p_{v2}$$

$$\Delta p_{12} = (6 \times 4)\,\text{Pa}$$

$$\Delta p_{12} = 24\ \text{Pa}$$

The 45° contraction in duct 2–3 after the filter has a velocity pressure loss factor of 0.04 and this is to be multiplied by the smaller duct velocity pressure p_{v3}.

$$p_{v3} = 0.5 \times 1.2 \times 3.5^2\ \text{Pa}$$

$$p_{v3} = 7\ \text{Pa}$$

$$\Delta p_{23} = (0.04 \times 7)\,\text{Pa}$$

$$\Delta p_{23} = 1\ \text{Pa}$$

Pressure drop through the fittings from node 2 to node 3 is 75 Pa for the filter and 20 Pa for the coil, 95 Pa ignoring decimal places.

Figure 6.17 is a simplified version and adequate for exercises in this book. To use the duct sizing chart, identify the air volume flow rate to be carried Q m^3/s, noticing that both horizontal and vertical scales are logarithmic, move horizontally to the right until the second coordinate limit is reached. This second limit may be either pressure drop rate Pa/m, air velocity *v* m/s or the duct diameter *d* m. If it is decided to choose duct diameters on the basis of a maximum pressure loss rate of, say, 1 Pa/m, then mark the point of this intersection. The diameter and velocity are estimated by interpolation between the lines, for example, at 0.3 m^3/s and 1 Pa/m, a 280 mm diameter duct has an air velocity of around 4.7 m/s, that is 4.87 m/s by calculation, demonstrating the danger of casually reading a logarithmic graph. Now find the real duct

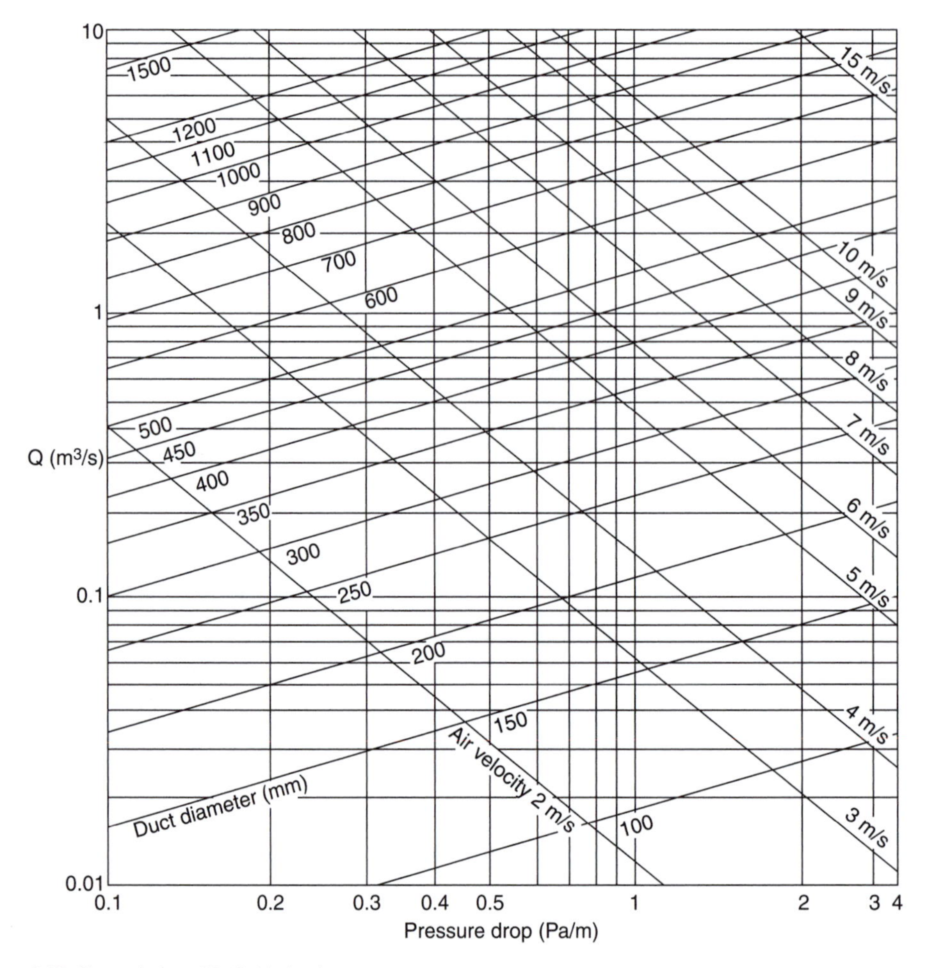

6.17 Flow of air at 20°C d.b. in ducts.

size to be employed by converting the 280 mm diameter into a standard circular, flat oval or rectangular equivalent and then finding the accurate diameter and velocity.

Airflow through sections 1 to 6:

$$Q_{16} = (1.25 + 0.5)\frac{m^3}{s}$$

$$Q_{16} = 1.75\frac{m^3}{s}$$

For duct 2–3 at a maximum air velocity of 3.5 m/s, $\Delta p/l$ in the duct will be 0.15 Pa/m:

$$\Delta p_{23} = 3\text{ m} \times 0.15\frac{Pa}{m}$$

$$\Delta p_{23} = 1\text{ Pa}$$

Air velocity through the fan is 10 m/s:

$$p_{v4} = 0.5 \times 1.2 \times 10^2 \text{ Pa}$$

$$p_{v4} = 60 \text{ Pa}$$

$$\Delta p_{34} = (0.04 \times 60)\,\text{Pa}$$

$$\Delta p_{34} = 2 \text{ Pa}$$

Before the k factor for the enlargement 4–5 can be found, the size of duct 5–6 is read from Figure 6.17. A 600 mm diameter duct carrying 1.75 m^3/s has a pressure loss rate of 0.6 Pa/m at a velocity of 6 m/s.

$$\text{Duct } \Delta p_{56} = 16 \text{ m} \times 0.6 \frac{\text{Pa}}{\text{m}}$$

$$\Delta p_{56} = 10 \text{ Pa}$$

A mitre bend with turning vanes, $k = 0.35$, is in the duct 5–6:

$$p_{v56} = 0.5 \times 1.2 \times 6^2 \text{ Pa}$$

$$p_{v56} = 22 \text{ Pa}$$

$$\Delta p_{56} = (0.35 \times 22)\,\text{Pa}$$

$$\Delta p_{56} = 8 \text{ Pa}$$

Enlarger 4–5 has an area ratio $\left(\frac{A_4}{A_5}\right)$:

$$A_4 = 1.75 \frac{\text{m}^3}{\text{s}} \times \frac{\text{s}}{10 \text{ m}}$$

$$A_4 = 0.175 \text{ m}^2$$

$$A_5 = \frac{\pi \times 0.6^2}{4} \text{m}^2$$

$$A_5 = 0.283 \text{ m}^2$$

$$\frac{A_4}{A_5} = \frac{0.175}{0.283} = 0.62$$

Sudden enlargement k factor is approximately 0.15; this concentric enlargement is at 30°, k 0.8:

$$\text{Enlargement } k_{45} = 0.15 \times 0.8$$

$$k_{45} = 0.12$$

$$\Delta p_{45} = (0.12 \times 60)\,\text{Pa}$$

$$\Delta p_{45} = 7 \text{ Pa}$$

Δp_{67} is due to a rectangular branch. Air velocity in each duct is expected to be 6 m/s so the velocity ratios for the branches 6–7 and 6–8 are unity when compared with the velocity in the main duct 5–6.

Take the branch bend to be a rectangular mitre bend k 1.25 and the discharge k as 1:

$$k_{67} = 1 \times 1.25 + 1$$

$$k_{67} = 2.25$$

$k_{68} = 0.09$ through the straight branch only.

$$\Delta p_{67} = (2.25 \times p_{v6} + \text{grille})\,\text{Pa}$$

$$\Delta p_{67} = (2.25 \times 22 + 50)\,\text{Pa}$$

$$\Delta p_{67} = 100\ \text{Pa}$$

Duct 6–8 carries 1.25 m^3/s at a maximum of 6 m/s and this requires continuation of the 600 mm diameter duct at a pressure loss rate of 0.3 Pa/m and v 4 m/s.

$$p_{v68} = 0.5 \times 1.2 \times 4^2\ \text{Pa}$$

$$p_{v68} = 10\ \text{Pa}$$

$$\text{Duct}\ \Delta p_{68} = 15\ \text{m} \times 0.3 \frac{\text{Pa}}{\text{m}}$$

$$\text{Duct}\ \Delta p_{68} = 5\ \text{Pa}$$

$$k_{68} = \text{straight through branch } 0.09 + \text{discharge } k$$

$$k_{68} = 0.09 + 1$$

$$k_{68} = 1$$

$$\Delta p_{68} = (\text{duct} + 1 \times p_{v68} + \text{grille})\,\text{Pa}$$

$$\Delta p_{68} = (5 + 1 \times 10 + 50)\,\text{Pa}$$

$$\Delta p_{68} = 65\ \text{Pa}$$

The fan has to provide a rise of total pressure *FTP* equal to the greatest frictional resistance. This is provided by the route 1–2–3–4–5–6–7. The branch 6–8 runs in parallel with 6–7 and has a lower resistance. When sufficient static pressure is provided at 6 to overcome the resistance of the branch 6–7, there will be more than enough to overcome the resistance to node 8. Each grille will be provided with a damper to regulate the air flow during commissioning to absorb the excess pressure in the non-index branches. It may be necessary to install balancing dampers in branches to absorb excess pressure that cannot be dropped across grille dampers.

The fan total pressure rise needed:

$$FTP = (24 + 96 + 2 + 7 + 18 + 100)\,\text{Pa}$$

$$FTP = 247\ \text{Pa}$$

$$\text{Excess pressure at node } 6 = (100 - 65)\,\text{Pa}$$

$$\text{Excess pressure at node } 6 = 35\ \text{Pa}$$

The grille damper should adequately balance the two branches and equalise them so that they both appear to the fan to be index routes.

Changing duct dimensions to fit the building, using oval or rectangular sizes, may require some recalculation of pressure drops.

Section 1–2 is likely to be square dimensions from intake grille to filter.

$$\text{Duct area} = \frac{\pi d^2}{4}$$

$$\text{Duct area} = \frac{\pi \times 1.4^2}{4} \text{m}^2$$

$$\text{Duct area} = 1.54 \text{ m}^2$$

$$\text{Square duct side} = \sqrt{1.54}\text{m}$$

$$\text{Square duct side} = 1.25 \text{ m}$$

$$\text{Square duct} = 1.25 \text{ m} \times 1.25 \text{ m}$$

$$\text{Square duct} = 1250 \text{ mm} \times 1250 \text{ mm}$$

Duct sizing workbook

Repetitive calculations can be written into a spreadsheet program and adapted for each new design. Dedicated programs are intended to handle large numbers of nodes and ducts but they require that each part of the system be fully specified with the same information that the user could enter into a spreadsheet. Download file *duct.xls* for use with these examples. Figure 6.18 is a schematic of the nodes used in *duct*.xls.

Follow a similar procedure to that already used:

1. Number all the system nodes and enter the section numbers.
2. For each section, enter the duct length l m, air flow Q m^3/s, maximum air velocity v m/s and the maximum allowable pressure loss rate $\Delta p/l$ Pa/m.
3. Duct diameter and the maximum carrying capacity of each duct are calculated.
4. The difference between the maximum carrying capacity and the required Q m^3/s is calculated and expressed as Q error %. If this is more than, say, 5%, enter a smaller value for maximum $\Delta p/l$ and repeat until satisfied.
5. Enter the likely duct diameter using increments of 50 mm.
6. Enter the dimensions of the rectangular duct to be used in mm. The circular equivalent diameter and actual air velocity are automatically calculated. Change the duct size until a satisfactory result is obtained.
7. The velocity pressure p_v Pa is calculated.
8. Enter the sum of the velocity pressure loss factors k for all the fittings in the duct section.
9. Enter the pressure drop through the air handling plant such as filters and heater batteries in the section, plant Δp Pa.
10. Enter the starting total pressure p_t Pa for node 1; this will normally be atmospheric pressure that is 0 Pa gauge pressure.
11. Each line is self-contained and calculations take place whenever new data are entered.
12. The total pressure at each node is displayed.
13. Enter a line of data for each section of duct.
14. Save your working spreadsheet into a unique file name. This preserves the original larger spreadsheet for future use.

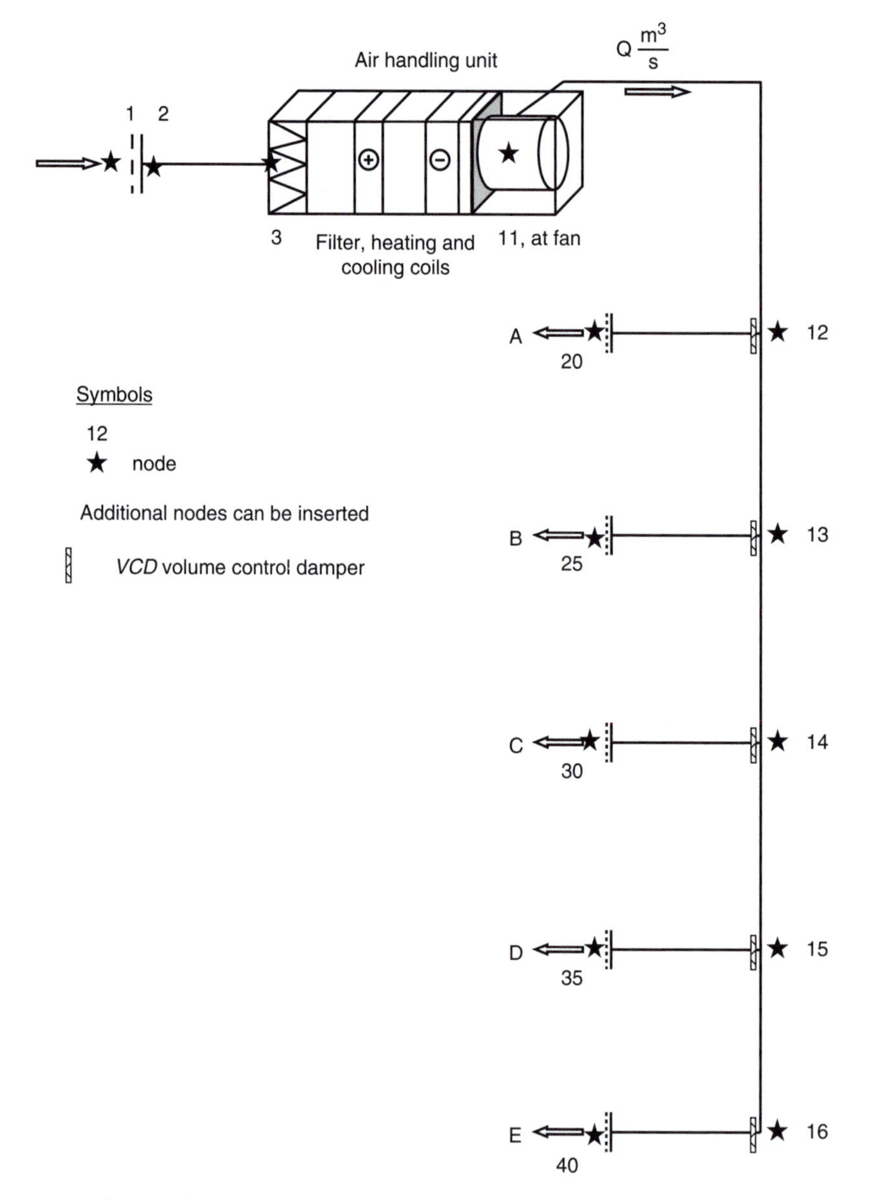

6.18 Schematic duct layouts for *duct.xls*.

15. If the original spreadsheet does not have enough sections, copy rows down the screen to make more. Edit the new lines to ensure that the correct cell reference for air density is read into the formula.
16. The column total Δp Pa summates the pressure losses in that row; p_t and p_s at each node are calculated.
17. The fan duty specification is shown.
18. Edit cells to correctly calculate pressure loss factors.
19. There are differences between the data used for the manual and spreadsheet calculations and these produce different fan total pressures due to manual truncation assumptions.
20. The format of the formulae may vary with different spreadsheet programs.

EXAMPLE 6.11

Use file *duct.xls* to design the supply air duct system shown in Figure 6.18. Section 1–11 and all other ducts are 10 m long. Supply air into each zone is 1 m^3/s. Limiting air velocities are 2.5 m/s in section 1–11, 10 m/s at the fan and 5 m/s in all other ducts. Section duct fitting factors are k_{12} 6, k_{23} 2, k_{11-12} 3, and each branch such as k_{12-A} 3. Plant Δp at each outlet grille is 50 Pa, outside air inlet grille 50 Pa, filter 100 Pa, heater 100 Pa and cooling coil 100 Pa. Duct air is at 20°C d.b. and 101325 Pa atmospheric pressure.

Correct data are provided in the file for the user to verify.
Fan specification becomes *FTP* 540 Pa, *FSP* 481 Pa, *FVP* 60 Pa with an air flow of 5 m^3/s.

Questions

1. State the three measurements made of air pressure within a duct, the direction they act, what each is used for and the scientist's name that is used to connect them. State the formulae connecting the three pressures.
2. Explain, with the aid of sketches, how the three airway pressures are measured. List all the equipment that would be needed. State how each item would be used.
3. A commissioning engineer needs to know the volume flow rate and mass flow rate of air through a 600 mm diameter duct. State all the measurements that are necessary and how they are to be acquired. Write all the formulae that would be needed and show the units of measurement used.
4. Sketch graphs of the three airway pressures changing along a duct of length *l* m and diameter *d* mm, showing the following cases.

   (a) Total and static pressures are above atmospheric.
   (b) Total pressure is above atmospheric but static pressure is below atmospheric.
   (c) The duct is on the suction side to a fan and air total pressure is below atmospheric pressure.
   (d) The duct tapers from 1 m diameter to 500 mm diameter along length *l* m. Pressures remain above atmospheric pressure.
   (e) The duct enlarges from 300 mm diameter to 600 mm diameter while the total pressure remains below atmospheric pressure.
   (f) A 500 mm diameter duct is above atmospheric pressure. The commissioning engineer omitted to seal a test hole halfway along the length of the duct.
   (g) Room air returns to the air handling plant through ducts that have inadequate joint sealing along their entire length, causing significant leakage. Total pressure at the commencement of the duct is above atmospheric. Static pressure within the duct starts at below atmospheric pressure.

5. Calculate the density of air for a temperature of 25°C d.b. when the atmospheric pressure is 101600 Pa.
6. Calculate the temperature of air that corresponds to a density of 1.1 kg/m^3 at standard atmospheric pressure.
7. Convert the following air pressures into pascals:

   25 mm H_2O, 50 mb, 125 mm H_2O, 0.3 m H_2O, 0.25 b.
   Note that 1 b = 1 bar = 10^5 Pa and 1 mb = 10^{-3} b
   Consequently, 1 mb = 100 Pa, also 1 kPa = 10^3 Pa

8. 3 m^3/s flows through a 1 m diameter duct. Calculate the air velocity.

9. Calculate the carrying capacities of air ducts of 400 mm, 600 mm, 1 m and 2 m diameters when the maximum allowable air velocity is 8m/s.
10. The temperature of air in a 400 mm diameter duct is 32°C d.b. on a day when the atmospheric pressure was 101105 Pa. The static pressure of the air in the duct was 45 mm water gauge below the atmosphere. The average air velocity was measured as 7 m/s. Calculate the air density, velocity pressure and total pressure.
11. 2 m^3/s are to flow through a 500 mm diameter duct at a static pressure of 300 Pa above the atmospheric pressure of 101500 Pa and at a temperature of 26°C d.b. Calculate the air density, velocity and total pressures.
12. Calculate the pressures that occur when 8 m^3/s flows through a 20 m long, 1300 mm diameter duct that then reduces to 1000 mm diameter and remains at 1000 mm for 20 m. The air total pressure at the commencement of the 1300 mm duct is 600 Pa above atmospheric. The reducer is the 60° concentric type. Air density is 1.2 kg/m^3. Frictional pressure loss rates are 0.23Pa/m in the 1300 mm diameter and 0.9 Pa/m in the 1000 mm diameter ducts.
13. Calculate the pressures that occur when 2 m^3/s flows through a 12 m long, 600 mm diameter duct that then reduces to 400 mm diameter and remains at 400 mm for 30 m. The air total pressure at the commencement of the 600 mm duct is 250 Pa above atmospheric. The reducer is the 30° concentric type. Air density is 1.15 kg/m^3. Frictional pressure loss rates are 0.85 Pa/m in the 600 mm diameter and 6.5 Pa/m in the 400 mm diameter ducts.
14. Calculate the static regain and pressure changes that occur when 3 m^3/s flow through a 10 m long, 450 mm diameter duct that then enlarges to 700 mm diameter and remains at 700 mm for 20 m. The air total pressure at the commencement of the 450 mm duct is 400 Pa above atmospheric. The enlarger is the 30° concentric type. Air density is 1.24 kg/m^3. Frictional pressure loss rates are 8Pa/m in the 450 mm diameter and 0.86 Pa/m in the 700 mm diameter ducts.
15. A 500 mm diameter duct supply air duct suddenly enlarges into a 1 m diameter plenum chamber containing filters. Calculate the static regain and pressure changes that occur when 1.5 m^3/s flows through the 35 m long, 500 mm diameter duct, enlarges, and then flows through the 1 m diameter plenum for 5 m. The air total pressure at the commencement of the 500 mm duct is 200 Pa above atmospheric. Refer to Figure 6.3 for the enlarger pressure loss factor. The air density is 1.22 kg/m^3. The frictional pressure loss rate is 1 Pa/m in the 500 mm diameter and can be assumed to be 0.05 Pa/m in the 1 m diameter ducts.
16. An air duct branch is similar to that shown in Figure 6.7. Use the data provided to calculate all the duct pressures at the nodes. The 800 mm diameter duct 1–2 is 22 m long and carries 3 m^3/s. The 600 mm diameter duct 3–4 is 12 m long and carries 2 m^3/s. The 450 mm diameter duct 5–6 is 10 m long and carries 1 m^3/s. Branch off-take is a short radius bend. Straight through contraction has a velocity pressure loss factor of 0.05. Air density is 1.2 kg/m^3. Pressure drop rates are, duct 1–2 0.43 Pa/m, duct 3–4 0.85 Pa/m, and duct 5–6 0.95 Pa/m. Total pressure at node 1 is 400 Pa.
17. An air duct branch is similar to that shown in Figure 6.7. Use the data provided to calculate all the duct pressures at the nodes. The 700 mm diameter duct 1–2 is 20 m long and carries 3 m^3/s. The 700 mm diameter duct 3–4 is 20 m long and carries 2 m^3/s. The 400 mm diameter duct 5–6 is 2 m long and carries 1 m^3/s. Branch off-take is a right angled bend having several turning vanes. Use circular duct data. Ignore the fact that turning vanes are not fitted to a circular branch. The vanes are in the branch duct. Air density is 1.2 kg/m^3. Pressure drop rates are, duct 1–2 0.85 Pa/m, duct 3–4 0.4 Pa/m, and duct 5–6 1.8 Pa/m. Total pressure at node 1 is 200 Pa.
18. A centrifugal fan delivers 3 m^3/s into an outlet duct of 600 mm × 400 mm. The inlet duct to the fan is 500 mm in diameter. Static pressure at the fan inlet was measured as −90 mm water gauge relative to standard atmospheric pressure. Ductwork system had a calculated resistance of 2000 Pa. Air density is 1.16 kg/m^3. Calculate the pressures either side of the fan.

19. The duct system shown in Figure 6.16 is to be installed in a false ceiling over offices and corridors. The air handling plant comprises the fresh air inlet, filter, chilled water cooling coil and an axial flow fan, and are located in a plant room. Office A is supplied with 1.5 m^3/s and Office B has 2.5 m^3/s. Find suitable sizes for the ducts and state the performance specification for the fan. Limiting air velocities are 2.5 m/s through the fresh air inlet grille and filter, 3 m/s through the cooling coil, 12 m/s through the fan and 5 m/s in the ducts. The fresh air inlet is constructed from 45° louvres having a free area of 60% and wire mesh. The filter, supply grille and cooling coil have air pressure drops of 65 Pa, 30 Pa and 40 Pa respectively. The contractions in the duct are at an angle of 60° and the enlargement is at 40°. Ducts are to be sized for an air temperature of 20°C d.b.
20. The ducts shown in Figure 6.16 are an extract system removing air from two workshops where low velocity air and heat reclaim are employed. The air flow direction arrows are to be reversed in Figure 6.16. Reverse the direction of node numbering starting with 1 at Workshop B and 2 at Workshop A. All the duct lengths are three times those shown. The air handling plant comprises a fan, heat reclaim cooling coil, a noise attenuator in place of the filter shown and an exhaust grille to outdoors. Workshop A has an extract rate of 3.5 m^3/s and Workshop B has extraction of 2.5 m^3/s. Find suitable sizes for the ducts and state the performance specification for the fan. Limiting air velocities are 2 m/s through the exhaust grille and attenuator, 3 m/s through the cooling coil, 15 m/s through the fan and 8 m/s in the ducts. The exhaust grille is constructed from 45° louvres having a free area of 70% and wire mesh. The attenuator, extract grilles and cooling coil have air pressure drops of 135 Pa, 70 Pa and 90 Pa, respectively. The contractions in the duct are at an angle of 45° and the enlargement is at 30°. Ducts are to be sized for an air temperature of 20°C d.b.

7 Controls

Learning objectives

Study of this chapter will enable the reader to:

1. know the component parts of control systems for air conditioning;
2. understand control terminology;
3. know the types and values of control signals;
4. understand the meaning and use of analogue signals;
5. know the voltages used for controls and actuators;
6. understand the use of thermistor temperature detectors;
7. know how digital communication takes place between computer and control equipment;
8. know how humidity is detected;
9. understand the use of enthalpy control;
10. know how pressure is detected;
11. know where and how air flow sensors are used;
12. understand weather compensation;
13. know the methods used for water temperature detection;
14. know the meaning and use of actuators;
15. know the principles of pneumatic actuators;
16. understand the operation of solenoids, relays and contactors;
17. understand the different modes of control operation;
18. know symbols for control diagrams;
19. realise the importance of the control system design to the air conditioning designer;
20. understand the operation of enthalpy control over fresh air ventilation;
21. draw plant operation graphs for all year control;
22. relate control signals to air conditions and plant status;
23. understand the use of optimum start control;
24. state the control equipment necessary for heating, ventilating and air conditioning applications;
25. list logical operating sequences;
26. know the starting sequence for heating plant;

27. know fan start procedures;
28. understand the control of single duct and variable air volume air conditioning systems;
29. identify the components in a controller;
30. understand direct digital control;
31. know how fan performance is regulated and matched to the air conditioning system requirement;
32. understand how chilled water refrigeration plant is controlled;
33. understand how a BMS is used in an air conditioned building;
34. understand the principles of electrical wiring diagrams.

Key terms and concepts

Actuator 220; analogue 220; BMS 240; capacity step 239; compressor load control 225; controlled condition 224; controlled variable 224; controller 224; controller action 225; dead time 224; desired value 224; detector 226; direct and reverse acting 226; direct digital control 244; enthalpy control 226; evaporator 238; optimum start time 227; pneumatic actuator 223; rectifier 220; refrigeration compressor 239; relay 243; set point 225; signal 224; solenoid 224; switch 228; thermistor 221; transducer 223; variable pitch vanes and blades 235; VAV 235; VFC 235; volt 221; weather compensator 231; wiring diagram 242.

Introduction

Automatic control of air conditioning aims to satisfy the temperature and humidity specification throughout the year. This can be for the thermal comfort of the occupants or for the benefit of a plant or process environment. Energy use minimisation is always a major factor. Each application is likely to be unique. The principles of operation and main features of commonly used systems are explained. The stages in control schemes are presented in logical order. Examples of practical control strategies are given. Data such as 20% minimum fresh air intake and a 5 volt control signal corresponding to a set point for air temperature are meant as examples, not as universal statements. Combinations such as fan starting methods, detector operation, and enthalpy control of dampers and temperature control requires the reader to bring together different sections of the book to formulate a complete system. The reader is challenged to propose suitable designs for control systems, operating diagrams, control signals, controller configuration and wiring diagrams. Principles are explained within the text. Questions that do not have a model solution are to build the reader's competence in preparation for tests and practical design work. The reader's answers to questions are for discussion with colleagues and tutors.

Further reference can be made to CIBSE publications Building Control Systems, Guide H (2001) Understanding Controls, and Commissioning Code C (2009) Automatic Controls. Also, manufacturers' publications such as Honeywell (1997) Engineering Manual of Automatic Control for Commercial Buildings, Johnson Controls (2008) Building Automation System over IP (BAS/IP) Design and Implementation Guide, and Siemens AG Download Center App (n.d.), and there may be others.

Components

Systems of automatic control consist of the principal components:

1. air humidity sensor;
2. air pressure modulating or balancing dampers;

3. air pressure or pressure difference sensor;
4. air temperature sensor;
5. air volume flow rate sensor;
6. analogue-digital module;
7. central processing unit, computer;
8. controller;
9. dedicated programmer;
10. disc data storage;
11. distributed processing unit, outstation;
12. electric or pneumatic valve and damper actuator motor;
13. electrical power cable;
14. electro-pneumatic and pneumatic-electric relay;
15. electro-pneumatic transducer;
16. flow modulating valve and damper;
17. fluid pressure modulating or balancing damper;
18. indicator instrument, analogue or digital, panel mounted;
19. instrument rack;
20. isolating valve and damper;
21. liquid crystal display;
22. low voltage control wiring, 0–10–24 volt;
23. mimic diagram on visual display unit;
24. motorised potentiometer;
25. motorised valve, two-, three- or four-way;
26. pressure switch;
27. printed circuit card, plug-in module;
28. printer;
29. programmable logic controller;
30. rectifier, alternating to direct current
31. relay;
32. standby electrical supply on mains failure;
33. switches, on/off, manual, selector, or electronic;
34. thermostat;
35. time clock, digital or motorised
36. transducer, detector signal to 0–10 volt signal;
37. transformer;
38. visual display unit, VDU;
39. water temperature sensor;
40. weather compensator.

Sensors and actuators

The standard electrical signal used between detector, controller and computer is 0–10 volt stabilised direct current from a rectifier. The current is kept constant at a value between 5 and 20 milliampere, mA. The variable voltage is an analogue of the condition being measured. Thus a 5 volt signal would represent 20°C if the range of temperature being measured is 10°C to 30°C. Electrical power to drive valve and damper motors is operated at 24 volts or 240 volts alternating current. Each controller has an alternating current to direct current rectifier to provide the signal voltage.

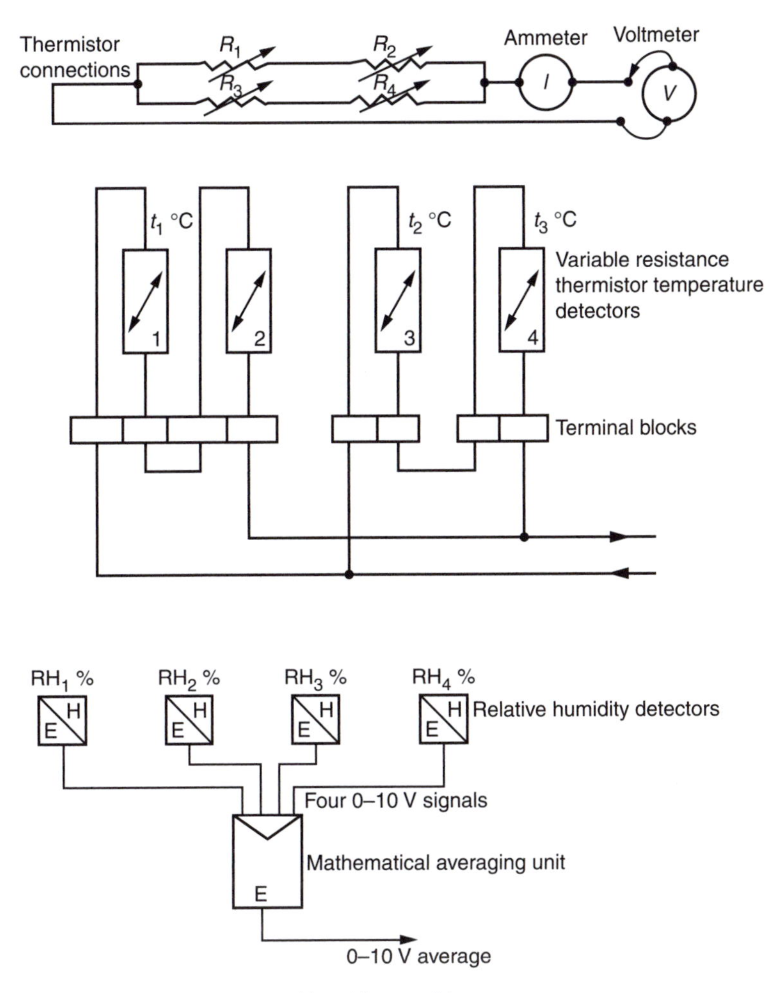

7.1 Averaging temperature and humidity conditions.

A temperature sensor is usually a thermistor. A fixed current of around 5 milliampere passes through the thermistor. Changes in detected temperature alter the electrical resistance of the thermistor. Ohm's law states that applied voltage equals current multiplied by resistance. The voltage drop through the thermistor changes with the detected temperature. The detector output signal varies from zero to ten volts, 0–10 V. This analogue output signal is either used directly by the controller or converted into a digital signal in binary code. Detected temperatures can be averaged by connecting two or more thermistor temperature detectors in series and parallel so that the whole circuit has the same resistance as only one detector. Figure 7.1 shows how this can be achieved. Space temperature t_1°C is measured with two thermistor detectors that are connected in series. The t_2°C and t_3°C temperatures are sensed by separate thermistors that are connected in series. The combined resistance of the four thermistors is equal to that of one thermistor.

Each thermistor produces outputs in the range 0.1 volt to 10 volt, V, at a current of 5 milliampere, m*A*. The variation of thermistor resistance, *R* ohm, with voltage is, from Ohm's law:

$$\text{Current } I \text{ ampere} = \frac{V \text{ volts}}{R \text{ ohms}}$$

$$R\,\Omega = \frac{V \text{ volts}}{I \text{ A}}$$

$$\text{At 10 V, } R = \frac{10 \text{ V}}{5 \text{ mA}}$$

$$\text{At 10 V, } R = \frac{10 \text{ V}}{5 \times 10^{-3} \text{ A}}$$

$$\text{At 10 V, } R = 2000\ \Omega$$

$$\text{At 0.1 V, } R = \frac{0.1 \text{ V}}{5 \times 10^{-3} \text{ A}}$$

$$\text{At 0.1 V, } R = 20\ \Omega$$

The electrical resistance circuit that is equivalent to the thermistor connections is shown in Figure 7.1. Series resistances are added, $(R_1 + R_2)$ and $(R_3 + R_4)$. The whole circuit resistance, R, is found by adding the reciprocals of parallel resistances.

$$\frac{1}{R} = \frac{1}{R_1 + R_2} + \frac{1}{R_3 + R_4}$$

When each thermistor has a resistance of 20 Ω:

$$\frac{1}{R} = \frac{1}{20 + 20} + \frac{1}{20 + 20}$$

$$\frac{1}{R} = \frac{1}{40} + \frac{1}{40}$$

$$\frac{1}{R} = \frac{2}{40}$$

$$R = 20\ \Omega, \text{ the same as one thermistor}$$

Temperature averaging is useful where there are significant variations within a room, zone or group of rooms or spaces. An averaging analogue unit contains amplifiers to process the mathematical calculation of the voltages. Figure 7.1 shows the use of a mathematical averaging unit with four humidity detector inputs. Digital controllers employ software programs for the mathematical processing.

Humidity sensors detect a change in the electrical resistance between two conductors. The conductors are fused onto plastic and coated with lithium chloride salt. The salt is hygroscopic, absorbing moisture from the air. The electrical resistance between the two conductors is related to the moisture absorbed. The 0–10 V output signal is calibrated to air percentage saturation.

Enthalpy detectors sense both air temperature and humidity, often in a combined sensor. The output 0–10 V signal is calibrated into a measure of the specific enthalpy of the air. They are used to compare the specific enthalpy of the incoming fresh air into a ducted air conditioning system, with that of the recirculated room air. The air mixing dampers or heat recovery equipment is operated from these enthalpy signals.

Air or gas pressure is sensed with a flexible diaphragm that separates two chambers. One chamber is connected to the air duct or space where the pressure is to be measured. The other chamber is either connected to atmospheric pressure or the other pressure space. A change in pressure between the two compartments causes the diaphragm to flex. The physical movement of the diaphragm operates a variable capacitor in a signal processor to generate a 0–10 V output. The sensor is a pressure transmitter that sends its output to a digital display, dial gauge or computer.

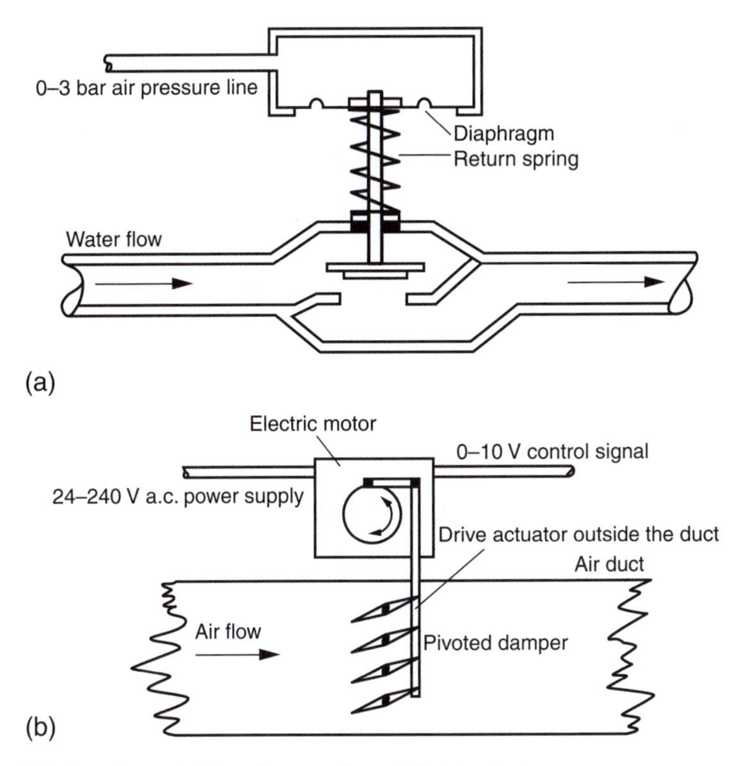

7.2 Operating principles of pneumatic and electric actuators.

An air flow rate transmitter is a pressure sensor that is connected to an averaging velocity pressure probe or flow grid as in Figures 6.1 and 6.13. The air flow measured is proportional to the square root of the pressure difference. The output signal has had the square root processed electronically. Pressure can be sensed with bellows or a bourdon tube that is expanded on rise of pressure and these are frequently used with instruments.

An external air temperature sensor is a thermistor within a weatherproof rigid plastic or cast alloy box. Temperature within the box is affected by solar radiation, wind and precipitation. The signal processor can be programmed to compensate the indoor plant operation for these weather conditions.

A water temperature sensor is a thermistor in a brass casing that is inserted into the waterway. An output signal of 0–10 V is proportional to the water temperature. Temperature sensing methods utilise a bimetal strip, bimetal rod and tube, liquid and vapour filled bellows, a thermocouple and an electrical resistance coil. They are for local indicators or used with automatic control. Temperature detectors take anything from 30 seconds to 5 minutes to change their output signal in response to a change in the measured variable.

An actuator is a device that converts the control signal into an action. Such actions are to operate valves and dampers that control fluid flow or to stop and start flows. Electric motor actuators can be operated on the line voltage of 240 V or extra low voltage of 24 V alternating current. The motor is either bi-directional, reversing, to drive the valve or damper in both directions, or spring return. The device may be normally closed or normally open depending upon the safe state when electric power is removed. The motor output shaft is driven through reduction gears or a linear screw thread. The motor run time can be 2 minutes from fully open to closed condition.

Pneumatic actuators use only air pressure for their motive power. They are spark free; require only a transducer to convert an electrical or electronic control signal into an air pressure and a 1–3 bar air pressure supply. It can become cheaper to install pneumatic actuators when many valves are needed. A control that combines electronic signals and pneumatic actuators is called a hybrid system. A pneumatic actuator comprises of a diaphragm that is pushed outward by increasing air pressure against a return spring to open the valve or damper controlling the fluid flow. On removal of the controlling pneumatic pressure, the spring returns the final control element, valve or damper, to its original position. This can be normally open or normally closed. A double acting actuator requires pneumatic pressure to move the output shaft in both directions and there may not be an automatic fail safe feature. Figure 7.2 shows the operating principles of electric and pneumatic actuators.

Electric solenoids are used to open and shut valves on fluid pipelines and to open and close electrical switches. They are usually employed in fail safe situations where power is required to keep them open.

Terminology

Terms used to describe control systems include the following.

Actuator	Valve or damper motor. The thing that does the work.
Analogue	Variable electrical voltage or current is transmitted between elements of the control system to initiate actions. 0–10 volts direct current of 0–5 milliampere. This is an electrical analogue model of, for example, a controlled room air temperature range of 18°C to 26°C.
Boost	Temporary operation of the heating or cooling plant at full output. Switch to 100% output for a while.
Cascade	The controller that receives two or more input 0–10 volt signals. The first signal is processed and then combined with the second signal for further processing to generate the output signal.
Controlled condition	Physical state of the controlled variable temperature, humidity, and flow rate or on/off run state.
Controlled variable	Temperature, pressure, speed or humidity that is being controlled.
Controller	Equipment that receives input data and produces an output signal that is used for control action. A direct digital control panel using software.
Controller action	Mathematical relationship between the input and output signals created by the controller. A formula.
Dead time or zone	Time between a change of input signal to a controller and the commencement of change in the controlled variable. This can be deliberately set to allow room air temperature or humidity to move within a comfortable range and minimise energy. The mechanical system is resting at its present activity level awaiting instruction to increase or decrease energy use. A long time or zone range saves energy but relies on users experiencing a variation in comfort standard.
Desired value	Condition of the controlled variable that is to be maintained (e.g. room air at 24°C).
Differential	Band of values of the controlled variable that is needed to initiate a response from the controller (e.g. room air at 24°C±1°C).
Direct digital control, DDC	Binary bytes are communicated between elements of the control system. A computer system.
Error	Difference between the actual value and the desired value of the controlled condition. Desired value 22°C, actual value of room air temperature at that moment 23°C, error 1°C.

Final control element Valve, damper or switch that operates upon the controlled variable. Modulates water or air flow rate or switches a fan off.

Inertia Resistance to change of the whole system. This includes the time taken to detect an error, send a signal to the controller, operate the final control element and change the state of the controlled variable. The thermal storage capacity of the building's surfaces may be a part of the overall inertia.

Input signal Signal received by the controller. This is normally the 0–10 volt signal from the detector.

Lag Time between a control signal being sent to a final control element and that element starting to respond. This may be less than dead time. The valve mechanism has to wake up.

Offset Controlled variable is maintained at a deviation from the set point or desired value. Desired value, set point, 24°C, actual value of room air temperature controlled to be 23°C, offset 1°C. This may be because solar radiation overheats a zone.

Output signal Mathematically processed output signal from a controller to the final control element. A formula. Room air temperature rises to 25°C, the controller sends 8 volts to open a cooling coil valve to 80% on the local fan cooled unit (FCU) chilled water (CHW).

Proportional Output signal from the controller is proportional to the deviation of the controlled variable from the set point. A small rise in room air temperature above set point produces a small increase in CHW valve opening.

Set point Setting of the controller that is necessary to achieve the desired value of the controlled variable. A desired value of room air temperature 24°C, the controlled variable, corresponds to 5 volts of the range 0–10 volts in the controller range.

Controllers operate according to a variety of methods, as follows.

On/off Controlled medium is switched on or off. A differential exists between the controlled conditions at the on and off switching points. The swing in the value of the controlled condition will be larger than the differential. This is due to the inertia of the whole system and the building.

Incremental Several fixed steps between on and off. Multiple boiler and refrigeration multi-cylinder compressor systems are controlled this way. The effect of each step is smoothed by the inertia of the fluid distribution system.

Proportional, P Output signal from the controller is proportional to the difference between the set point for the controlled variable and the sensed condition. The response is in proportion to the problem.

Floating Valve or damper actuator is moved at a fixed speed. The controller gives pulses of movement to bring the controlled condition within a neutral band where no pulses are made. The controlled condition varies continuously and the actuator is used frequently.

Integral, I Output signal from the controller is time dependent. The valve or damper actuator is moved at a speed that is in proportion to the error signal from the controlled condition. The detection of a large error causes the maximum speed response of the actuator. This is analogous to an emergency stop when driving a vehicle.

Derivative, D	Valve or damper actuator is moved at a speed that is in proportion to the rate of change in the controlled variable. This is to match fast load changes. It is usually employed with another mode. Not used for air handling unit (AHU) controls as it is too fast. A car engine ECU needs it, so does anti-lock braking.
Proportional plus integral, P+I	The combination provides stable proportional control and the ability of the integral action to minimise offset. Provides an accurate control signal to valve or damper motor.
Proportional plus integral plus derivative, P+I+D	Ability to match fast load changes is added to P+I. Not used in HVAC.
Direct acting	Controller causes the valve or damper to close on increase of detected temperature, pressure or humidity. This is the normal action for heating system control. The room air temperature rises, the LPHW valve closes a bit.
Reverse acting	Controller causes the valve or damper to open on increase of detected temperature, pressure or humidity. This is the normal action for cooling system control. The room air temperature rises, the CHW valve opens a bit.

Symbols used for control diagrams are shown in Figures 7.3, 7.4 and 7.5.

Control system diagram

The air conditioning design engineer makes an initial analysis of the requirements and likely systems of control. These are discussed with other interested parties to ensure integration with the other services. A full description of the system logic is made. Schematic diagrams are used to demonstrate the location of all the components. Diagrams of the electrical wiring are needed. It is particularly important to specify all the connection points between the wiring and the elements of the control system.

Figure 7.6 shows the automatic control schematic diagram for the variation of outdoor air during the year. When the specific enthalpy of the outdoor air h_o exceeds that of the extracted room air h_r in summer, it is reduced in volume flow. This is to avoid overheating the room or to minimise the cooling plant load. Air temperature and humidity detectors are located in the return air duct from the room and in the fresh air inlet duct. Display devices allow manual reading of the data. The enthalpy controller compares the two sets of data. The output signal from the controller is 0–10 volt. A 5 volt signal corresponds to equal specific enthalpy of the two air streams h_o/h_r of 1. In winter, when h_o is much less than h_r and h_o/h_r is 0.25, the controller output signal is 10 volt. In summer, when h_o is much greater than h_r and h_o/h_r is 2, the controller output signal is 0 volt.

An output of 0 volt corresponds to the minimum fresh air inlet quantity during the summer. This is typically 12 litres per second per person occupying the building. An output signal of 10 volt corresponds to the minimum fresh air inlet quantity in winter. 5 volts occurs when the maximum quantity of outdoor air can be passed through the building, perhaps 100%. The heating and cooling plant are switched off when h_o/h_r is 1. All the winter outdoor air conditions lie within the range 0–5 volt. Between these limits, the outdoor air intake is gradually reduced. 5 volt may correspond to 100% fresh air and 0 volt to 20% fresh air. A controller output of 2.5 volt will always produce 60% fresh air, that is, half way between 20% and 100%. All the summer conditions are in the 5–10 volt range. A 10 volt signal will produce the minimum summer fresh air inlet quantity, say 20%. A controller output of 7.5 volt will always produce 60% fresh air, that is, half way between 100% and 20% on the summer programme. The controller will always know the difference between summer and winter by the voltage signal.

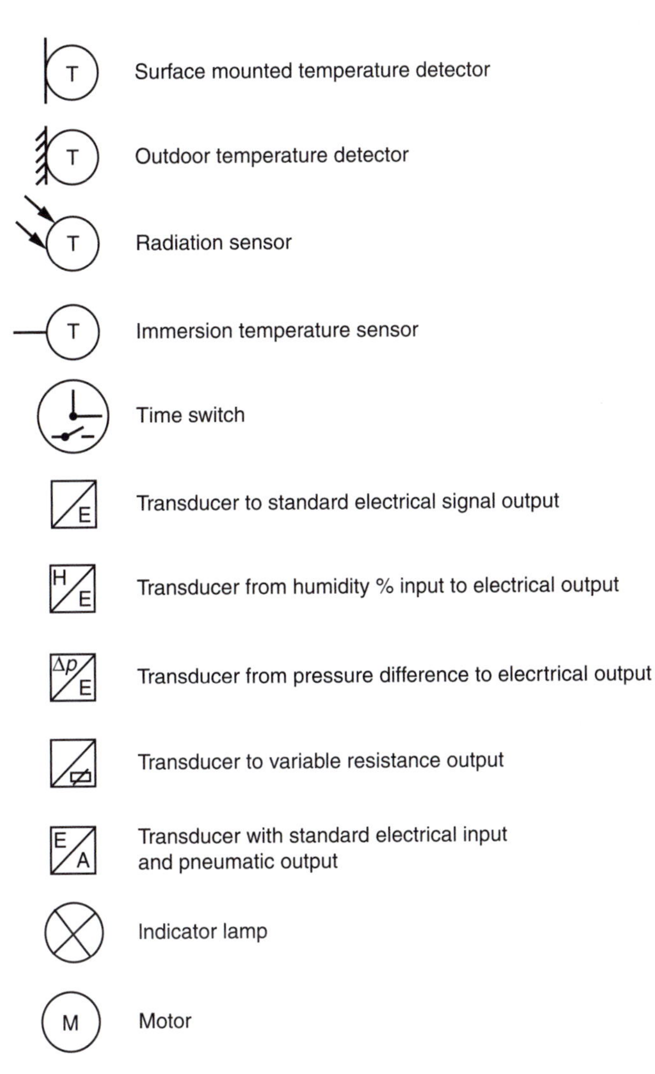

7.3 Control symbols.

Heating and ventilating control

A typical heating and ventilating control schematic is shown in Figure 7.7.

Cooling is not provided. The boiler plant has additional controls for frost protection, ignition cycle, flue induced draught fan and other connected services such as hot water storage. The ventilation plant has an optimum start and stop time switch controller that decides when to activate the plant. The user inputs the desired time of occupancy. An algorithm within the controller calculates when it is necessary to activate the heating to raise the occupied space to the desired value at the time of occupancy. The calculation includes the current space air temperature, outdoor temperature and the length of the off period. A frost temperature detector within the building is used to override the overnight and weekend off instruction. A minimum indoor air temperature of 10°C will prevent internal freezing and serious condensation. The optimum start software can be adapted by the user to allow variation of the system performance.

Controller plus mode symbol: P, proportional; I, integral; D, derivative; E, electrical output; A, pneumatic; ↡, incremental; ↑ direct acting, ↓ reverse acting.

Proportional plus integral controller, electrical output, direct acting.

Switch: T, thermostat; H, hygrostat; Δp; pressure

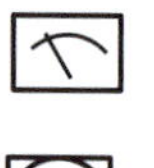

Analogue indicator

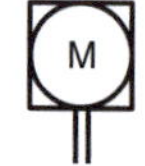

Electric motor driven actuator

Pneumatic membrane actuator

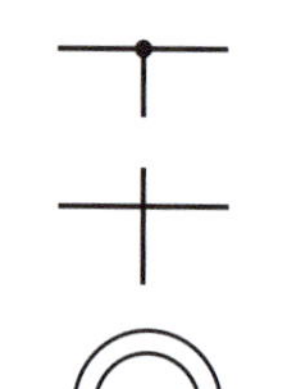

Electrical wire junction

Crossed lines with no connection

Heat user, general

Alternating current to direct current rectifier

7.4 Control symbols.

Motor driven air damper

Set point adjustment

Electrical wire junction

Humidity detector

Velocity detector

Pressure difference detector

7.5 Control symbols.

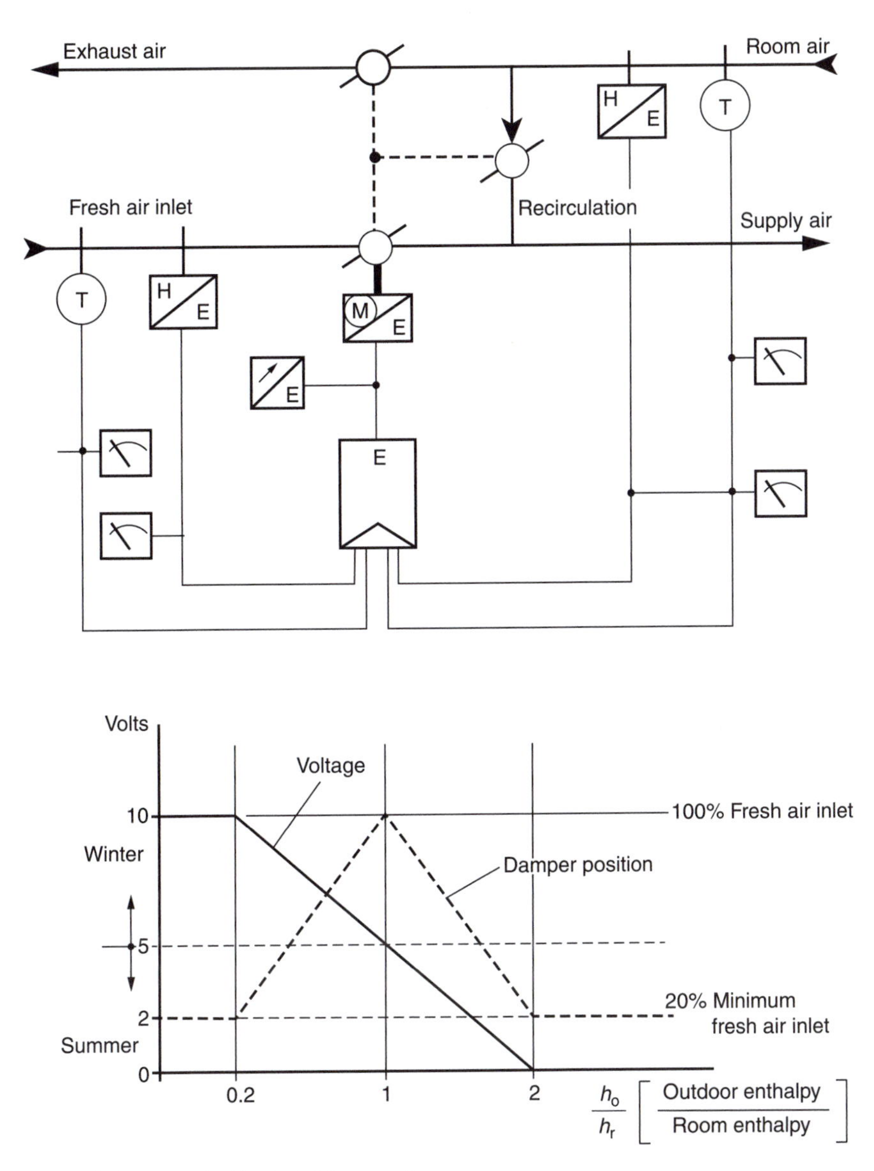

7.6 Enthalpy control of fresh air inlet quantity.

The starting sequence may be as follows.

1. Overnight frost protection.
2. Fresh air inlet and exhaust air dampers remain shut overnight.
3. Optimum start time.
4. Start signal sent to boiler plant.
5. Boiler plant start sequence is initiated:

   (i) combustion air and flue fans start;
   (ii) air flow detectors confirm fan operation;

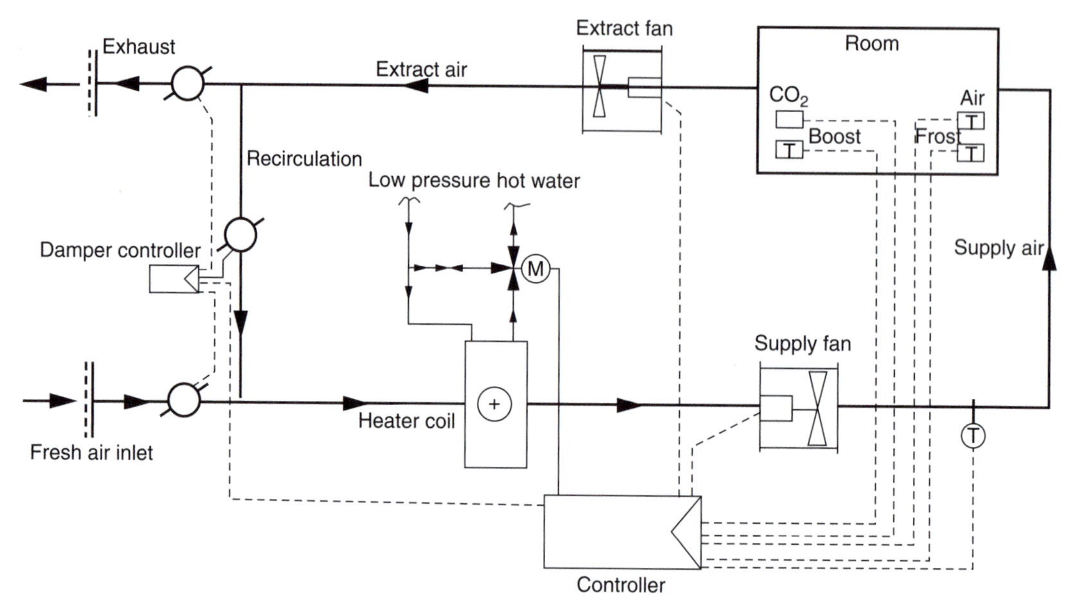

7.7 Heating and ventilating system control.

(iii) time delay of 1 minute to purge combustion passage;
(iv) water circulation pumps started;
(v) water flow sensors confirm circulation;
(vi) ignition transformer and spark activated;
(vii) time delay to confirm ignition device;
(viii) gas valve or oil pump and valve opened;
(ix) fuel is ignited; ignition remains active;
(x) photo-cell registers presence of flame;
(xi) ignition spark maintained for 1 minute or longer;
(xii) heat output controlled from flow water temperature detector;
(xiii) multiple boilers start and are controlled by a multi-step controller.

6. Two- or three-port flow control valve on the low pressure hot water flow or return at the air heater coil is held in its normally open position.
7. Heater coil reaches working temperature.
8. Water temperature detector in the return pipe from the coil can be used to signal correct performance.
9. Frost detector is switched out of circuit.
10. Boost temperature detector is activated.
11. Supply air fan is started.
12. Time delay of 30 seconds prior to starting the extract fan.
13. Flow switches in the supply and extract ducts confirm air flow.
14. Boost limit temperature detector signals the end of the boost period.
15. Room air temperature detector is activated and assumes its normal control function over the water flow control valve.
16. Time of commencement of occupancy is used to open the fresh air inlet and exhaust dampers to their minimum settings.

17. Three damper motors are interlocked to operate in synchronisation.
18. External air temperature sensor, or weather compensator, resets the room air temperature set point or the supply air duct temperature to minimise energy use.
19. Occupancy detector measuring room CO_2, may be used to increase the fresh air inlet quantity.
20. At the end of the occupancy period, fresh air and exhaust dampers are shut and the heating plant is switched off.
21. Recirculation damper and heater coil flow control valve are run to their open positions.
22. Heating water pump continues to run for 10 minutes to distribute remaining heat into the building before being switched off.
23. Supply and extract fans run for 15 minutes to distribute heat from the heater coil before being switched off.
24. Plant reverts to control from the frost detector.

Single duct variable air temperature control

A single duct air conditioning system is shown in Figure 7.8.

It may incorporate features previously described that are not repeated here. The main aspects of its operation are as follows.

1. Temperature of the air supplied to the conditioned space and that of the extracted air are both detected.
2. Supply air temperature has minimum and maximum limits.
3. A cascade controller resets the set point of the supply air temperature according to the return air duct temperature. Reset schedules and sequence graphs are shown in Figure 7.9.
4. 0–10 volt output from the cascade controller operates the heating valve, mixing dampers and cooling coil valve in sequence.

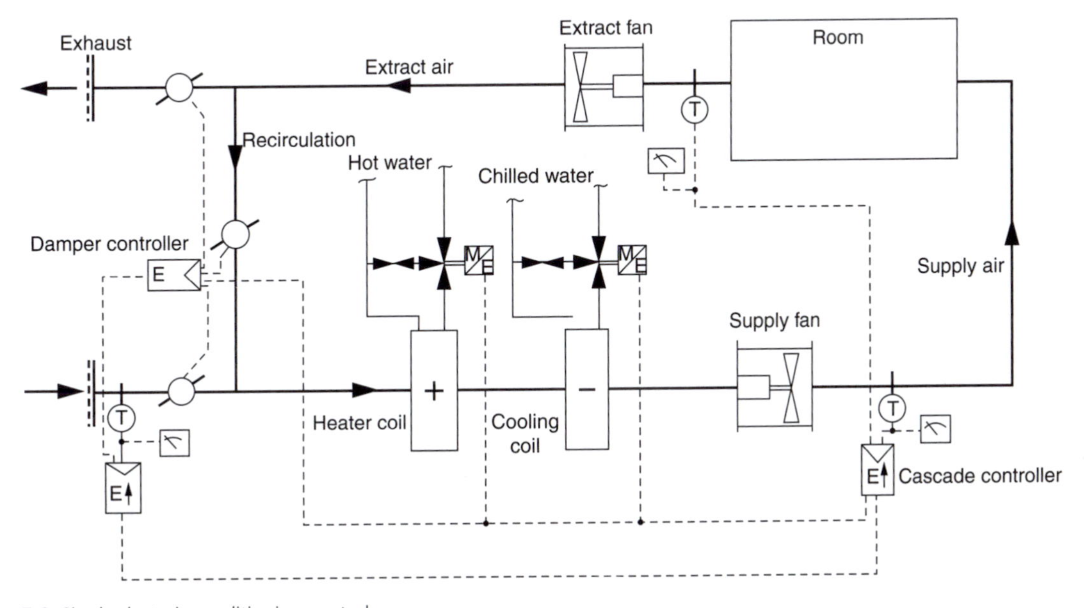

7.8 Single duct air conditioning control.

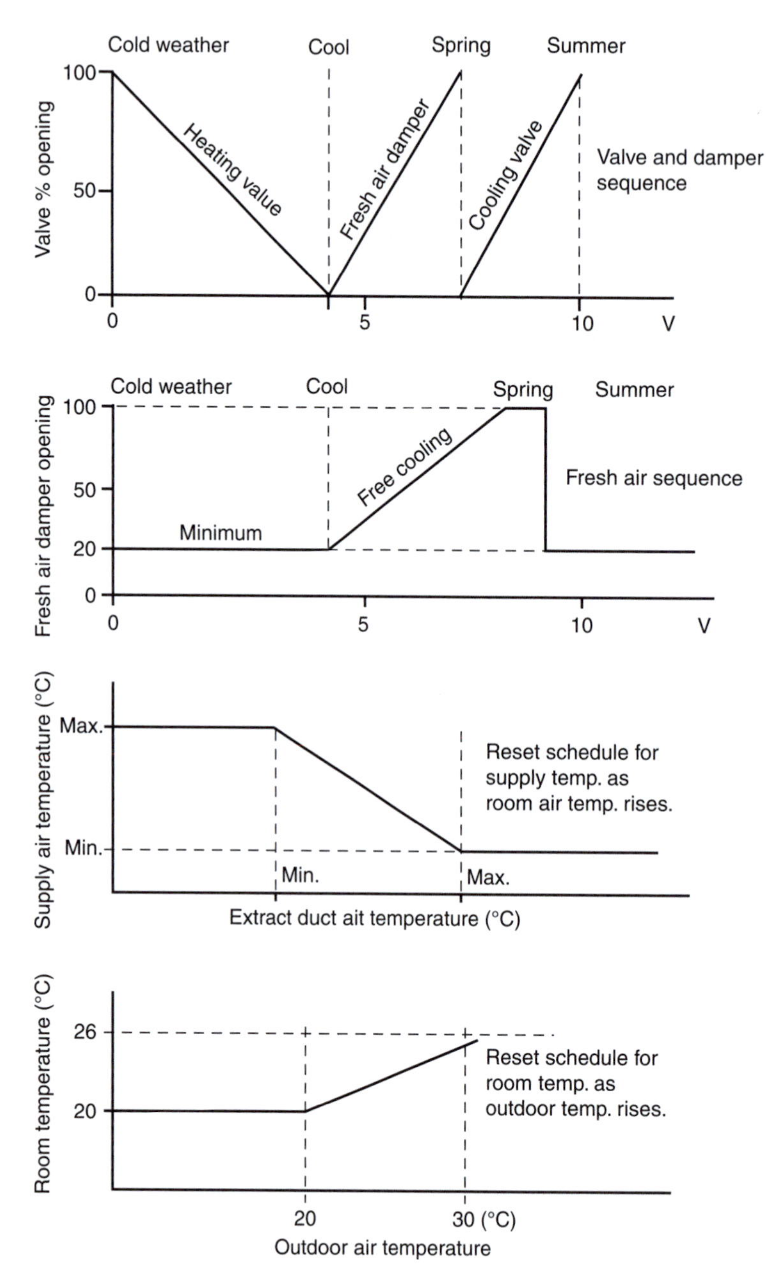

7.9 Control sequences for single duct air conditioning.

5. Three mixing dampers have a controller that provides free cooling from outdoor air and extract room air temperature detectors.
6. When the outdoor air temperature is higher than the air temperature extracted from the room, the damper controller closes the fresh and exhaust air dampers to their minimum opening. The recirculation damper is opened.
7. Minimum fresh and extract air damper setting is adjustable.

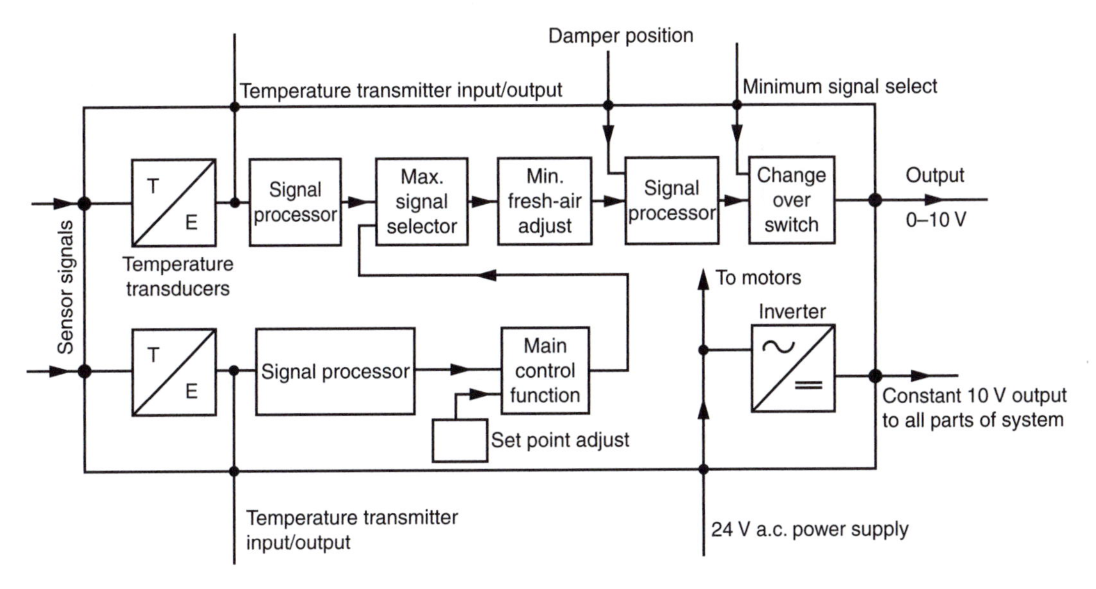

7.10 Internal components of a mixed air controller.

8. Outdoor air temperature is used by a controller to reset the room extract air temperature set point. This allows increases in the room temperature as outdoor temperature rises. This can be permissible for comfort cooling. It reduces refrigeration plant load and running costs.

A simplified block diagram of the internal components of the controller for the three dampers is shown in Figure 7.10.

The thermistor circuit electrical resistance signals from the outdoor air and room air temperature detectors enter the controller. Each signal is converted into 0–10 volt by a transducer. The voltage is processed and its set point and proportional band can be adjusted. The two signals are then compared and the higher voltage prevails. When the outdoor air temperature signal is higher than that from the room air temperature, the higher value is used to close the damper to its minimum opening. A damper position indicator signal is compared with the chosen voltage. If the damper is not at its minimum opening, the signal operates a changeover switch to drive the damper motors. An output voltage signal to the damper motors causes the 24 volt alternating current power circuit to close the dampers.

The building energy management computer system can be connected into any controller that uses the standard 0–10 volt range. The voltage is an analogue of the control system's status, that is, if 10 volt represents 100% relative humidity, 5 volt corresponds to 50%. The computer has an analogue to digital conversion card in its data bus. Digital bytes representing the voltage signal are processed by the software. The server computer communicates in both directions with the controller. Several controllers can be accessed, interrogated and reset from the remote server computer. A local intelligent microprocessor outstation may be used between a group of controllers and the supervising computer. Each controller and outstation has a digital address. The supervisory computer scans the control and data system every few seconds or minutes as required.

Single duct variable air volume control

This air conditioning system supplies occupied spaces with a single supply duct air temperature. It is a cooling system. Terminal reheating can be provided by a low pressure hot water heating coil or electric

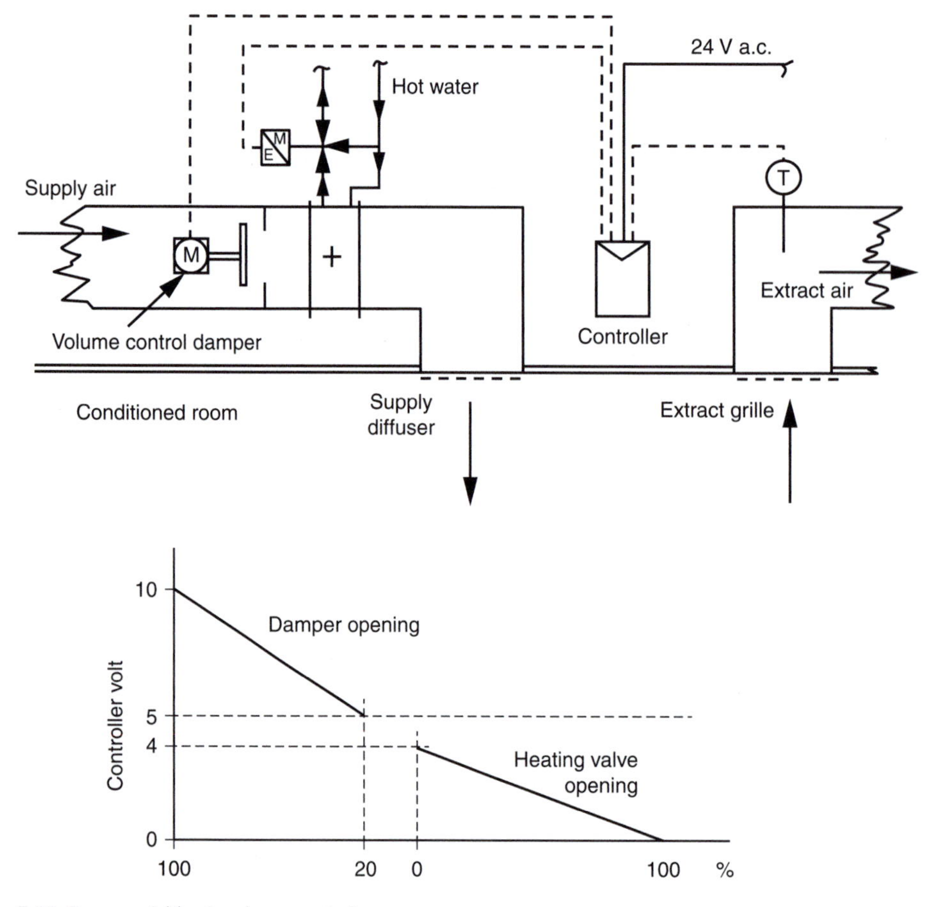

7.11 Room variable air volume control.

resistance element. Supply air duct temperature is scheduled against the outdoor air temperature. A weather compensator may be used to account for sunshine, wind and rain. Different orientations of the buildings may be grouped in different zones. Enthalpy control of the fresh air intake may be used to minimise refrigeration use. The part load performance needed for most of the year is met by reducing the supply air into each room. An arrangement for the room variable air volume controller is shown in Figure 7.11.

A room air temperature detector is located in the room or within the extract air duct or false ceiling. A 24 volt alternating current power supply provides the 0–10 volt stabilised and rectified control signal circuit. A 10 volt room air temperature signal corresponds to maximum cooling. The volume control damper is open and the heating coil valve is shut. The controller output from 10 volt down to 5 volt is caused by a reducing cooling load from the room. This is found from a fall in the room air temperature. The volume control damper is scheduled to close to its minimum setting at 5 volt. The minimum air flow may be 20% of the full load condition. It is necessary to maintain the Coanda effect of the supply air entering the space at reduced flows. The cooling requirement within the room has now ceased. A further fall of the room air temperature means that heating is called for. At 4 volt, the controller commences to open the two- or three-port heating valve. At 0 volt the heating valve is at full load and is open. The controller directs 24 volt alternating current power to the valve and damper motors. The reverse of the control process occurs as the space moves from a heating load to that requiring cooling.

The air handling plant supply and extract fans interface with a duct system requiring fluctuating air flow rates. It is wasteful of electrical energy to run fans against volume control dampers. Noise could arise from the excess pressure difference across room air terminal units. The fan delivery pressure is reduced by one of several possible methods:

1. variable frequency control, 0–50 hertz, of the fan motor;
2. variable pitch inlet guide vanes on the inlet to a centrifugal fan;
3. variable pitch blades of an axial flow fan;
4. variable resistance damper on the fan inlet or outlet.

It is preferable to reduce the speed of the fan to lower its discharge air pressure. This results in considerable savings in electrical energy consumption, lower duct pressure, reduced air leakage from the ducts, lower air velocity, less noise and less wear on the electric motor, drive belts and bearings. This is achieved with a variable frequency controller that takes 0–10 volt signals. The cost of this approach is up to a 10% increase in electrical power consumption to run the frequency inverter. This 10% electrical loss produces an equal heat gain into the plant room.

Use of variable pitch inlet guide vanes and axial fan blades reduces the efficiency of the fan to lower its performance. The fan operates at a constant speed and motor power consumption is reduced. Hydraulic actuators are used to move the blades and vanes. Inserting a damper to reduce the air volume flow rate through the duct system while running the fan at full load is not good engineering practice. It is analogous to driving a car on full throttle and using only the brake to moderate the speed.

Figure 7.12 shows how the air handling plant supply and extract fans are controlled in a variable air volume system. A variable frequency control (VFC) is designed to maintain a constant duct air pressure. Partial closure of the room terminal unit VAV damper increases the resistance of the ductwork system. The fan output is constrained to the shape of the curve shown. An increase in the duct system resistance causes the fan to supply less air at the elevated pressure. The duct air pressure transmitter detects the rise by comparing the duct air with the atmosphere. The air pressure set point corresponds to 5 volt. An increase in duct pressure is scheduled to the 5–10 volt signal band. A signal above 5 volt causes the fan speed controller to reduce speed until 5 volt is established. A reduction in duct air pressure occurs when the variable volume terminal dampers open in response to rising room air temperature. Duct air pressure that falls below the set point is scheduled to the 0–5 volt range. A signal below 5 volt causes the fan speed controller to increase speed until 5 volt is established. A duplicate pressure transmitter, pressure controller and variable frequency controller are fitted to the extract fan and duct. The supply and extract fans will be different in size and power consumption. The controlled pressure in the extract system is not identical to that in the supply system. An alternative to duct pressure control for the extract fan is to detect the volume flow rate in the supply and extract ducts. Reduction in the supply air flow triggers a corresponding lowering of the extract flow with a fan performance controller.

A variable volume air conditioning plant room control system is shown in Figure 7.13. Enthalpy control of the fresh air and exhaust dampers provides low cost cooling. The supply air fan speed is controlled from duct air pressure. The extract fan is controlled from air velocity detectors in the supply and extract air ducts to balance the flows. The plant only needs a chilled water cooling coil. Terminal hot water coils satisfy the heating demand. A heating coil in the plant room may be needed for frost protection and boost heating on cold starts during the winter. The building is protected from cold weather by closure of the external air dampers and recirculating the room air. The fresh and exhaust air dampers remain at their 20%, 2 volt, minimum opening during the cooler weather until a room cooling need is identified. Increasing amounts of fresh air are admitted until the fresh air enthalpy equals the extracted room air enthalpy. The dampers are moved to their minimum fresh air intake positions. The controller signal is reset to 0 volt. The controller commences opening the chilled water valve on the cooling coil. The controller signal varies from 0 to 10 volt

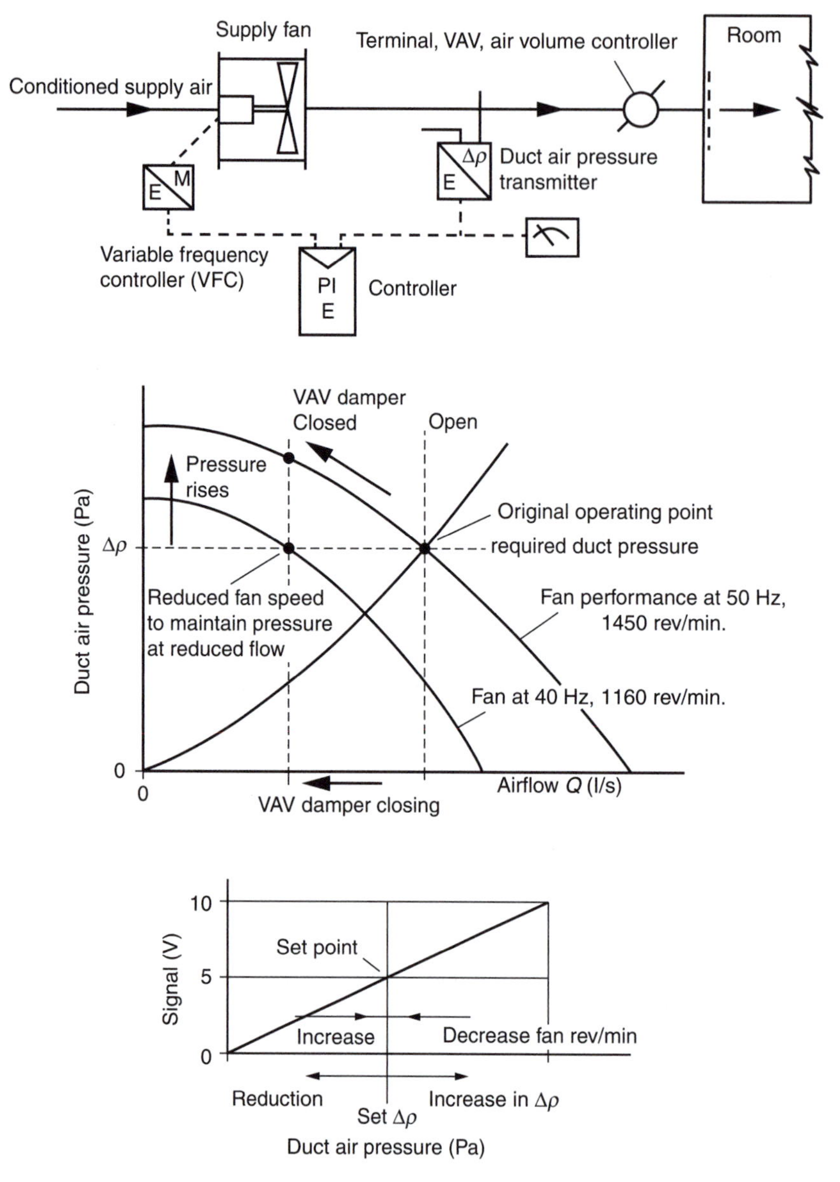

7.12 Fan output control for a variable air volume system.

over the 0–100% chilled water flow through the cooling coil. The supply air temperature is scheduled to the external air temperature to avoid excessive cooling and to minimise refrigeration plant load.

Chilled water plant

Chilled water is provided from a variable output refrigeration plant. A chilled water flow temperature of 6°C and return of 10°C is often suitable for air conditioning. The plant serves several cooling coils. The coils are in the zone air handling units and room terminal fan coil or induction units. Controlling the cooling output from the refrigeration machine and matching it to the current load, is a similar problem to that with

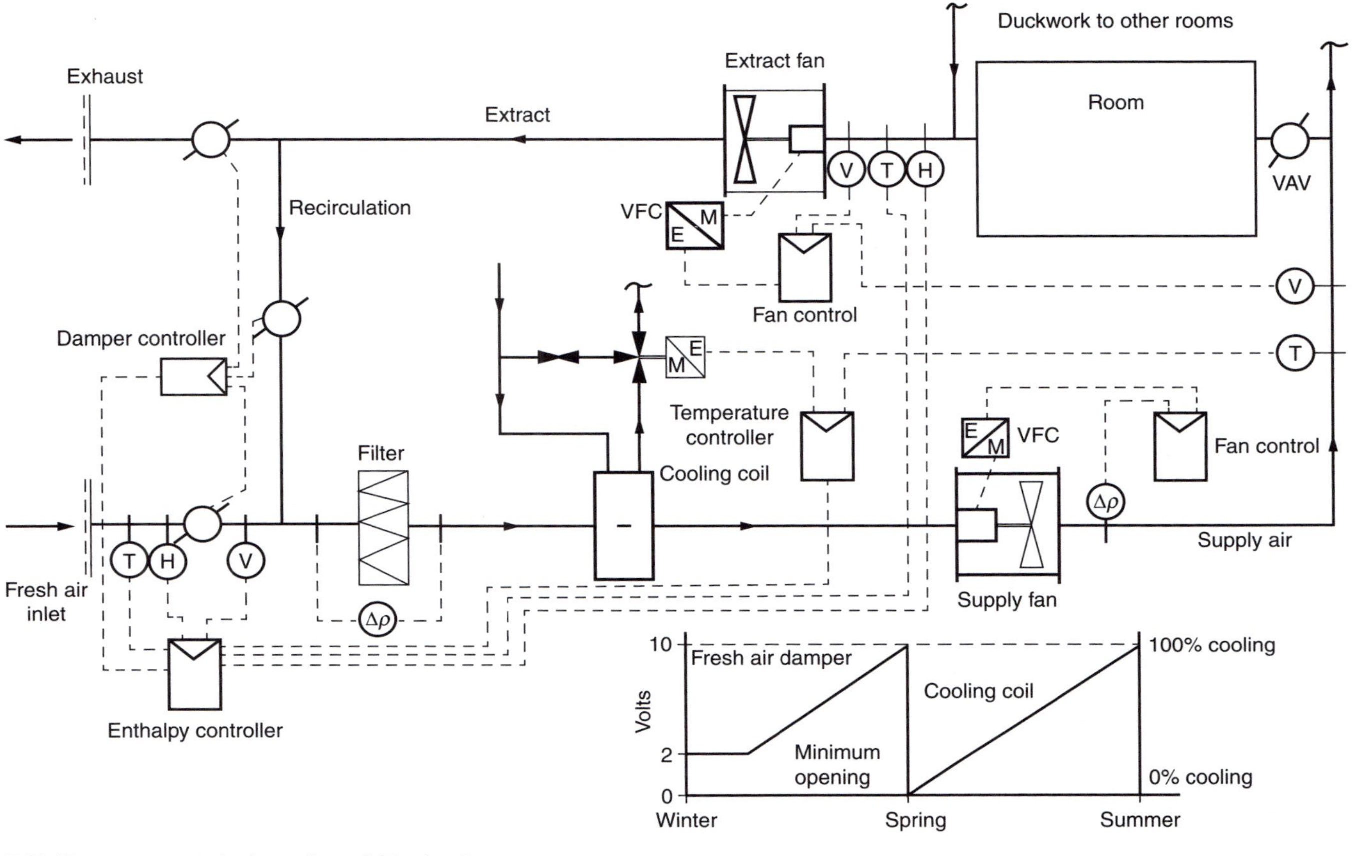

7.13 Plant room control scheme for variable air volume.

a boiler and a heating system. Heating or cooling load varies with the outdoor air temperature while the heat supply is generated in steps.

The cooling output of rotary compressors and absorption refrigeration machines can be smoothly variable. Reciprocating compressor refrigeration machines are single- or multi-cylinder. Each cylinder has a reciprocating piston and spring steel valves. To reduce the compressor performance, the suction valve is held open with hydraulic pressure. The valve actuator is operated by the oil pressure from the lubrication pump. All the pistons reciprocate whether the valves are open or not. Cylinders which are unloaded in this way do not pump refrigerant. This reduces the mass flow rate of refrigerant vapour through the evaporator. The evaporator lowers the temperature of the water circulating through the chilled water circuit. The cooling performance of the plant is reduced in steps corresponding to the number of cylinders that are unloaded.

To achieve close control over the chilled water temperature, two multi-cylinder compressors are used. One compressor leads the other. The lag compressor becomes activated when the lead machine is fully operational. The lag machine is the first to be switched off. The result of the capacity control steps is a smoothed control of the chilled water temperature.

Figure 7.14 shows the arrangement of a chilled water refrigeration plant. Figure 7.15 shows the resulting mixed flow water temperature leaving the compressors.

The design flow and return water temperatures are 6°C and 10°C at full cooling load. The compressor control maintains a leaving water temperature of 6°C plus or minus a tolerance of 0.5°C. A cooling load dead band of 1.5°C around the set point is the fluctuation in the flow water that corresponds to one

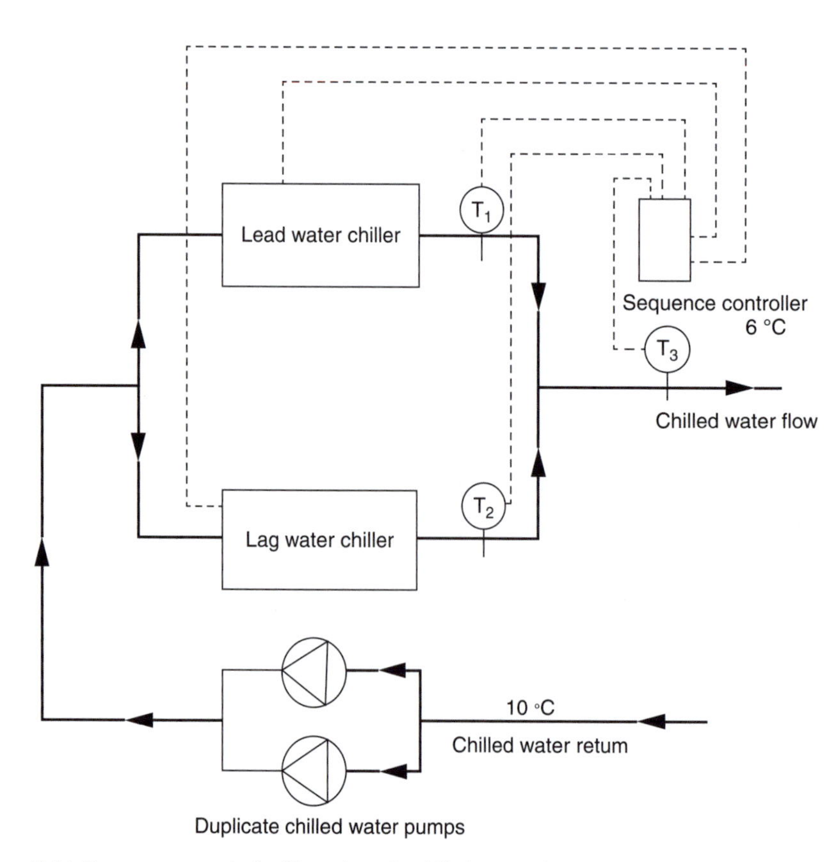

7.14 Sequence control of refrigeration of a chilled water plant.

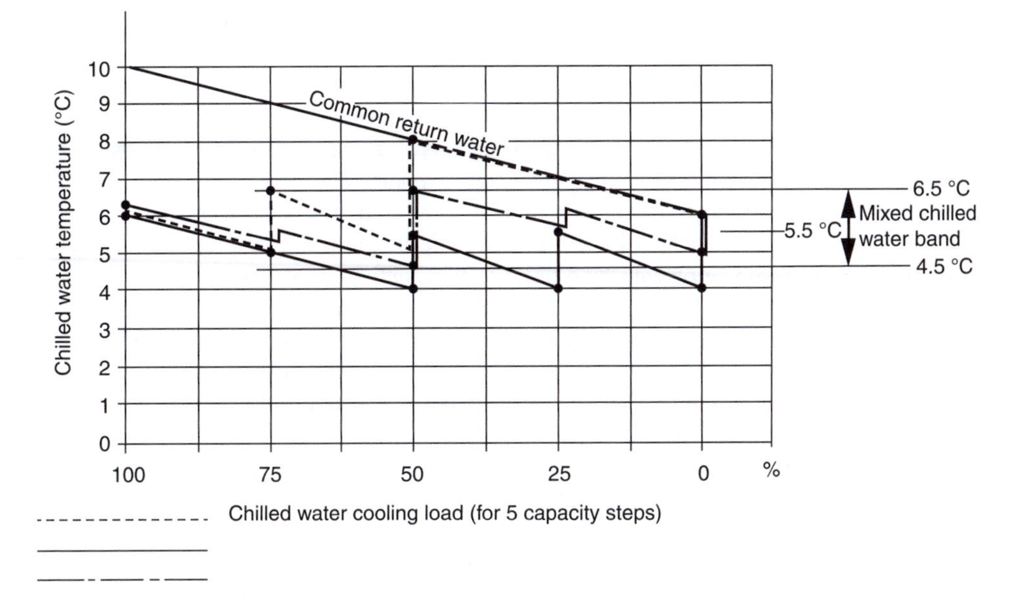

7.15 Temperature schedule for a chilled water plant.

capacity step. 6°C is produced for all cooling load conditions within the building. A reducing demand for cooling within the building causes the return water temperature to fall to 6°C when the cooling load is removed.

While one refrigeration machine is unloaded, return water is still circulated through its evaporator. The combined flow water leaving the plant is the mixture of two streams, one chilled, the other not. The two compressor machines are of equal size and chilled water flow rate. To achieve a leaving mixed flow temperature of 6°C from a chilling machine and an unloaded machine, the chiller set point is 4°C. Refrigerant evaporates at 3°C. The chilled water leaving the evaporator will approach 3°C on occasions. Freezing of the water must be avoided. Lower temperature systems for industrial process cooling utilise brine at sub-zero temperatures. The mixing of water as low as 3°C with uncooled water at 10°C produces the desired value of 6°C ±0.5°C. When it does not, another capacity step is loaded or unloaded as appropriate.

Building energy management systems

Air conditioning control is a major part of the building energy management system (BEMS) for almost any size of building above residences and tower blocks of apartments; these are individually controlled by the owner or tenant and often with packaged direct expansion reverse cycle air conditioning units. BEMSs comprise:

1. The programmable logic controller (PLC) is a dedicated microprocessor programmed to operate a particular plant item such as a boiler, refrigeration compressor or passenger lift.
2. The energy management system (EMS) is a dedicated microprocessor that is linked to all the energy and power using systems such as heating, air conditioning, electrical power, lighting, lifts, diesel generators and air compressors. It may appear as a metal box on a wall of the plant room, having numbered buttons and a single line of screen display for maintenance staff to use for carrying out a limited range

of routine changes. Such a unit may serve as an outstation that is either intelligent, having its own microprocessor, or dumb, merely passing data elsewhere.

3. A building energy management system (BEMS) is a supervisory computer that is networked to microprocessor outstations, which control particular plant such as heating and refrigeration equipment. All the energy-using systems within one building are accessed from the supervisor computer, which has hard disk data storage, a display unit, a keyboard, a printer and mimic diagrams of all the services. Additional buildings on the same site can be wired into the same BEMS by a low-cost cable. Remote buildings or sites are linked to the supervisor through the internet, telephone modem, fibre optic cable, microwave line of sight long distance transmitters or satellite communications. A modem is a modulator–demodulator box, which converts the digital signals used by the computer into telecommunication signals suitable for transmission by the network to anywhere in the world.
4. The plant management system (PMS) is used to control a large plant room such as an electrical power generator or district heating station. A PMS can be anything from a small dedicated PLC on a water chiller to an extensive supervisory computer system.
5. The building management system (BMS) is used for all the functions carried out by the building including the energy services, security monitoring, fire and smoke detection, alarms, maintenance scheduling, status reporting and communications. Types of BMS range from systems serving one small office, shop or factory to systems serving government departments and international shopping chains. Systems may carry out financial audits, stocktaking and ordering of supplies each night utilising telecommunications. Suppliers of, say, refrigeration equipment, maintain links with all their installations in clients' buildings and are informed of faults as they occur, and often prior to clients being aware of the problem.

An outstation is a microprocessor located close to the plant that it is controlling and is a channel of communication with the supervisory computer server. An intelligent outstation has a memory and processing capability that enables it to make decisions regarding control and to store status information. A dumb outstation is a convenient point for collecting local data such as the room air temperature and whether a boiler is running or not, which are then packaged into signals for transmission to the supervisor computer in digital code. Each outstation has its own numbered address so that the supervisor can read the data from that source only at a discrete time. The supervisor is the main computer, which oversees all the outstations, PLCs and modems, contacting them through a dedicated wiring system using up to 10 V and handling only digital code. Such communication can be made every few seconds, and accessing the data can take seconds or minutes depending on the quantity of data and the complexity of the whole system, i.e. the number of measurements and control signals transmitted. The engineer in charge of the BEMS receives displayed and printed reports from the supervisor and has mimic diagrams of the plant, which enable identification of each pump, fan, valve and sensor together with the set points of the controllers that should be maintained. Alarm or warning status is indicated by means of flashing symbols and buzzers, indicating that corrective action by the engineer is required. Plant status, such as the percentage opening of a control valve and whether a fan is running, is recorded, but only the engineer can ascertain whether such information is correct, as some other component may have been manually switched off in the plant room by maintenance staff, or may have failed through fan belt breakage. Therefore a telephone line between the supervising engineer and the plant room staff is desirable to aid quick checking of facts. Connections are made to an outstation by means of a pair of low-voltage cables.

Data which are sent to an outstation include measurement sensor data such as temperature, humidity, pressure, flow rate and boiler flue gas oxygen content, control signals to or from valve or damper actuators, plant operating status, which can be determined from the position of electrical switches (open or closed), for example to check whether a pump is operational.

EXAMPLE 7.1

Describe how the BEMS of a large commercial building interacts with its variable air volume air conditioning system, gas-fired water heating, water chillers, cooling towers, fans and pumps. There are two gas-fired heating boilers, one centrifugal refrigeration water chiller sized for 60% of the AHU's full load, a multi-stage reciprocating water chiller that can meet 40% of the imposed load, and two cooling towers. There are 10 VAV boxes on each floor. All pumps have duplicates on standby. Refer to the figures provided in this chapter.

Large spaces such as foyers, circulation corridors, meeting and conference rooms have multiple room air temperature and humidity sensors, Figure 7.1, transmitting averaged 0–10 V to a P+I outstation controller on each floor. Each controller has its own control software on an EPROM memory chip, initially downloaded from the control engineer's laptop computer during commissioning. Each controller communicates with the supervisory computer to transmit zone information and receive updated control instructions and functions, such as scheduled room usage changes and revised settings. Alarm signals are sent from the controller when any variable moves beyond its allowed operating range, such as zone air temperature moving outside the range of 22°C to 24.8°C.

Each floor has an AHU, Figure 7.13, serving 10 VAV boxes, each box serves a zone. Each VAV box and reheat coil has its own dedicated PLC controller, Figure 7.11, maintaining room conditions from a return air duct temperature sensor.

Humidity is controlled in each floor AHU with face and bypass dampers on the chilled water cooling coil from the floor outstation.

Outdoor air economy is provided with enthalpy sensing and comparison for AHU damper modulation, Figure 7.6, to minimise outside air heating and cooling.

Each floor AHU supply and exhaust air fan has a VFC controlled from duct air pressure sensors, Figure 7.12, to minimise fan energy as the VAV terminal unit dampers reduce supply airflow.

The basement plant room has a dedicated outstation controller that switches the boilers and chillers on and off, Figures 7.14 and 7.15. This controller has a sequence control algorithm that compares the actual demand for heating and cooling from external air and solar radiation sensors with the building time of use schedules, to switch on the appropriate plant, such as the small chiller first, then switch it off when the 60% chiller is called on; the small chiller comes back on when the centrifugal machine cannot maintain 6°C CHW temperature on its own, and also the cooling towers. Boilers and chillers have their own manufacturer's onboard PLC for their specific control functions. The BMS only gives start and stop instructions. CHW and cooling tower condenser water pumps are switched as scheduled with monthly pump changeovers. Condenser water temperature returning to the chillers is controlled to 29°C with a three-way bypass valve alongside the towers to modulate heat rejection. Cooling tower fans are scheduled to outside air dry bulb, condenser water temperature and time schedule. This controller receives electric, gas, water and heat meter data from transmitters on those meters.

The building's maintenance engineer physically supervises the front end server computer on the desk, noticing red flag alarm conditions that arise, checking what the issue is and taking corrective action, probably before the floor occupants realise there was a problem. All alarm signals are automatically relayed to the engineer's mobile phone or pager and recorded in a continuous data log on the server. The server is remotely accessible by the maintenance engineer and other relevant supervisors through pagers and mobile phones at all times. Reset instructions are sent by mobile phone to the server computer through a dedicated phone number and access code system, so that errors can be resolved remotely. An event log is maintained and continuously printed for record keeping, showing plant switching, all alarms, alarm resets, remote log-ins and plant status. Energy audit report data are collected in meter data files for regular access.

Electrical wiring diagram

A diagram of the electrical wiring shows all the components of the power equipment and control system, how they are connected and the logic of their operation. A simple wiring diagram for part of a ventilation system control is shown in Figure 7.16.

A single-phase 240 volt alternating current circuit is indicated. The protective conductor, earth, has not been shown. The making of switches completes the circuit between the line and neutral conductors and

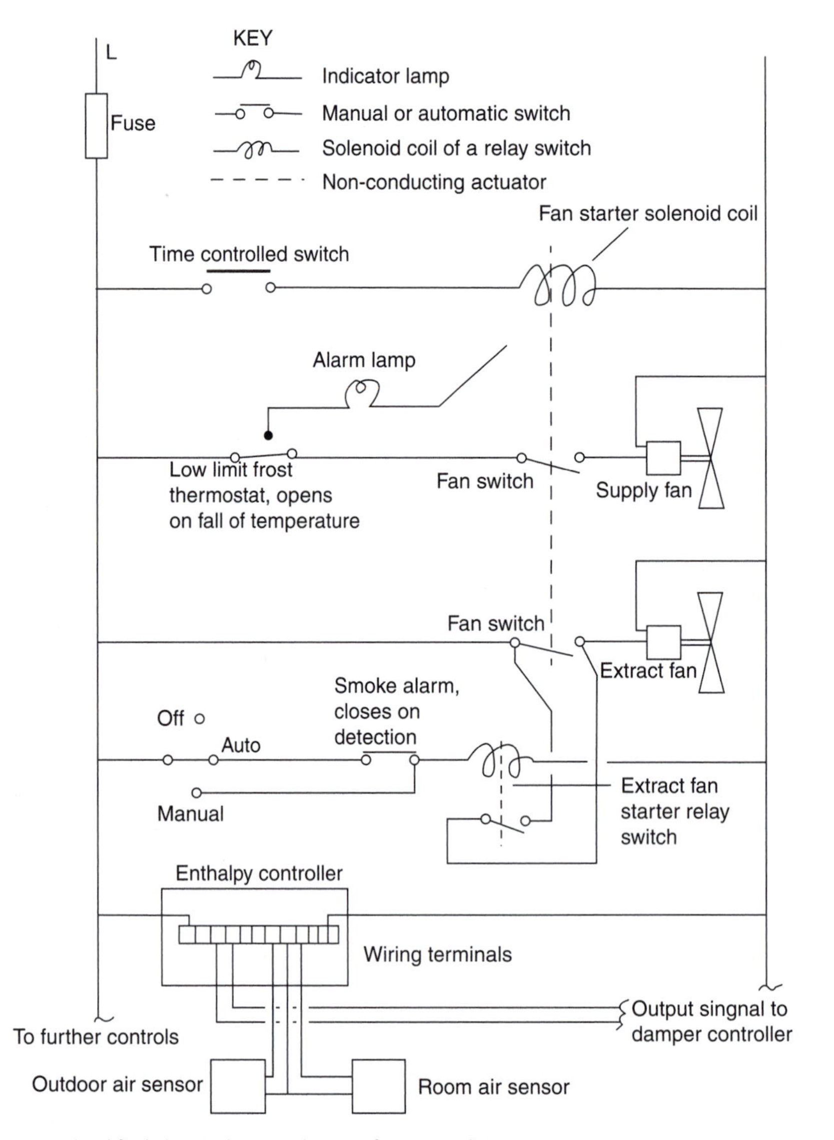

7.16 Simplified electrical wiring diagram for air conditioning equipment.

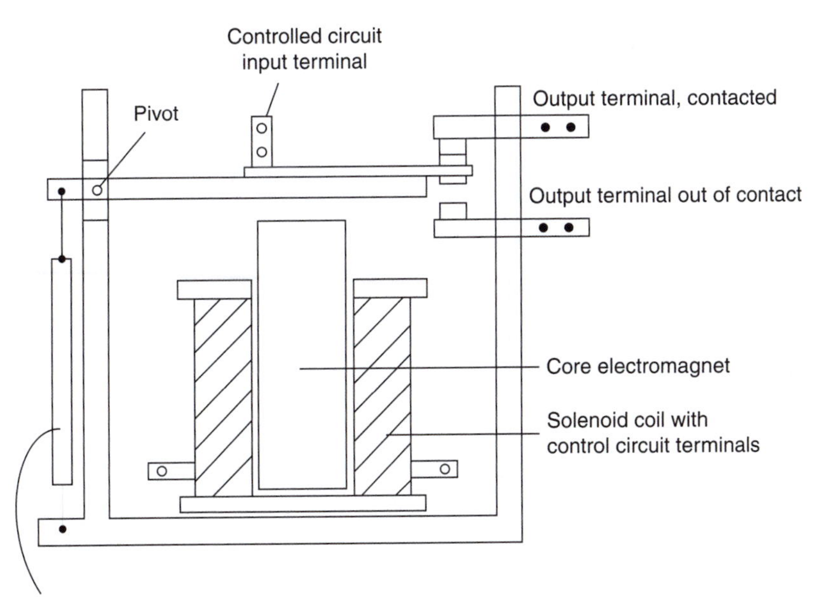

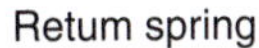

7.17 Solenoid relay.

current flows. When a solenoid coil is energised in a circuit a second switch contact is made. The metal actuator is attracted toward the solenoid coil by magnetic attraction. The movement of the actuator pulls the switch into the closed position. This energises the controlled circuit that may be at a higher voltage and current. A 10 volt control signal can switch on a 415 volt three-phase circuit to a fan motor. The actuator does not conduct electricity between the low and higher voltage circuits. This arrangement of solenoid, actuator and switch is called a relay or contactor.

A relay is shown in Figure 7.17; energizing the solenoid coil with an external electrical current causes the metal core to become magnetised. The magnetic field attracts the metal actuator towards the core. Movement of the actuator closes the switch contacts in the controlled circuit. Removal of the control circuit energising current switches off the magnetic field. The return spring pulls the actuator back to its off position.

Figure 7.16 shows a 240 volt line conductor serving a time controlled switch. The line conductor has a fuse or overload micro circuit breaker. The time switch closes at the designated time and energises the fan starter solenoid coil. The fan starter relay closes the power switch to the supply air fan. The extract fan relay switch is closed simultaneously. If the low limit frost thermostat switch is closed, electrical power flows to the fan starters or direct on line motors. Each fan may have a starter control such as star-delta, time delay, soft start or variable frequency speed control. If the low limit thermostat detects frost, the thermostat switch contact opens and the alarm circuit is illuminated. The fans cannot be started. Additional lamps may be in the fan wiring to indicate their operation. When the fans are switched off, the smoke alarm closes the automatic detection circuit. This energises the relay that provides power to the extract fan. This bypasses the time switch and low limit thermostat. The operation of the smoke alarm circuit can be manually tested or switched off. The enthalpy controller for the fresh air damper is shown. A row of numbered terminals is provided for installation work. Temperature and humidity detectors have a two-wire cable for a common input voltage and an output signal. The enthalpy controller processes the signals and sends its output to the damper controller.

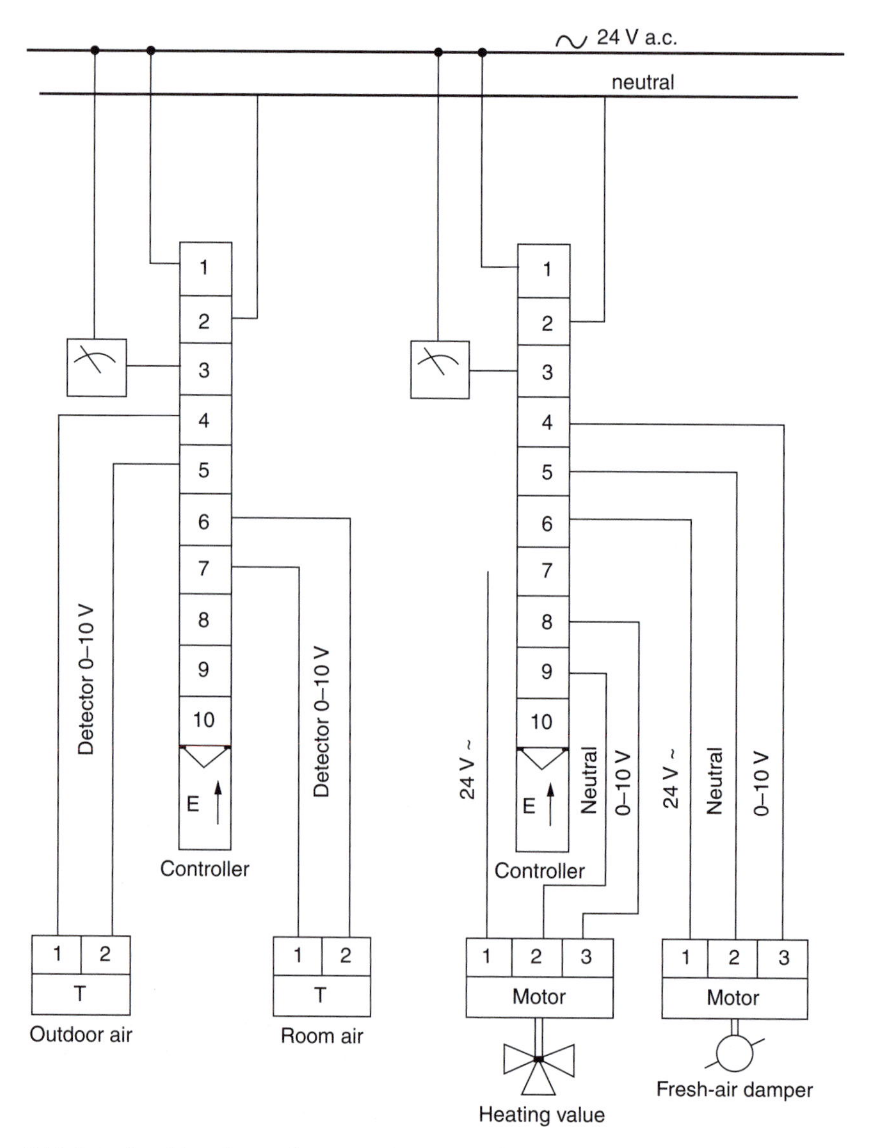

7.18 Part of a wiring diagram for a control system.

The wiring diagram of a control system is shown in Figure 7.18. Each controller has a numbered terminal block. A common 24 volt alternating current and neutral cable is connected to each controller and power operated actuator. The controller has an electronic rectifier to produce a stabilised 10 volt direct current. The 10 volt is used as the control signal for the detectors and signal processors. The analogue outputs from the controller to actuators are 0–10 volt. An analogue to digital circuit board may be within the controller. This enables direct digital control (DDC) from the supervisor computer. Status and qualitative digital output data from the controller are transferred through the BEMS and computer network cable. These data can be transferred through the 240 volt alternating current cables, mains borne signals, to the supervisor. Each package of data is transmitted by its controller at a discrete frequency in the MHz range. Each controller has its own digital address. Signal interference with the 50 Hz system or other possible sources of electronic noise, is unlikely. The MHz harmonic frequencies produced by variable frequency controllers on electric motors need to be checked for possible interference.

A switch is used for control. For direct current and single phase it is single pole and only opens the line conductor. An isolating switch on single phase is double pole. Both the line and neutral conductors are switched open. Three-phase wiring has three line conductors and a neutral. Each phase is a different colour. Yellow, blue and red are used. The phase currents are equal. Each phase is switched open or connected simultaneously. A three-phase wiring diagram shows the three line and neutral conductors. Isolating switches are triple pole and neutral (TP&N). Electric motors of 1 kW and above power consumption are normally on three phase.

Questions

Discussion questions require the use of the text and may benefit from additional information. It is expected that references, manufacturers' literature, colleagues and tutors will be utilised to increase the reader's breadth of knowledge. It may be necessary to assume source data to answer some questions.

1. List the components of an automatic control system for:
   (a) domestic gas-fired central heating;
   (b) a ducted air heating and ventilation system in a single-storey factory used for assembly of electronic components;
   (c) a single duct variable air temperature air conditioning system for a lecture theatre for 120 people;
   (d) a single duct fan coil air conditioning system serving a 12-storey office building. Chilled water refrigeration compressors and oil-fired boilers are used.
2. Explain what is meant by analogue and digital values in control.
3. Explain how an analogue signal can be created and used to represent room air temperature.
4. Draw a graph of the output signal from a thermistor temperature sensor that has an operating range of 0°C to 40°C and a constant current of 5 milliampere. A linear voltage output of 0–10 volt is produced by the thermistor. 10 volt corresponds to 40°C air temperature. Calculate the voltage that corresponds to an air temperature of 25°C and the resistance of the thermistor at this value.
5. Explain the sensing and operating principles of:
   (a) a thermistor temperature detector;
   (b) a humidity detector;
   (c) an enthalpy detector;
   (d) an air pressure transmitter;
   (e) a detector to measure the air volume flow rate in a duct and use its value for automatic monitoring and control;
   (f) a weather compensator;
   (g) a water temperature detector.
6. An open plan occupied space is served by a single duct variable temperature air conditioning system. Explain how the average air temperature and humidity of the space can be sensed and used for the control of the hot and chilled water diverting control valves. Sketch the arrangement of the detectors, controllers and actuators on a plant schematic.
7. A thermistor produces a 5 volt direct current output when passing a control current of 10 milliampere. Four thermistors are used to find the average of four air temperatures in a lecture theatre. Draw a suitable wiring circuit that would produce an average voltage for use by the controller. Calculate the

resistance of the circuit. Validate your connection design by calculating the circuit resistance from first principles.

8. Explain the use of enthalpy control of the fresh air intake to a ducted ventilation or air conditioning system. State why it is used and what limitations may arise.
9. State the types of actuators used to control heating, ventilating and air conditioning equipment. Sketch and describe their operating principles.
10. Discuss the use of pneumatic actuators. Include in your discussion their operating principles, the plant necessary, their interaction with electrical, electronic and digital signals, their advantages and limitations.
11. Explain, with the aid of sketches, what an electric solenoid, relay and contactor is. State what the device is used for.
12. Explain how a personal computer is used to monitor and control an air conditioning system. State how 24 volt, 240 volt and 415 volt alternating current electric powered actuators and plant are interfaced with the binary code used by the computer and network cables. Give examples of the software used by the supervising computer and the engineering management personnel in dealing with the data accessed.
13. Discuss, with examples, how lags occur in the detection and control of air conditioning within an occupied building.
14. Explain the following terms: proportional, integral, derivative, proportional plus integral, proportional plus integral plus derivative, offset, differential, boost, controlled variable, controller, dead time, set point.
15. State the ten controller operation methods used. Briefly describe the principle of each method.
16. A public entertainment theatre and conference centre seats 500 people. The basic layout of the air conditioning system is shown in Figure 7.7. Add a chilled water cooling coil and three-port diverting valve on the chilled water circulation. The room condition is to be maintained at 20°C d.b. ±2°C and 50% percentage saturation ±10%. The outdoor air temperature varies from −10°C d.b. to 32°C d.b. during the year. Design an automatic control system that will maintain thermal comfort throughout the year. Specify the types of detection and control equipment necessary. Draw a schematic diagram of the air handling plant and control system to describe its components, locations and modes of control. Draw operating graphs for the controls to demonstrate the voltage signals that correspond to plant status. Describe the logical operation of the control sequence. Chilled water is available at a flow temperature of 6°C and hot water is at 82°C. Ignore the boiler and refrigeration plant operations. Frost protection of the building and air handling plant is needed due to the low external air temperature. Make any assumptions that may be considered necessary. The design may be discussed with colleagues, tutors and suppliers of control systems.
17. Explain the logical operation of the control of a variable air volume air conditioning system. Draw the plant and control equipment schematic diagram.
18. Discuss the four methods used to control the output performance of centrifugal and axial flow fans. Illustrate the methods to show their application. Sketch the effect of each method on a fan performance graph. Each graph is to identify the duct system resistance, the control effect and the combined performance point. State the energy savings, advantages and limitations of each method.
19. Explain the methods used to control the chilled water output performance of refrigeration plant. Include the various types of refrigeration system and compressors.
20. Two three-cylinder refrigeration reciprocating compressor and water chilling evaporator sets are to produce a flow temperature of 5°C when the return water is at 11°C. The minimum chilled water temperature is 3°C. The tolerance of the chilled water temperature control is ±0.5°C. Each cylinder corresponds to a cooling load dead band of 1°C around the set point. Design a control schematic for

the refrigeration plant. Draw a capacity control graph to show the control steps and the mean chilled water flow temperature that will be produced.

21. Draw an electric wiring diagram, similar to Figure 7.16, for the heating and ventilating system in a shop. The supply and extract fan motors are single phase. The fresh air inlet fan has an electric resistance heater that raises the supply air to 20°C. A low temperature limit thermostat in the supply duct switches the supply fan off after a 2 minute time delay. The extract fan is started from a 30 second time delay unit after the supply fan has started. A time controlled switch activates the fans and two fan powered single phase electric resistance heaters in the shop. An air thermostat for each heater switches them on and off. Each fan and heater has an 'on' status indicator lamp. A frost thermostat is set at 8°C in the shop and it overrides the time switch. Three smoke detectors switch on a smoke extract fan. The smoke extract system is always operational and it has a ready status indicator lamp, a smoke alarm indicator lamp and an audible smoke alarm.
22. Draw the wiring diagram for a three-phase 415 volt power supply to two air conditioning fans, a refrigeration compressor and a steam humidifier. Each item has a triple pole and neutral isolating switch. The humidifier is controlled from a 10 volt control signal from a humidity detector within the air duct. The supply duct air temperature controls a hot water valve. There is an isolating switch for all the air conditioning circuits. An overload circuit breaker protects the whole system. The plant room has a three-phase distribution board. Single phase is used by the temperature and humidity controller.
23. Water chillers are a major source of cooling for air conditioning. List the types of packaged water chillers available; provide sketches to show their principle of operation and typical application.
24. Multiple chillers are used in large buildings to match capacity with demand. Sketch and describe the following means of providing a satisfactory cooling service while minimising energy use: parallel and series connected chilled water evaporators; parallel and series connected condenser water cooling heat exchangers; chilled water primary circuits; chilled water secondary circuits; chilled water common headers; constant pump speed primary chilled water circuits; variable speed secondary chilled water circuits; the use of chilled water pressure difference sensors and control valves.
25. Heat rejection is where indoor cooling is transferred to the external environment. Explain with the aid of sketches and manufacturers' literature how the following systems function and where they can be employed: direct air cooled condensers; water cooled and evaporative condensers; open circuit cooling towers; closed circuit cooling towers; forced draught cooling towers; induced draught cooling towers; cross draught cooling towers; evaporatively precooled dry heat exchangers.
26. Controlling water chilling plant plays a big part in economising on energy use. Explain with the aid of sketches and resource material the meaning of the following control methods: each chiller in a multi-chiller installation as a capacity step; multi-compressor step control; compressor cylinder unloading; variable speed control; hot gas bypass; evaporator pressure regulator.
27. Explain how centrifugal, screw, gear and scroll refrigeration compressors are controlled to match refrigeration demand.
28. Find a BMS company's single chiller control graphic and explain the control functions. Show typical temperature set points.
29. Four centrifugal water chillers are connected in parallel to a common chilled water flow and return header. Sketch the CHW pipe and chiller schematic. Show one typical pumped secondary CHW circuit connected to the common header. The secondary circuit supplies AHU coils through diverting control valves. Annotate the schematic to demonstrate how the control system functions to minimise energy use and show typical temperature set points.
30. A roof mounted cooling tower is to have a condenser water diverting valve to control heat rejection capacity from a reciprocating water chiller located in the basement. Sketch and describe the control schematic to demonstrate functionality and show typical temperature set points.

31. Find a BMS manufacturer's digital controller configuration schematic for multiple water chillers and cooling towers. Explain the functionality of the system and show typical temperature set points.
32. Sketch and describe a typical digital control schematic for multiple cooling towers and show typical temperature set points.
33. Find a BMS manufacturer's control graphic for multiple water chillers, show typical temperature set points and explain its operation.
34. Decide on a climate region for your answer. Give reasons for sequencing unequally sized water chillers in a large commercial building; for example, when each chiller capacity is 15%, 25%, 60% of the design's cooling load. Quote a specific installation if one is known to you.

8 Commissioning and maintenance

Learning objectives

Study of this chapter will enable the reader to:

1. understand the purposes and importance of commissioning;
2. know the range of information needed to commission an air conditioning system;
3. recognise the plant and systems that are connected with the commissioning of air conditioning;
4. identify the visual data checks needed;
5. know the safety criteria for the safe activation of electrical systems;
6. know when to commission plant and systems;
7. understand how fluid pressure tests are conducted;
8. know the cleanliness and hygiene criteria for systems;
9. understand fan and pump starting and speed control methods;
10. recognise how the stages of commissioning relate to the other members of the construction team;
11. know the checks and data needed when commissioning fans;
12. understand the meaning of variable frequency control of fans;
13. know how to conduct a duct air leakage test;
14. use practical duct air leakage test data;
15. devise suitable test, commissioning and maintenance data record forms and logs;
16. know how to set to work and regulate an air duct system;
17. understand what is meant by proportional balancing;
18. know the information needed by the commissioning engineer;
19. identify the instruments used during commissioning;
20. know how air flow rates are measured;
21. know the principles of operation of gas detectors;
22. know the application of gas detectors;
23. recognise why gas detectors are used;
24. understand the use of tolerance in measurement;

25. know the methods used in the measurement of ventilation rates;
26. know what is needed to commission automatic control systems;
27. recognise the need for a commissioning sequence document;
28. know the items requiring maintenance;
29. identify maintenance schedules;
30. understand the use of standby plant;
31. know how and why systems are cleaned and disinfected;
32. know how systems are maintained in clean and disinfected condition;
33. recognise the possible sources of airborne health hazards;
34. know the principal maintenance items.

Key terms and concepts

Aerosols 264; anemometer 260; automatic controls 251; biocide 265; catalyst 261; chlorine 265; commissioning 264; cooling tower 264; drive belt 274; electrochemical 262; gas detector 261; inspection 264; legionnaires' disease 264; maintenance 249; manometer 256; pneumatic controls 263; pressurisation 253; proportional balancing 268; recording 250; sick building syndrome 264; sodium hypochlorite 265; soft start 254; switching 253; thermistor 262; tracer gas 262; trend log 251; variable frequency control 255; vibration 255.

Introduction

The meaning and importance of commissioning and the scope of work that is normally carried out are introduced. While not an academic subject, commissioning and maintaining air conditioning systems makes and keeps them functioning. Refer to the extensive CIBSE PROBE reports and observe the ways in which designed technical functions of buildings can sometimes fall short of expectations.

An air conditioning system may be allied to other mechanical services systems such as heating plant. Maintenance of air conditioning has energy consumption implications and also health and safety at work legislative obligations. The main elements of commissioning and maintenance are included but the topics and lists are not exhaustive. The commissioning and maintenance of the air handling system is prominent. The information provided for water systems, automatic control, refrigeration and heat generation are general rather than specific and detailed. Items are listed alphabetically and not in their order of importance or timing. The reader's answers to descriptive questions should be discussed with colleagues or tutors.

Questions require the use of the information within this chapter, references, manufacturers' recommendations and details of specific applications.

Further reference may be made to CIBSE Guide M (2008); CIBSE TM26 (2000); CIBSE KS02 (2005); CIBSE TM31 (2006), and BSRIA (2010).

Commissioning

The purpose of commissioning an air conditioning system is to set it into operation, regulate flows and verify that it performs according to the specified design. The completed installation is visually inspected and has certain measurements taken. The measured data will be unique to the installation rather than the test figures supplied by manufacturers under standard conditions. The accurate and logical recording of data is essential. Commissioning work can extend over months or years in large applications. Several groups of engineers and clients may have access to the data. The client needs records of performance for future use during maintenance, repair, refurbishment, and for continual energy and plant management.

Commissioning codes, CIBSE Commissioning Code A (2004), and others, are in use for air distribution, boiler plant, automatic control, refrigerating, and water distribution systems. These represent a standard of works and detail that competent engineers will hand over to the client upon completion of the commissioning process. Commissioning procedures extend over a wide range of specialist areas such as refrigeration, computers, electrical power and electronics. It is recommended that a single authority should have overall responsibility for commissioning. This may be one contractor or consultant, with subcontracted specialists.

Some data reading will be witnessed by the client to ensure that satisfactory performance has been agreed. Output numeric and graphical data from supervisory computer systems will form part of the information. Such graphical representation of system execution is termed a trend log. Plant status is recorded as timed events such as the starting of a fan or the opening of a motorised valve. Each event and time are recorded and often printed at line printers as they happen. Computer records of trends and status are stored on hard disc or magnetic tape for a predetermined interval upward of 24 hours.

Commissioning documentation may be incomplete when the contractors walk away from the site, having achieved 'practical completion', that is, the commissioning is practically (almost) complete. The owner may re-commission the plant to prepare records of the operational parameters of the entire system and commence energy and building data logging for performance certification.

Information requirement

Commissioning information includes:

1. description of the plant and its function;
2. design data on temperature, pressure, humidity and flows;
3. description of how to start and stop the systems;
4. location of all plant and distribution items;
5. fault diagnosis data, alarm limits and corrective action;
6. manufacturers' instructions and literature;
7. spare parts and sources of supply, stock list;
8. plant operational instructions;
9. maintenance schedule with frequency;
10. record drawings of all services;
11. plant schematic logic diagrams, framed and displayed;
12. lubrication charts, lubricants list, frequency;
13. valve and damper lists, number, location and setting;
14. log book of work carried out, data, date and signature.

Plant and systems directly connected with air conditioning are:

1. air distribution ductwork, fans and terminal units;
2. artificial lighting;
3. building energy management computer-based control;
4. dedicated automatic controls;
5. electrical power;
6. fire and smoke detection plus active smoke control;
7. fire-fighting systems;
8. heat generation;
9. hot and chilled water distribution;

10. interaction with the manufacturing process;
11. refrigeration, heat reclaim and heat rejection;
12. security and alarm systems;
13. telecommunications;
14. transportation systems;

Visual data

The system is visually inspected prior to being started to ensure that it is operational and safe. Some items that are checked are:

1. access hatches and test holes in air ducts are closed;
2. air and water pressure sensors;
3. air and water temperature sensors;
4. air duct joints are sealed;
5. air filters, access panels and driving motors;
6. air intake and exhaust louvres;
7. air volume control dampers, electric motors and linkages;
8. all component bolts, fixings and supports are secure;
9. anti-vibration mountings;
10. boiler and burner equipment;
11. builders' works ducts are clear of debris and dust;
12. cooling coil condensate tray and drain;
13. correct polarity of electrical connections to motors;
14. drain water seals are filled and drains functional;
15. drive guards in place and secure;
16. electrical switches, circuit breakers and fuses;
17. electrical wiring, earth conductor and insulation;
18. external cleanliness of all plant and distribution;
19. fan blades, bearings, drive belts and mounting bolts;
20. fan drive belt tension;
21. fire dampers in air ducts;
22. flue system;
23. fresh and correct grade of lubricant to motor and fan bearings;
24. refrigeration compressor lubrication;
25. fuel supply;
26. grilles and diffusers in rooms are fully open;
27. heater and cooler coils;
28. hot and chilled water pumps and valves;
29. interior of air ducts for cleanliness;
30. noise attenuators in ducts;
31. steam or water spray humidifier;
32. switch and circuit breaker operation;
33. thermal insulation securely in place;
34. water quality;
35. water storage tanks and float valves;
36. water supplies, tanks and float valves.

Electrical items

Before switching on the electrical power to any circuit, the following checks are made:

1. access equipment, ladders are available;
2. all wire terminal connections are tightly connected;
3. cables are electrically insulated;
4. carbon dioxide fire extinguisher is available;
5. equipment and floors are clean and dry;
6. competent and qualified electrical engineer;
7. correct current rating of fuses and circuit breakers;
8. correct isolating switches are in place and operative;
9. correct voltage is available for each item;
10. cover plates and cubicle doors are closed;
11. earth cables and bonds to utilities are in place;
12. electrical insulation tests are satisfactory;
13. equipment is undamaged;
14. lighting equipment is operational;
15. line and neutral wires are correctly connected;
16. mechanical equipment, such as fans, is ready for use;
17. no uninsulated wires within control panels;
18. packing is removed from switches, contactors and motors;
19. services are electrically bonded to the earth terminal;
20. spare fuses and lamps are available;
21. test certificates are ready;
22. test equipment is available;
23. thermal cut outs are operational;
24. wiring of control equipment is complete.

Setting to work

Parts of the installation are tested during the construction phase. Tests are to ensure that the components are safe and will retain their containment and distribution function when fully operational. Large plant such as boilers, refrigeration machines and fans cannot be tested without having their connecting distribution services filled and usable. Pipework, air duct systems and cables are tested in sections as each is completed. The test fluid and electrical current may not be the same as will be finally used, but is compatible with the system materials. Such tests are as follows:

1. Pressurisation of air ducts. The leakage of air from ductwork is measured by sealing the ends of a section and blowing air into it with a test fan. The static air pressure in the duct is raised to between 200 Pa and 2000 Pa depending upon the pressure class of the duct system.
2. Pressurisation of pipework. Prior to filling with water, gas or refrigerant, an air compressor pressurises the network up to 5 b depending upon the design working pressure. The test pressure is held for an hour. Loss of pressure is investigated, with a soap solution being applied to joints.
3. Pressurisation of drainage systems. The ends of above and below ground drain pipework are temporarily sealed and water seals are filled. A static air pressure of 100 mm water gauge is applied with a hand pump. The test pressure is not to fall below 75 mm water gauge during a 5 minute period without further pumping for the system to pass the test.

4. Water fill test. The completed hot or chilled water pipework system is filled with water. It is visually inspected for leaks. The water pressure is increased either by connecting the water main or by connecting an air compressor to increase the test pressure to 5 b or more, depending upon the design working pressure. Leak inspection is conducted for an hour.

When the water distribution system is proven to be reliable and has been pressure tested, it is drained of test water, flushed with clean water to remove debris, metal swarf and surplus jointing material and filled. Large plant is de-scaled and cleaned internally by the circulation of chemicals in water under cold and hot conditions. The filled system is ready for boiler or refrigeration plant operation.

The air distribution system is ready for use when all the components are installed, grilles and dampers are fully open and the building is sufficiently clean to allow air circulation. This will usually be when the internal decorations, floor and ceiling finishes are in place. The initial air circulation will contain dust and debris from within the air ducts, air handling plant, rooms and false ceilings that avoided the cleaning operation. High efficiency air filters, terminal mixing boxes, induction units, duct attenuators, finned coils, velocity, pressure and humidity sensors and variable air volume boxes are susceptible to dirt contamination. Air filters used for commissioning may be replaced prior to handover to client. The discharge of dust from supply grilles into conditioned spaces can cause secondary damage or contamination. The initial starting of fans is done when appropriate temporary protective measures have been taken.

Tests and commissioning that is conducted during cold weather requires attention to possible frost damage. Hot water air heating coils are to be either drained or be passing the correct hot water flow rate prior to starting the fresh air inlet flow. Automatic control systems and automatic fabric air filters are set to manual control. Boiler and refrigeration plant remain off load.

The supply air fan is started. Switching methods are a direct on line switch starter, timed soft start that provides a reduced initial voltage, or a variable frequency inverter control (VFC). A VFC start allows the fan to be accelerated from zero to maximum revolutions per minute during an adjustable time. Run the fan up to speed as gradually as available from the method installed. The extract or recirculation fan is started 30 seconds or longer after switching on the supply air fan. This is to avoid overloading the electrical supply with the simultaneous start of two high current motors. It may be unsafe to run only one fan due to the creation of high or low duct or room air static pressures. Damage might be caused to ductwork, room doors or partition walls. The building contractor is kept informed of test operations and advised if structural faults could be caused. Where excessive pressure variations could be generated, the supply and extract air fans are interlocked with a short time delay. The second fan is automatically started after the first fan. Fans should always be started at light load with reduced supply voltage or frequency to reduce wear on the drive belts and bearings, and minimise starting current and duct air pressure surge. Closure of the main supply air duct damper can be used with a direct on line starter as this minimises the electrical power demand.

Activation of the fans is accompanied by checks:

1. anti-vibration mountings and duct air connectors are secure;
2. belt drive security;
3. dynamic balance of the rotating items is acceptable;
4. electric motor speed and current are correct;
5. fan and motor shaft rotation direction;
6. fan rotation speed is correct;
7. full load start
8. harmonic frequencies from VFC do not cause interference;
9. light load start
10. lubricant seals on bearings are not leaking;

11. motor and fan bearing temperatures are correct;
12. running in period may be several days;
13. spark free motor operation;
14. starting and stopping controls are operational;
15. three phase circuits have equal current;
16. vibration and noise are acceptable.

Frequent starting of electric motors on fans, pumps and refrigeration compressors is to be avoided. The high initial current required, over double the running current, will cause overheating of the motor, belt drive, switch, fuse and circuit breaker, leading to premature failure. Up to six starts per hour may be allowable, if cooling intervals are long enough.

Variable frequency control of fan motor speed works by digitally reproducing alternating current from the input 50 Hz supply. The output from a VFC is controllable from 0 Hz up to more than 50 Hz. The rotation speed of an alternating current motor is directly proportional to the applied frequency. 50 Hz is 3000 revolutions per minute (RPM), but an induction motor has slip, so a maximum of 2900 RPM is obtainable. 25 Hz produces 50% full speed, 1450 RPM. The VFC can accelerate and decelerate the fan during a variable time and hold the speed at any value. In doing this, it generates a wide range of frequencies into the kHz and MHz ranges. These harmonic frequencies are injected into the neutral electrical cable and distributed to other parts of the building. These high frequency electromagnetic emissions can cause radio and electrical interference with other systems. The commissioning process should include their measurement and analysis for secondary affects.

Acceleration meters can be attached to fan and motor bases to measure the vibration frequency and movement. Anti-vibration mountings are designed to absorb a particular frequency and limit the deflection of the fan. The correct functioning of the mountings is checked. Excess vibration that is transmitted to the building structure may produce unacceptable low frequency noise at some distance from the source. Fans have a critical, resonant, rotation frequency that may be as low as 100 RPM. Constant rotation at this speed will equalise the forcing and natural vibration frequencies of the rotating mass. The amplitude of vibration can become infinite at resonance, leading to rapid failure of the mountings. It is essential to accelerate the fan through its resonant frequency without pausing. Vibration measurements are particularly important near critical RPM values.

When the running in period has been satisfactorily completed, the fan is started at full load condition with dampers fully open and the regulation activity is commenced.

Duct air leakage test

Low pressure class ducts have an internal static pressure limit of ±500 Pa and a maximum design air velocity of 10 m/s. The test arrangement is shown in Figure 8.1.

A test fan supplies air into a completed duct section and maintains the test pressure. Ductwork is inspected for leaks. The leakage rate is equal to the rate of air flowing into the duct through the test fan. An air flow meter, Figures 6.10, 6.11 and 6.12, records the leakage rate. The allowable leakage rate for low pressure category ducts Q l/m^{2}s of duct surface area is found from:

$$Q = 0.027 p_S^{0.65} \frac{\text{l}}{\text{s}}$$

p_S = static air pressure maintained during the test, Pa

Leakage air flow can be measured by a venturi, orifice or conical inlet flow meter from the equation:

$$Q = \frac{C\pi d^2}{4}(2\rho\Delta p)^{0.5} \frac{\text{m}^3}{\text{s}}$$

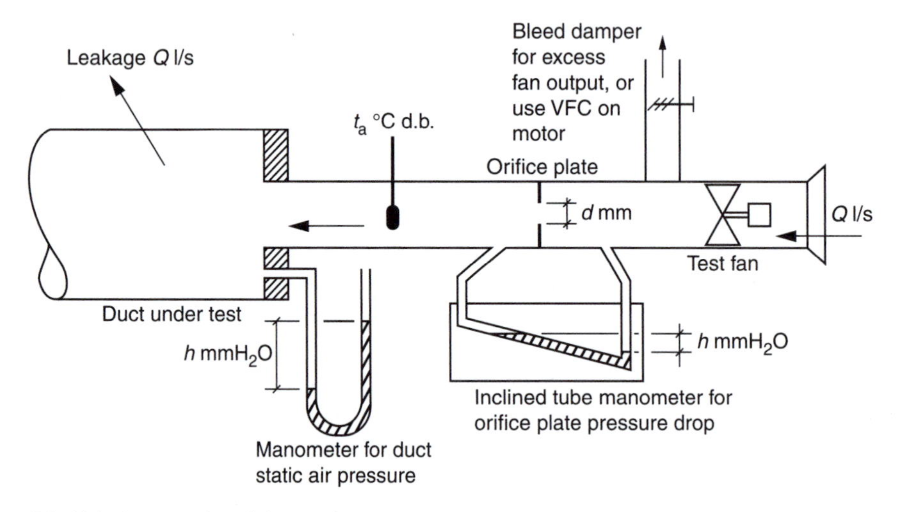

8.1 Air leakage testing of ductwork.

C is the flow coefficient, approximate values are 1 for a venturi, 0.65 for an orifice and 0.95 for a conical inlet, *d* is the orifice or venturi throat internal diameter, m

$$\rho = \text{air density}, \ \frac{\text{kg}}{\text{m}^3}$$

$$\Delta p = \text{pressure drop through meter, Pa}$$

$$\rho = 1.205 \times \frac{273 + 20}{273 + t} \times \frac{101325 + p_s}{101325} \frac{\text{kg}}{\text{m}^3}$$

$$t = \text{temperature of air in the test duct, °C}$$

The test procedure is as follows:

1. Seal the air duct to be tested with inflatable bags, polythene sheets or blank plates inserted at flanged joints.
2. Allow the joint sealant to cure.
3. Air handling plant such as filters, heater and cooler coils are not included in leakage tests.
4. Tests may be witnessed by the client's representative.
5. Record all test data on a standard form.
6. Install test equipment and run until stable conditions are found.
7. The leakage rate should be stable for 15 minutes.
8. Switch the test fan off.
9. Wait until duct static pressure drops to zero.
10. Immediately switch on the test fan.
11. Verify that the same test results are found.
12. Record all the data; sign, date and witness the forms.

A sample test data sheet is shown in Table 8.1.

Table 8.1 Data sheet for sample duct air leakage test

Air leakage test data		
Test number	1001	
Date	8 February 2014	
Client	Watt Air PLC	
Job	South wing VAV, Chilworth office	
Contract number	TC/150144	
Part 1, Installation		
(a)	duct section	nodes 34–88
(b)	ducts shown on drawings	HV/4007
(c)	surface area of duct	65 m^2
(d)	test static air pressure	300 Pa
(e)	leakage equation	$Q = 0.027 p_s^{0.65} \frac{l}{s}$
Part 2, Test		
(a)	duct air static pressure reading	280 Pa
(b)	duct air temperature	22°C
(c)	maximum allowable leakage	68 $\frac{l}{s}$
(d)	air flow measuring device	venturi
(e)	meter flow coefficient *C*	0.98
(f)	flow meter range, l/s	10–200
(g)	flow meter serial number	D342
(h)	flow meter throat diameter	50 mm
(i)	flow meter calibration certificate	attached
(j)	pressure drop at flow meter	25 mm H_2O
(k)	interpreted duct leakage l/s	47
(l)	switch fan off and repeat	done
(m)	test duration	46 minutes
Part 3, Conclusion		
(a)	result of test	satisfactory
(b)	test engineer	A.N. Smith, Airsure Commissioning Ltd
(c)	witness	I. Care, Watt Air PLC

EXAMPLE 8.1

Use the air flow test data shown in Table 8.1 and calculate the results.

Allowable duct leakage is found from,

$$Q = 0.027 p_s^{0.65} \frac{l}{m^2 s}$$

$$p_s = 280 \text{ Pa}$$

$$\text{Duct surface area} = 65 \text{ m}^2$$

$$Q = 0.027 \times 280^{0.65} \times 65\ \text{m}^2 \frac{\text{l}}{\text{m}^2\text{s}}$$

$$Q = 68.38 \frac{\text{l}}{\text{s}}$$

$$\rho = 1.205 \times \frac{273+20}{273+t} \times \frac{101325 + p_S}{101325} \frac{\text{kg}}{\text{m}^3}$$

$$\rho = 1.205 \times \frac{273+20}{273+22} \times \frac{101325+280}{101325} \frac{\text{kg}}{\text{m}^3}$$

$$\rho = 1.2 \frac{\text{kg}}{\text{m}^3}$$

$$C = 0.98$$

$$d = 0.05\ \text{m}$$

$$\Delta p = 25\ \text{mm}\ H_2O$$

$$\Delta p = (9.807 \times 25)\,\text{Pa}$$

$$\Delta p = 245\ \text{Pa}$$

$$Q = \frac{C\pi d^2}{4} (2\rho\Delta p)^{0.5} \frac{\text{m}^3}{\text{s}}$$

$$Q = \frac{0.98 \times \pi \times 0.05^2}{4} (2 \times 1.2 \times 245)^{0.5} \frac{\text{m}^3}{\text{s}}$$

$$Q = 0.047 \frac{\text{m}^3}{\text{s}}$$

$$Q = 47 \frac{\text{l}}{\text{s}}$$

Air flow regulation

Figure 8.2 shows a typical air duct system with the information needed by the commissioning engineer. The system commences with the balancing dampers and grille deflectors and dampers fully open. The design calls for a specific air flow of 1 m^3/s through duct *B* and 2 m^3/s through duct *C*. The balancing dampers at *B* and *C* need to be set to the position that ensures the correct proportion of the total air supply from duct *A* passes into each branch. When this is done for each branch, the duct system will be proportionally balanced.

The proportions are,

$$Q_A = 3 \frac{\text{m}^3}{\text{s}}, \text{ or} 100\%$$

$$Q_B = \frac{1}{3} \frac{\text{m}^3}{\text{s}}, \text{ or } 33\%$$

$$Q_C = \frac{2}{3} \frac{\text{m}^3}{\text{s}}, \text{ or } 67\%$$

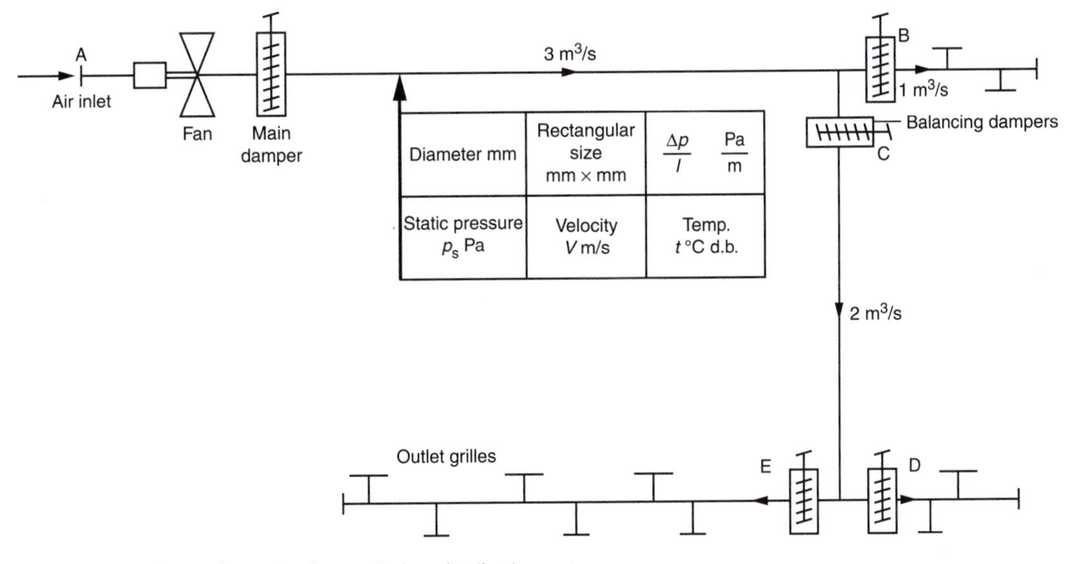

8.2 Commissioning information for an air duct distribution system.

Whatever air flow is found in duct *A*, 67% of it must be directed into duct *C*. Throttling the air flow with dampers *B* and *C* will increase the overall system resistance and reduce the air flow at *A*. Repeated flow measurements and damper adjustments are made until the desired proportions are stabilised. The same operation is carried out at successive locations in an upstream direction.

The correct system air flow can then be achieved by throttling the main damper or changing the fan speed, revolutions per minute (RPM). Fan and motor combinations are normally selected to deliver slightly greater air flow than called for at the design operating point. Fan speed is changed by using different pulley diameters on the fan or motor shaft. The slack in the belt drive is taken up with a sliding motor support. The electric motor runs at a constant 2900 RPM. If the fan pulley is the same diameter as the motor pulley, the fan also runs at 2900 RPM. This is generally undesirable due to the aerodynamic noise that would be generated. Fan speeds of up to half the motor speed, 1450 RPM, are preferred. If the correct duct system air flow cannot be provided by the fan as installed, then a smaller fan shaft pulley will increase the fan speed and allow a higher flow rate. If the fan is delivering too much air flow, as is expected, a larger fan shaft pulley will reduce fan RPM and the air flow delivered. This is generally the most economical solution. A variable frequency control would simply be set to a frequency lower than 50 Hz until the correct air flow is achieved. Either method of fan speed reduction saves operational energy consumption and is preferable to throttling the air flow with the main damper. A damper introduces additional frictional resistance into the system that the fan motor has to overcome. This is paid for at the electricity kWh meter.

Air ductwork varies in size from 100 mm diameter thin sheet galvanised metal up to walk in room dimensions constructed from builders, work concrete blocks. Specialist applications may have thick stainless steel ducts for toxic substances in air. The information needed by the commissioning engineer includes:

1. air duct internal dimensions;
2. air ductwork and plant locations;
3. air static pressure at controlled locations;
4. air temperature in ducts under operational conditions;
5. air volume flow grid locations and type;
6. damper design flow rate and pressure drop;

7. damper locations and type;
8. design air flow rate in each duct;
9. design air velocity in each duct;
10. drive motor speed;
11. duct material thickness;
12. duct test pressure;
13. ductwork materials and method of jointing;
14. electrical supply voltage and frequency;
15. fan design rotation speed;
16. fan type and characteristic curve;
17. filter pressure drop when clean and dirty;
18. location of access hatches into the ductwork;
19. location of test holes;
20. pressure drop through heating and cooling coils.

Instruments

1. A rotating vane anemometer, Figure 8.3, gives a stable mean velocity due to its large diameter, 100 mm, and the inertia of the blades. It can be used at the throat of a venturi, or hood, which

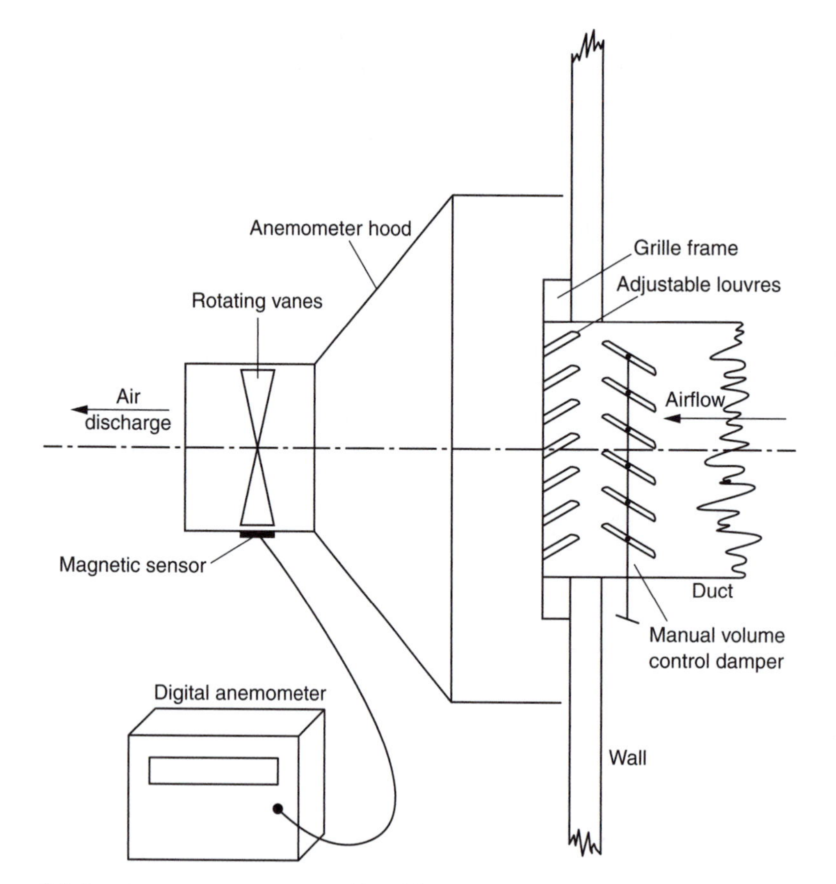

8.3 Rotating vane anemometer and hood for measuring air discharge from a grille.

collects all the air issuing from a grille. This enables one reading of air velocity to be multiplied by the cross-sectional area of the throat to find the total air volume flow rate. Calibration is achieved by placing the anemometer in a jet of air from a test duct. The mean air velocity of the jet is known from pitot-static tube readings within the duct.

2. Pitot static tube and manometer, Figure 6.1, is used for measuring the air velocity pressure and static pressure in ducts. An inclined tube liquid manometer can be used and this requires no external power supply or calibration. An electronic micro manometer can provide a wider range of pressure readings, it can have greater accuracy, its output can be sent to a data logger, chart recorder or remote computer, and it can be left in place unattended. Static pressure measurements are needed at strategic locations within ductwork, for example either side of fans, Figure 6.8, at controlled pressure locations and at air inlets to induction and variable air volume units and dual duct mixing boxes.
3. Gas detectors are used in the commissioning of boilers to measure the constituents of the flue gas. Adjustments are made to the combustion air supply to optimise the boiler efficiency. They also may be employed in leak detection to trace the presence of methane or other combustible or explosive gas. Manual inspection of the insides of large oil or water tanks and underground service tunnels or drains is accompanied by gas detection.

It is unreasonable to expect that air flow can be measured with absolute accuracy by the commissioning team, or that it is necessary. The equipment has tolerance limits. Air flows rarely produce steady velocity readings. The internal surfaces of the ductwork will be far from smooth. The air flows are turbulent and fluctuating. A skilled operator will produce carefully obtained data, but the time necessary to balance out all supposed errors is likely to be uneconomic. It is important that the design air flow is achieved within the conditioned space at the commissioning stage. The minimum is −0% of the design air flow.

Commissioning is conducted with the air filters in their clean state. Soiling of the filters will provide increased resistance and air pressure drop. The filters are changed at a predetermined pressure drop. The ductwork system has a greater air flow when the filters are clean. This initial flow gradually reduces as the filter clogs. Some air conditioning systems have a motor driven damper to introduce a variable resistance that is compatible with the changes in the filter. This equalises the filter resistance at an externally constant pressure drop.

The accumulation of errors, which is inherent while acquiring air flow data, leads to an overall tolerance on the desired flow. It is reasonable to be satisfied with increased air flows in steps of 5%, 10%, 15% and 20% depending upon the designers' criteria and the need to provide an overall satisfactory installation at reasonable cost. Excessive air flow may cause undesirable noise or draught that is noticeable by the users of the conditioned space. A balance needs to be achieved between these factors and agreed at an early stage. The designer specifies the air flow tolerance as design Q l/s + 20% − 0%.

Gas detectors

Oxygen content of air is 20.9% by volume, most of the remainder being nitrogen. The presence of unwanted gas may displace or consume oxygen to a dangerously low level. An alarm condition is usually 19% oxygen in air. Leaking oxygen cylinders used for welding in confined spaces can increase the oxygen content of air. This increases the flammability of materials with consequent danger. Detection instruments include the following types.

Pellistor sensors

These detect flammable gas and vapours. Two coils of platinum wire are each embedded in a bead of alumina. One bead is impregnated with a catalyst that promotes oxidation. An electrical current passes through each coil that raises the bead temperature. When the target gas is present, the heated catalyst causes the gas to be oxidised, that is, combustion takes place. The local combustion further increases the temperature of the coil. Increasing the wire temperature increases its resistance. The voltage applied to the coil circuit remains constant, so the current flowing through the coil reduces. The coil without the catalyst remains unaffected and has a constant current flow. The imbalance between the two electrical circuits is detected in a Wheatstone bridge circuit and an output reading of gas percentage is calibrated on the output display. Pellistor sensors can last in service for years. They can be poisoned by halogens, lead or silicone substances.

Thermal conductivity sensors

Two thermistor beads have a current passing through them. A thermistor is a semiconducting material made from oxides of manganese and nickel. Their resistance varies with temperature. One thermistor is exposed to the gas/air atmosphere while the other is sealed within an air container. They are connected in a Wheatstone bridge circuit. The rate of heat loss from the detector depends upon the thermal conductivity of the gas/air mixture. Imbalance between the two thermistor circuits is a measure of the gas in the atmosphere. Carbon dioxide, helium, hydrogen and methane can be detected by this method.

Electrochemical sensors

These are used to measure toxic gas and oxygen. An electrochemical sensor is a fuel cell having two electrodes and a liquid electrolyte. Its sensing head has a permeable membrane covering the electrolyte. Gas diffuses from the atmosphere through the membrane and reacts on the surface of the electrodes to produce a flow of ions and electrons. Ions flow through the electrolyte to the other electrode. A reaction produces a flow of electrical current through the external wiring circuit. Thermal efficiency of fuel cells is around 90%. Electrical current produced is directly proportional to the gas concentration being measured and is displayed as percentage gas in air. Oxygen detectors are replaced annually but hydrogen sulphide and other sensors last for up to three years.

Ventilation rate measurement

The rate of natural ventilation in a building can be measured using the technique described for the leak testing of air ducts. Figure 8.1 shows how a space can be pressurised and the air flow through it measured. A typical test kit for a complete house would include a replacement door with the test fan and air flow meter attached. The building is pressurised to between 25 Pa and 50 Pa above the atmosphere while leakage rate is measured. Very leaky buildings are tested at lower pressures depending on what can be maintained. This flow represents the natural ventilation rate under normal conditions. Results for 50 Pa standard are calculated from the lower maintained pressure.

An alternative is to release a tracer gas such as nitrous oxide or helium into the room to achieve a measurable percentage concentration and then switch on the mechanical ventilation system. The concentration of the tracer gas is measured at suitable intervals. The rate of decay of the gas is used to calculate the ventilation rate in air changes per hour or air flow rate.

Commissioning control systems

Activation of the air and water distribution and electrical systems allow the automatic control equipment and BMS to be commissioned CIBSE Commissioning Code C (2009). The work necessary includes checking the following.

1. Actuating motors for dampers and valves have full movement.
2. Air and water temperature and pressures are correct.
3. Air flow dampers are lubricated and move fully.
4. Air flow meters are ready for use.
5. All detectors, valves and dampers are numbered and listed.
6. Automatic sequence of switching is correct.
7. Calibration of detectors and flow meters is completed.
8. Controls respond in intended direction and scale.
9. Client is informed of commissioning work.
10. Commissioning drawings and specifications are provided.
11. Commissioning timetable or sequence is published.
12. Control valves and dampers have full movement.
13. Control valves are correctly located.
14. Control valves have the correct pipe connections.
15. Controllers operate over their correct range of values.
16. Correct control is maintained at low flow rates.
17. Data display and storage meet design criteria.
18. Desired states of measured variables can be stabilised.
19. Detector locations are marked on drawings.
20. Detectors are not subject to external influences.
21. Electrical interlock, circuit breaker and switches are in place.
22. Electrical items are operational and safe.
23. Electrical systems are operational and safe.
24. Fan pressure rise is within design tolerance.
25. Full air flow is within design tolerance.
26. Gas, water, electricity and drainage services are functional.
27. Isolating gate valves are open.
28. Main contractor is informed of commissioning activity.
29. Major building operations are completed.
30. Manufacturers' instructions are complied with.
31. Measuring instruments have calibration certificates.
32. Occupants are informed of commissioning work.
33. Pipework systems are tested and operational.
34. Plant switches on and off at correct settings.
35. Pressure balancing valves are ready for use.
36. Pressure sensors are operative.
37. Pump pressure rise is within design tolerance.
38. Schematic system logic drawings are current edition.
39. Sensing elements are sampling representative conditions.
40. Signal transmitters and outstations are functioning.
41. Specification of equipment is correct.
42. Steady conditions within the building can exist for tests.
43. Supervisory computer system is fully functioning.
44. Temperature and humidity detectors are correct range.
45. Temperature, pressure and flow settings are listed.
46. Water flow is within design tolerance.

Where pneumatic control systems are in use, the additional commissioning operations include the following.

47. Air compressor is tested and operational.
48. Air compressors are clean and free of leakage.
49. Air drying systems are functional.
50. Air pressure reducing set is clean and operational.
51. Air volume flow rate is within design tolerance.
52. Branch pipe air pressures are within design tolerance.
53. Compressed air receiver tank is pressure tested.
54. Dual compressor changeover is operational.
55. Electrical power to air compressor is complete and tested.
56. Pneumatic control actuators are functional.
57. Pneumatic pipelines are clear and pressure tested.
58. Pneumatic pressure is maintained within design tolerance.
59. Pressure switch operates compressor correctly.
60. Safety pressure relief valves are set and functioning.

Commissioning sequence

Commissioning work can take from days to months depending upon the system size, complexity and detail needed. A sequence of activity is published by the design team itemising each step to be undertaken. Designers need to initiate the work, as only they are in a knowledgeable position at the time. Formalisation of a plan of action will ensure that all interested parties are included.

Maintenance schedule

Maintenance is organised for regular work at specified intervals and at the expiry of the running hours of specific plant. Scheduled maintenance is the planned execution of replacement and servicing operations. This takes place prior to plant breakdown and is the product of knowledge and experience. Lamps, drive belts, electric motors and lubrication fluids are replaced as their service period finishes. Although they remain operational, their efficiency will have deteriorated and replacement is desirable or essential. Unscheduled maintenance is due to breakdown. When the building has essential functions such as 24 hour computing equipment, manufacturing or health care, standby plant is switched in automatically. In crucial applications, a third plant item is at standby. The first standby is a duplicate fan, pump, boiler or refrigeration compressor that is fitted in parallel with the running item. Changeover between the running and standby plant is made at frequent intervals as part of the maintenance schedule. The third standby requires some manual intervention to bring it into service. It might be an item kept in storage on the site or permanently connected into the circulation system.

Air conditioning systems are susceptible to the accumulation of dust, dirt, bacteria and rust within air ducts, air handling plant and cooling towers. Inspection of the internal surfaces is necessary to ascertain the need for cleaning, repair and disinfection. Contamination of the air passing through plant and ducts can lead to illness, outbreaks of legionnaires' disease and cases of sick building syndrome.

Water aerosols cause the dispersal of micro-organisms and they come from:

1. dehumidifying cooling coils;
2. drainage systems;
3. evaporative humidifiers;
4. fountains and garden sprinklers;
5. piped hot and cold water systems;
6. showers;

7. spas, whirlpool baths and therapy pools;
8. spray washing equipment such as for vehicles and processes;
9. sprayed chilled water cooling coils;
10. water spray humidifiers.

Cleaning and disinfection of equipment is carried out:

1. prior to commissioning;
2. after a shutdown of 5 days or more;
3. after system alterations;
4. when the cleanliness of the system is in doubt.

Internal surfaces of air ducts, cooling towers and water storage tanks are cleaned with sprayed detergent solution. Slime, rust and debris are loosened by brushing or grit blasting and then vacuumed. Cross-contamination from other items is avoided by cleaning all parts of the system simultaneously. Pipework is cleaned with chemical dispersant and then disinfected by circulating chlorine solution for 6 hours. Water storage tanks can have their internal surfaces coated with epoxy resin that is compatible with potable water supplies.

Chlorine is used as the biocide disinfectant, but it has no detergent cleansing property. Chlorine reacts with organic matter, ferrous salts and hydrogen sulphide that may be present in water or on wetted surfaces. Its disinfecting ability can be neutralised by them and other biocides. Residual chlorine must be present in the disinfecting solution after such reactions have taken place. Sodium hypochlorite solution is used and it contains up to 15% free chlorine. The disinfectant solution is made from sodium hypochlorite to obtain a minimum free residual chlorine concentration of 5 mg/l (ppm).

A cleaned and disinfected system is drained, flushed and filled with treated water for continued use. The water authority is informed before the discharge of chlorinated water into the public sewer. Cooling tower water is dosed with 30 mg/l of sodium hypochlorite; water condition is regularly monitored so that it remains unfavourable to the proliferation of micro-organisms.

Maintenance of air conditioning systems includes the following:

1. air filter material cleaning or replacement;
2. air filter pressure drop monitoring;
3. air intake louvres clear and clean;
4. bearing wear assessed on rotary plant by vibration spectrum measurement and records;
5. building energy management computer system diagnostic check;
6. chemical dosing of water measured;
7. clean and disinfect cooling towers;
8. clean and disinfect water storage tanks
9. clean surfaces of air and water distribution systems;
10. clean surfaces of electric motors and fans;
11. combustion efficiency measurements on boiler plant;
12. compressor pressure and temperature controls working;
13. cooling tower packing in good condition;
14. corrosion check made on all metalwork;
15. disinfection of air and water distribution systems;
16. electric motor carbon brush replacement;
17. electric motor full load current measured;
18. electrical insulation resistance inspected and tested;

19. electrical switches and circuit breakers operational;
20. fan belt drive tension adjustment;
21. float valves operational;
22. flow meters calibrated;
23. hours run meter readings taken;
24. investigate vibration and noise sources;
25. leakage from pipes or plant assessed for repair;
26. lubricate and test air duct dampers;
27. lubrication of electric motor bearings;
28. lubrication of fan bearings;
29. manufacturers' maintenance instructions followed;
30. no loose bolts, supports or plant connections;
31. operation and maintenance documents completed;
32. outstation telecommunications functioning;
33. overflow and drain systems clear;
34. paintwork and corrosion protective finishes in place;
35. pipeline strainer cleaning;
36. pressure sensors calibrated;
37. pressures of air, water and refrigerant measured;
38. records of maintenance and alterations up to date;
39. refrigerant pressures and temperatures correct;
40. room terminal grilles clean and quiet;
41. standby pump and fan changeover to running operation;
42. standby refrigeration, boiler, fan and pump tested;
43. steam humidifier de-scaled;
44. storage of spare parts and chemicals checked;
45. temperature and humidity sensors calibrated;
46. temperature, pressure and humidity transmitters operational;
47. valve operation verified, motor driven and manual;
48. valve packing gland leakage checked;
49. visual inspection of thermal insulation.
50. temperature scanning of motors, bearings and thermal insulation with surface temperature thermistor and remote infra red scanning.

Questions

1. A leakage test on completed ductwork revealed the following data: duct static pressure 45 mm H_2O duct surface area 120 m^2, duct air temperature 12°C, orifice plate flow coefficient 0.67, orifice throat diameter 70 mm, orifice pressure drop 160 mm H_2O. Calculate whether the duct system meets the maximum leakage criteria when it is calculated from $0.027p_s^{0.65}$ 1/m^2s duct surface area.
2. A commissioning engineer is to carry out a leakage test on a section of completed ductwork. On the day of test, the duct surface area is measured as 93 m^2, duct air temperature 14°C, orifice plate flow coefficient 0.68, orifice throat diameter 50 mm and orifice pressure drop 230 mm H_2O. Calculate the duct static pressure h mm H_2O that must be maintained by the test fan in order for the duct system to meet the maximum leakage criteria when it is calculated from $Q = 0.027p_s^{0.65}$ 1/m^2s duct surface area.
3. A leakage test on a section of completed ductwork with a surface area of 260 m^2, duct air temperature 18°C, venturi flow coefficient 0.97, throat diameter 60 mm and venturi pressure drop is 190 mm H_2O. The duct static pressure is held at 62 mm H_2O by the test fan. Calculate whether the duct

system meets the maximum leakage criteria when it is calculated from $Q = 0.009p_s^{0.65}$ 1/m^{2}s duct surface area.

4. Write a complete schedule for the commissioning work necessary for the air handling equipment for a single duct air conditioning system serving a lecture theatre. Heat is provided from a low pressure hot water two-pipe system from a boiler house. Cooling is provided by chilled water two-pipe systems.
5. Explain the difference between scheduled and unscheduled maintenance work on an air conditioning system. State the items that will require replacement, their likely length of normal service and whether they should be held in storage on the site.
6. Describe how failure of plant during normal use is overcome to maintain the air conditioning service.
7. Discuss the approach to a suitable maintenance programme for the air conditioning in the following applications:

   (a) a large general hospital;
   (b) a university;
   (c) residential buildings in the UK;
   (d) office accommodation in Brisbane, Australia;
   (e) a manufacturing building including the containment of biological and radioactive materials;
   (f) comfort control for the workplace in the UK.

8. A refrigeration condenser is cooled with water that is circulated to an evaporative cooling tower on the roof of a hospital in a city. State the actions that are taken to ensure that the tower operation will not cause health hazards to patients and the public.
9. Design a maintenance log for a city office building that has low pressure hot water heating and a single duct heating and ventilation system. There is no refrigeration plant. Maintenance work is to be carried out by contractors. The office users will allow some interruption of the heating and ventilating systems during 09.00-17.00 hours for repairs.
10. Discuss the reasons for using standby plant in an air conditioned building. Include the implications for plant room size, storage commitment, stock control, capital and recurring costs for the user.
11. List the frequency of visual checks, physical measurement, changeover of running plant and planned replacement of the items and systems in an air conditioned building of your choice from the types given in question 7.
12. Find the maintenance records for a building that is accessible to you. This may be a residence, office, factory, warehouse, college or university. Request the help and cooperation of the professionally employed maintenance engineering staff where this is applicable. Write a report on the comprehensiveness of the records that are kept. Make appropriate recommendations as to improvements that should be made.
13. Compare the quality of maintenance work and its recording: (a) when it is conducted by employees of the same organisation that uses the site, and (b) by contract companies. This may involve the acquisition of information from several sources that are dissimilar. Discuss the advantages and the budgetary control implications of each method.
14. Draw a schematic diagram of part of an air conditioning system. Show the location of all the necessary valves, dampers and controls. Number all the controls and test points and produce a schedule of their data. Write on the diagram all the data that will be needed by the commissioning and maintenance engineers.
15. Design a maintenance log for the refrigeration plant of an air conditioning system that serves a 10000 m^2 floor area complex of offices and computer manufacturing facilities. Evaporative cooling towers are located on the roof. Refrigeration plant is in a ground floor plant room. Each room has a chilled water cooling coil and local temperature control. Make any assumptions about the systems that are necessary. State the intervals between servicing and replacement work.

16. Acquire manufacturers' literature that demonstrates the use of computer screen system logic diagrams for air conditioning systems. Note how the air circulation, detectors and control systems are represented. Choose a different type of air conditioning system and create an equivalent diagram. List all the data points to be used, with sample data.
17. Use the solution to question 4, 14 or 16 and list all the points that are to be connected into the automatic control system. A point is where a detector, switch, control panel, outstation, modem or other item is wired into the automatic control system. The control, commissioning and maintenance engineers need to know this information.
18. Explain the advantages gained when the original designer of an air conditioning system publishes the schedules for commissioning and maintenance work. State the information that is included and the form that the documentation should take. State how a computer-based system can aid good maintenance practice.
19. List the order in which each part of an air conditioning system will be commissioned. State the condition required of the building works for each stage of commissioning.
20. Explain the following:

    (a) why frequent starting of fans and pumps is to be avoided;
    (b) why internal surfaces of air ducts, cooling towers, water storage tanks and water circulation pipework systems need to be cleaned and disinfected;
    (c) how cleaning and disinfection work is conducted;
    (d) methods available for starting the electrical motor drive on a fan;
    (e) why the first installed air filters will be temporary;
    (f) proportional balancing;
    (g) the need for vibration measurements;
    (h) harmonic interference from electrical equipment;
    (i) fan speed regulation;
    (j) full load start;
    (k) why and where gas detectors are used;
    (l) tolerance limits;
    (m) how cross-contamination between services is avoided;
    (n) where water aerosols originate;
    (o) what reduces the effectiveness of chlorine dosing and disinfection.

21. List the instruments that may be needed to commission an air conditioning system. Describe, with the aid of sketches, the operating principle and method of use of each instrument.

9 Fans

Learning objectives

Study of this chapter will enable the reader to:

1. know the types of fans used in building services;
2. apply fan types correctly;
3. understand the use of fan plenums;
4. calculate the opening force on a plenum door;
5. understand the use of fan characteristic performance curves;
6. calculate fan pressure and power curves manually and with a spreadsheet;
7. generate fan characteristic curves manually and with a spreadsheet;
8. understand and calculate the fan motor power over a range of air flows;
9. know why the control of fan pressure may be needed;
10. know the methods that are used to control the performance of a fan;
11. use the fan laws to predict the performance of fans at different conditions;
12. calculate the characteristic resistance of a ductwork system and plot it on a fan performance graph;
13. find the fan and system operating point;
14. know the fixed motor speeds that are used;
15. calculate and plot fan performance data for different fan speeds;
16. know the results of using fans that are connected in series;
17. know the components of the electrical power used by a fan;
18. know the methods of starting and controlling fans;
19. analyse the energy costs for the different methods of controlling the air flow in a duct system, manually and with a spreadsheet;
20. know the methods used to protect electric motors;
21. know the advantage of running electric motors at varying speeds;
22. know the commissioning and maintenance procedures for fan and air ductwork installations.

Key terms and concepts

Aerodynamic forces 286; aerofoil 271; air contaminants 279; air density 280; air filter 273; air flow control 289; apparent power 285; atmospheric contaminants 289; atmospheric pressure 284; attenuators 273; auto transformer control 288; axial flow 272; backward curved 272; balancing damper 279; belt drive 273; bi-metallic cut-out 287; centrifugal 272; characteristic curves 276; delta connection 288; fan laws 281; fan motor power 276; fan static pressure 277; fan testing 283; fan total pressure 276; fan velocity pressure 280; fans in series 284; forward curved 271; frequency inverter 289; impeller diameter 281; impeller efficiency 279; mixed flow 273; motor contactor 287; motor efficiency 286; motor poles 279; motor speed 279; non-overloading 279; opening force 273; operating point 292; overheat protection 287; overloading 279; power factor corrector 286; propeller 270; soft start 288; star/delta controller 288; thermal protection 279; triac motor speed controller 288; useful power 285; variable speed drive 279.

Introduction

Types of fans and their applications are introduced. Fan and system characteristic curves are demonstrated through the use of manual and spreadsheet calculations. Spreadsheet calculation files are provided for downloading. Fan performance calculations are explained with worked examples. Fan starting and the methods of controlling the air delivery from fan and ductwork systems are discussed. Energy consumption is calculated for the different methods of control. Commissioning and maintenance procedures for fan installations are listed.

Fan types

Fans are the prime movers for air conditioning and ventilation systems. They are used to exhaust vitiated air and flue gases, as well as to generate the required air movement within occupied spaces of the building. They are a source of acoustic energy and the possible creation of noise for the occupants of the treated building and for those outside it. Fans are rotodynamic machines that are used to generate low to medium increases in the pressure of air or other gases. Passage of each blade of the fan impeller imparts a pulse of energy to the air flow. The number and geometry of the blades, and the rotational speed of the impeller, determine the frequency of the pulsations in the air stream and the noise that is produced. Linear speed of the outer tip of the fan impeller is a factor in the material design of the rotating components.

Impeller blade shape, complexity, dimensions, constructional material, rotational speed and the shape of the enclosure depend upon the performance that is required and the application of the fan. Low cost fans are formed from cut sheets of flat steel. Their blades may be flat or slightly curved. These appear in domestic appliances and in vehicle cooling and ventilation. High cost fans have aerofoil cross-section blades, a high pressure volute casing and they can be run at high speeds and are used in high air pressure and air flow applications. Blade types are shown in Figure 9.1.

Types and applications are:

Propeller

These are used where there is a low frictional resistance and high air volume flow rate system requirement, such as exhaust fans and outdoor air cooled heat exchangers; see Figure 9.2.

Axial

These are used: in ducted air and hot gas exhaust systems; where the fan has to be in line with the ductwork due to space restriction; where pressure rise is limited; see Figure 9.3.

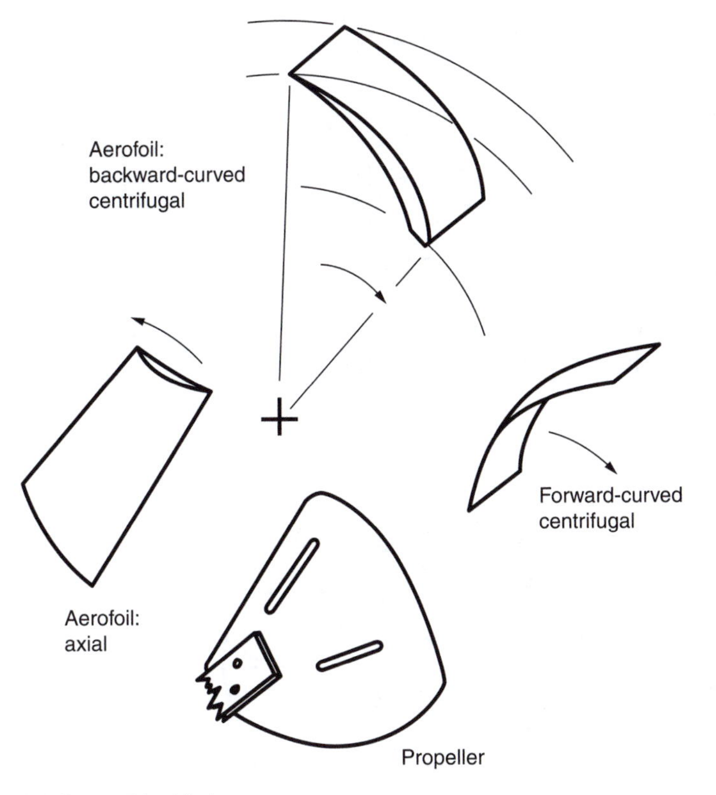

9.1 Types of fan blade.

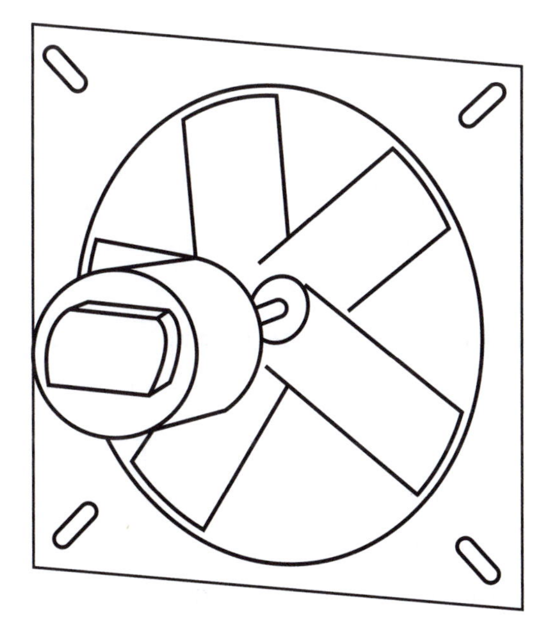

9.2 Propeller fan.

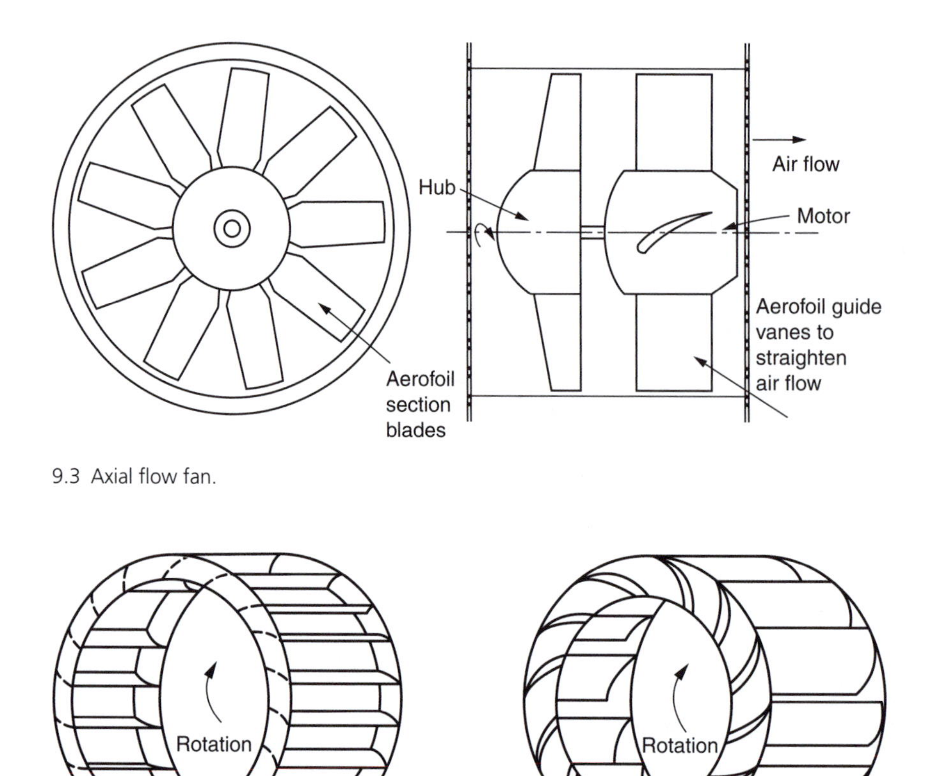

9.3 Axial flow fan.

9.4 Forward and backward curved centrifugal fan impeller.

Forward curved centrifugal

These are used in: packaged fan coil, air handling and air conditioning units; roof mounted exhausts; low cost ducted supply and exhaust ventilation, air conditioning; see Figure 9.4.

Backward curved centrifugal

These are used in large duty ventilation and air conditioning with high frictional resistance and air flow; see Figure 9.4.

Mixed flow

These are used where the fan casing is to be in line with the ductwork but a higher fan pressure than can be produced by an axial fan is required; see Figure 9.5.

Centrifugal fans are installed as part of packaged ventilation and air conditioning equipment such as room air conditioners, heating and cooling fan coil units, prefabricated air handling units or outdoor dry and evaporative heat exchangers. Where a high pressure and air flow is needed in extensive ducted

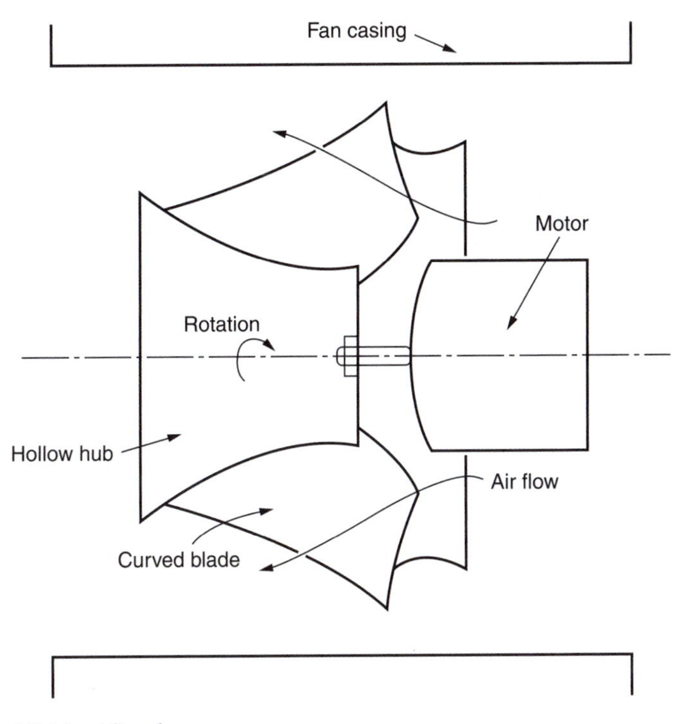

9.5 Mixed flow fan.

systems, the centrifugal fan is usually installed free-standing within a plant room. In large air flow applications the fan plant room also serves as a path for the air flow. In this case, the fan plant room becomes a plenum chamber that is maintained at a different static pressure from that of the atmosphere. A fan plenum chamber may be constructed from concrete and brickwork as part of the building structure. The building materials that are used, the quality of the surface finishes and airtightness of the plenum are carefully controlled to ensure reliable long term performance and low maintenance requirements. Large builders' work plenums and air ducts for filters and coils have been used for television studios and swimming pool ventilation plants. The centrifugal fan, motor and belt drive for metal ducted systems may be installed within an acoustic plenum. The inlet and outlet air ducts in the plenum are fitted with noise attenuators. Manual access into a fan plenum is usually barred while the fan is in operation. The plant control system will include the facility of switching the fan, or fans, off while the plant operator and the maintenance engineer are still outside the plenum. An airtight access door is provided into a fan plenum. The door is designed to maintain the desired attenuation, air sealing and a safe means of access.

When manual access is allowed into a fan plenum while the fan is running, safeguards are required. These include interior illumination, an interior light switch, emergency and exit lighting, warning notices, ear defenders, eye protection, protective overalls, covers to all rotating components, a local fan motor isolating switch, a fan motor interlock contactor to switch the fan off if the plenum door remains open for, say, 30 seconds and a limit on the manual force that is required to open, or close, the door against the air pressure difference. A plenum door that is being sucked shut by a negative air static pressure within the enclosure, can require a significant opening force, will not remain naturally open while the maintenance person is carrying tools, replacement drive belts and air filter panels through the doorway, can slam shut and entrap the person, or parts of the person, as well as being difficult to open from the inside. There is a danger that loose items will be sucked into the fan, drive belt, air filter or coil during maintenance work

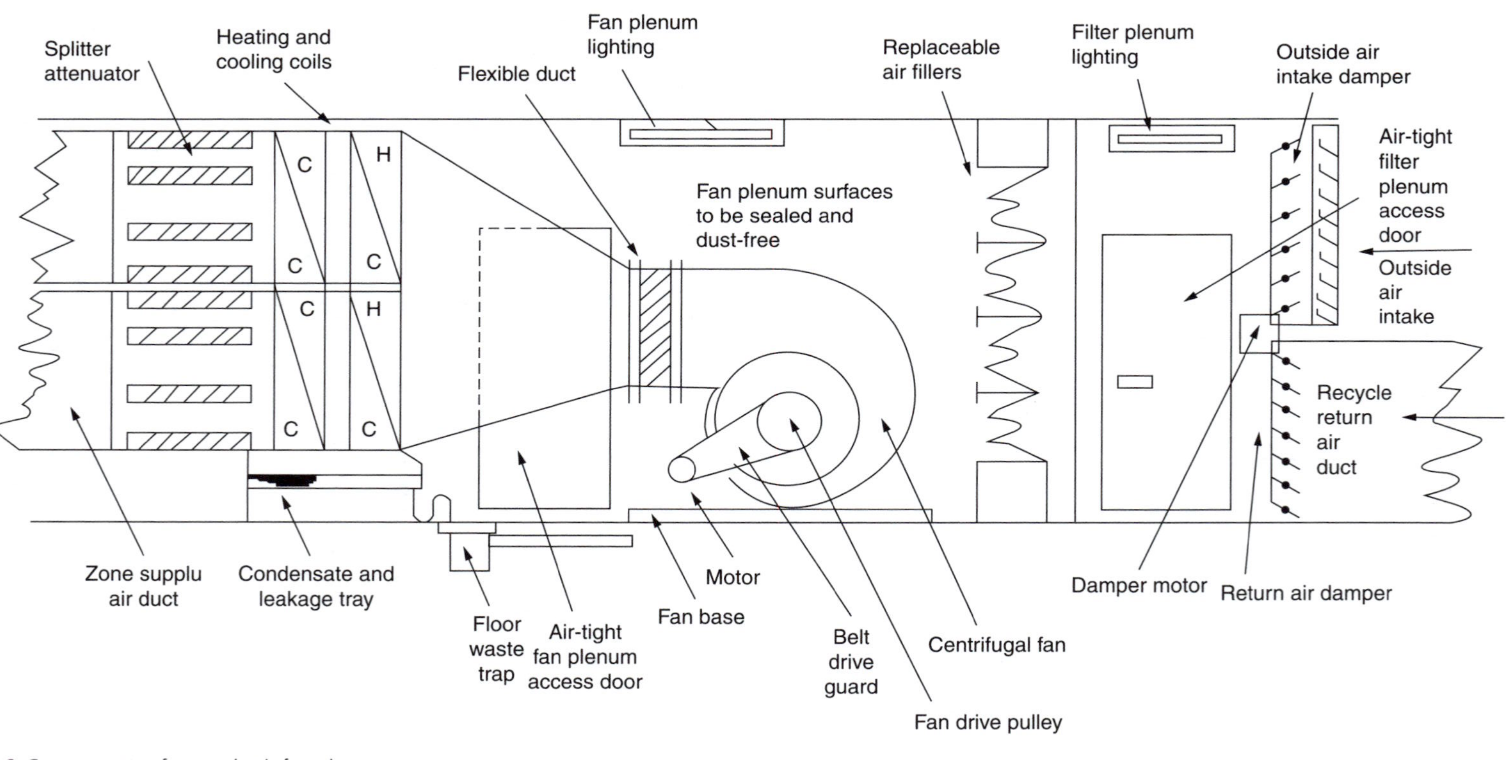

9.6 Components of a supply air fan plenum.

or inspections. Good, safe practice is to only allow access to pressurised plant rooms and plenums when the plant can be switched off. Figure 9.6 shows the components of a fan plenum.

EXAMPLE 9.1

A backward curved centrifugal fan maintains a static air pressure of 100 Pa below that of the atmosphere, within the fan acoustic plenum. An access door of 750 mm width and 2000 mm height allows entry for regular maintenance of the air filters and drive belt. The door is hinged along one vertical side and has a handle on the opposite side. Calculate the manual force that is necessary to open the door.

Air pressure difference creates a force on the door.

$$\text{Force } F_1 \text{ N} = \text{pressure } p \text{ Pa} \times A \text{ m}^2$$

$$A = 0.75 \text{ m} \times 2 \text{ m}$$

$$A = 1.5 \text{ m}^2$$

$$p = 100 \text{ Pa}$$

$$1 \text{ Pa} = 1 \frac{\text{N}}{\text{m}^2}$$

$$F_1 = p \frac{\text{N}}{\text{m}^2} \times A \text{ m}^2$$

$$F_1 = 100 \frac{\text{N}}{\text{m}^2} \times 1.5 \text{ m}^2$$

$$F_1 = 150 \text{ N}$$

A force F_1 on the door that is created by the difference of air pressure acts equally over the surface area of the door. F_1 will be considered to act at the centre of the door. When an opening force F_2 is applied to the door handle, the person is applying an opening torque, T_o Nm that acts on the door hinge. This opening torque is:

$$\text{Opening torque} = \text{manual force } F_2 \text{ N} \times \text{distance } L_2 \text{ m}$$

$$L_2 = 0.75 \text{ m}$$

The opening torque must equal the closing torque, overcome the frictional resistance of the hinge, move the door away from its perimeter airtight seals and move the door against the wind induced drag force. The simple closing torque represents the minimum torque that is needed to open the door.

$$\text{Closing torque } T_c = \text{closing force } F_1 \text{ N} \times \text{distance } L_1 \text{ m}$$

$$T_c = T_o$$

$$L_1 = 0.375 \text{ m}$$

$$T_c = 150 \text{ N} \times 0.375 \text{ m}$$

$$T_C = 56.25 \text{ Nm}$$

$$T_O = F_2 \text{N} \times 0.75 \text{ m}$$

$$56.25 \text{ Nm} = F_2 \text{ N} \times 0.75 \text{ m}$$

$$F_2 = \frac{56.25 \text{ Nm}}{0.75 \text{ m}}$$

$$F_2 = 75 \text{ N}$$

$$F_2 = 75 \text{ N} \times \frac{1 \text{ kg}}{9.81 \text{ N}}$$

$$F_2 = 7.65 \text{ kg}$$

This is equivalent to the force required to lift an 8 litre bucket that is full of water, but pulling horizontally, while the inward rush of air is sucking the door shut, the engineer is attempting to hold the door open, switch on the lights, carry equipment into the plenum and, once inside, hold the door ajar to stop it slamming shut and trapping a hand, foot or filter frame. The designer is obligated to avoid such unsafe practice.

Fan characteristics

Characteristic performance curves are used in graphical form as the result of tests on geometrically similar fans. The performance of a fan is specified by the pressure rise that it generates at the air volume flow rate that is passing through it. Typical curves of fan total, static and velocity pressures, fan motor power and a duct system resistance, for propeller and axial flow fans are shown in Figures 9.7 and 9.8.

The downloadable file *axial.xls* is used for these propeller and axial fan curves.

Figures 9.9–9.11 show typical fan and system performance curves for forward curved centrifugal, backward curved centrifugal and mixed flow fans. The file *prop.xls* is for the propeller fan, *forcent.xls* is for the forward curved centrifugal fan, *backcent.xls* is for the backward curved centrifugal fan and *mixed.xls* is

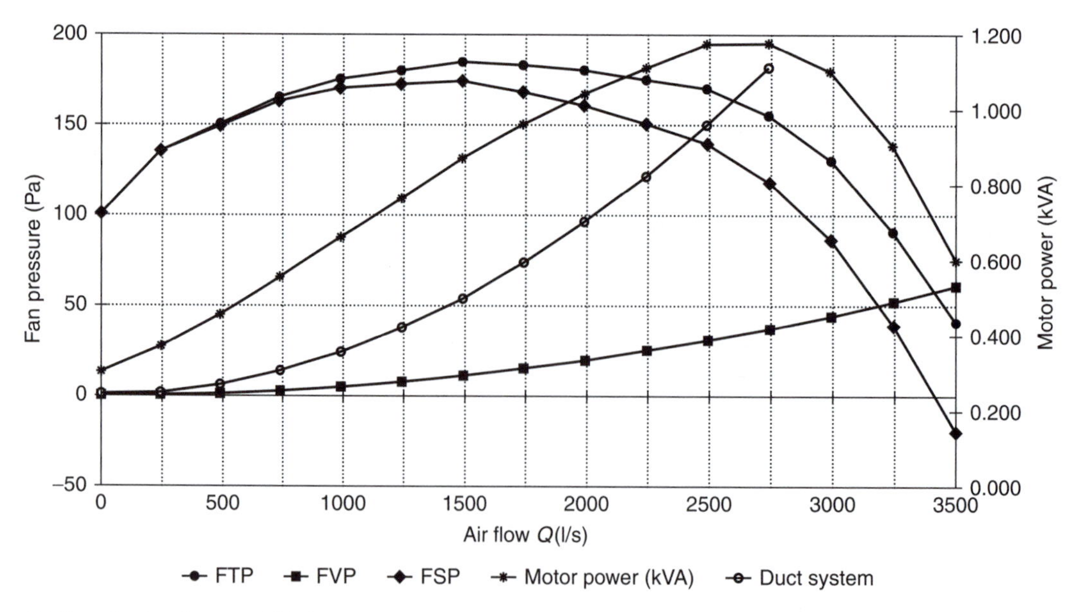

9.7 Propeller fan, fan speed 24 Hz.

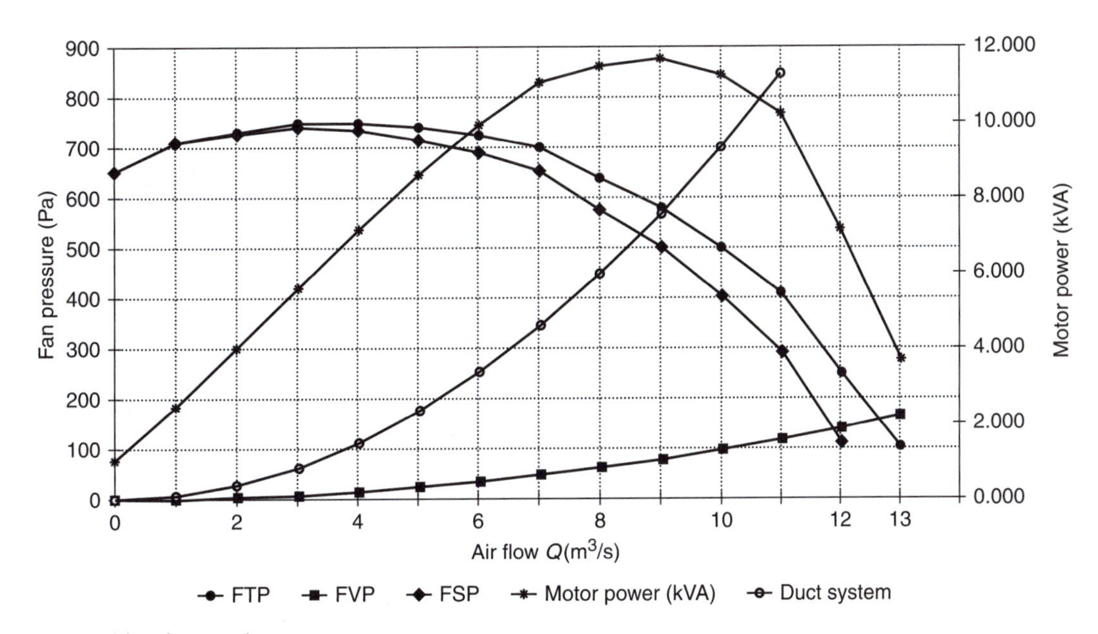

9.8 Axial fan, fan speed 24 Hz.

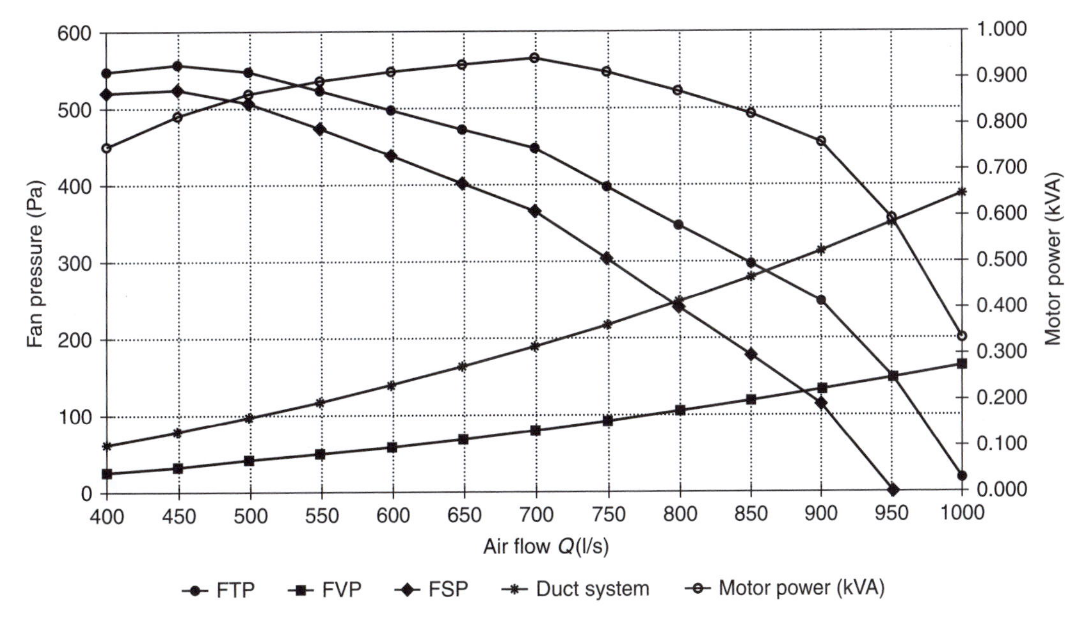

9.9 Forward curved centrifugal, fan speed 12 Hz.

for the mixed flow fan. Each of these files can then be edited with the data for new fan types for examples and questions within this book, assignments or design office work.

Notice on the charts that when the fan total pressure coincides with the fan velocity pressure, the fan static pressure is zero. This point defines the end of the fan performance curve. All the fan pressure development is being used to generate kinetic energy and there is no pressure available to overcome duct resistance. The commencement of the fan total pressure curve is not so easy to define. When a fan is

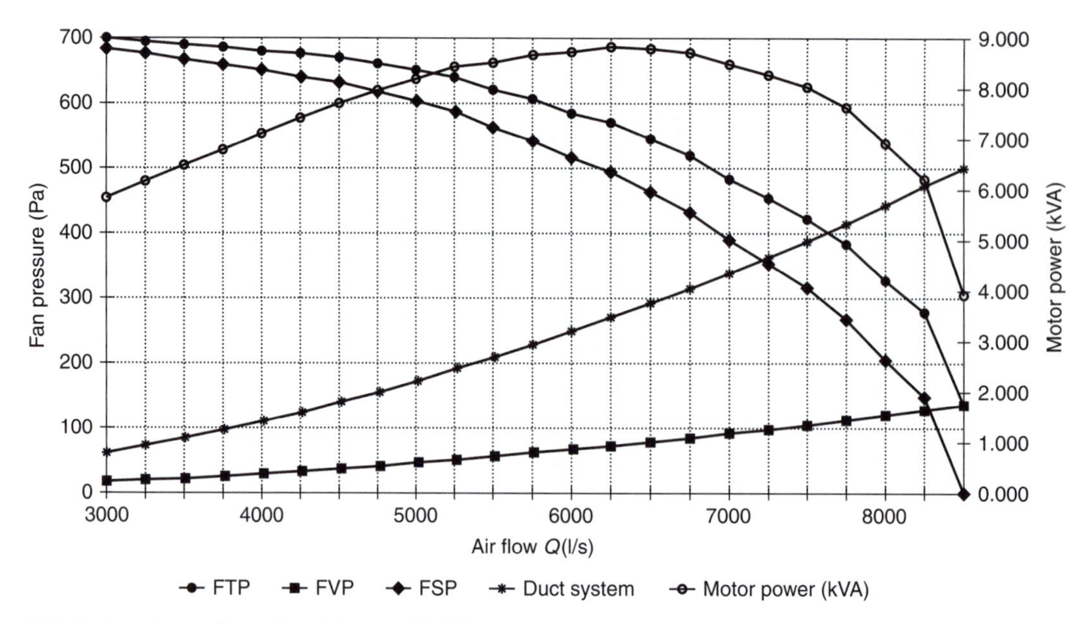

9.10 Backward curved centrifugal, fan speed 16 Hz.

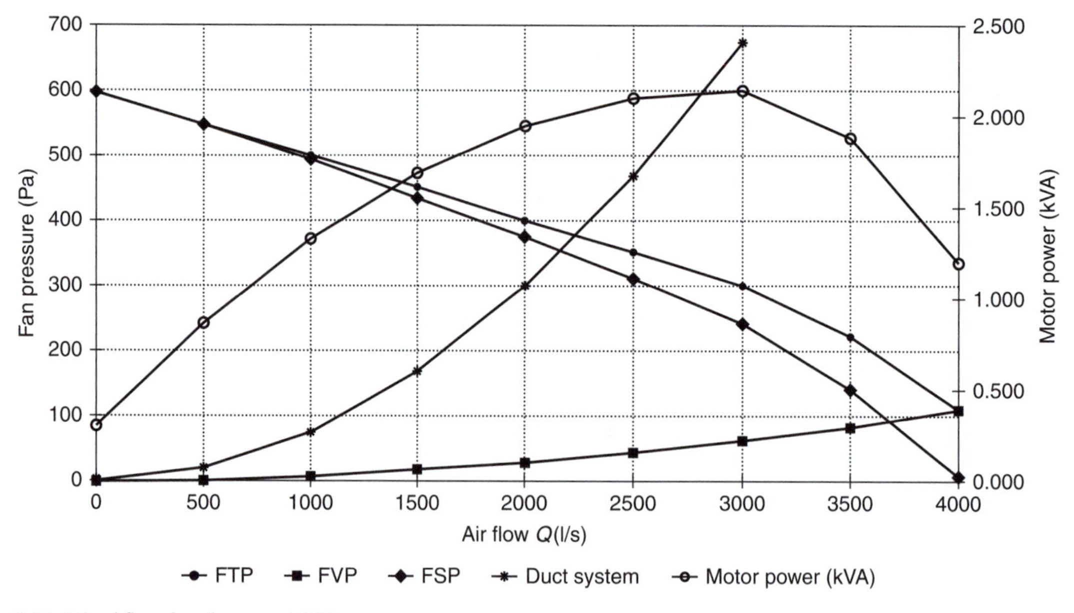

9.11 Mixed flow fan, fan speed 16 Hz.

operating against a closed and airtight dampened duct, there is no air flow delivered by the fan. Air only recirculates within the fan casing. No power is being delivered to the air system as the product of Q l/s and FTP Pa is zero. The driving motor is consuming power to overcome the overhead losses, as described later. Heat that is generated by the electric motor when it is within the air stream in a direct drive arrangement, the air compression by the impeller and the friction within the shaft bearings, is not dissipated, and the temperature within the fan casing will rise. Operating a fan against a closed damper could be used as a means of controlling the air flow and static pressure within the duct system during start up for short time periods. It may be necessary to control static pressure within the ventilated room during the starting of high pressure supply and exhaust fans, for example, where room air contaminants must be contained within the ventilated spaces through the room air static pressures. However, for the cost of a modulating damper, duct air pressure detector, damper motor and automatic controller, an electrical soft starter or a variable speed drive provide a higher quality performance and energy savings, when lower speeds are used.

At zero fan air delivery the fan velocity pressure is zero while the fan total and static pressures coincide. Increasing the system air flow by opening the fan outlet damper or, more likely, the system balancing dampers, produces the general performance curves shown in Figure 9.9. The fan supply air flow region between zero and that at the peak of the fan total pressure curve is not normally used for fan selection. The impeller efficiency is low in this region and fan operation can be unstable. The central portion of the highest fan total pressure curve is used for the selection of a fan to meet the duct system design. The fan motor power curve is termed as being non-overloading when peak power consumption is reached within the operational range of system air flow, as shown in Figure 9.9. An overloading fan motor power characteristic is where the power continues to increase as air flow increases. When a fan has an overloading power curve it may be possible to overload the driving motor and cause the excess current circuit breakers or the motor thermal protection to operate.

Manufacturers predict the pressure output of a new size of fan from knowledge of the performance of geometrically similar models that have been tested. Curves of pressure against volume flow are scaled upwards and downwards depending upon the rotational speed of the new fan. New fan models are subsequently tested to confirm their predicted performance.

Selection of a suitable fan is made by plotting the ductwork system characteristic curve onto the fan performance curve. The intersection of these curves shows where the fan and system will operate together. A graphical enlargement of the cross-over region allows the designer to make any tolerance adjustments that are desired. The operating point is selected to maximise the fan impeller efficiency.

Curves of fan total pressure *FTP* against air volume flow rate *Q* l/s correspond to a smooth polynomial curve. *FTP* is the fan total pressure rise in Pascal and *Q* is the air volume flow rate delivered by the fan in litres per second. The use of mathematical models of fan characteristic curves is demonstrated on spreadsheets and charts. These allow fan speed to be changed. The system characteristic curve is used to locate the operating point of the installed fan and duct system.

The highest practical motor rotational speed on a 50 Hz electrical power supply is normally 2900 RPM. 50 Hz corresponds to 3000 RPM. The 60 Hz supply that is used in some countries, corresponds to 3600 RPM. There is always a slip between the alternating current frequency in the stator of the motor and the rotor shaft speed. Fans are either directly driven from the motor shaft or with a v-belt drive running on pulleys. A larger pulley on the fan shaft than on the motor shaft produces a reduction gearing effect. Fan rotational speeds are usually within the range of 350 to 1450 RPM in order to minimise the generation of noise and mechanical stress within the rotating components. Some manufacturers show curves for fan speeds up to 4000 RPM. These have a smaller diameter pulley on the fan shaft than on the motor shaft in order to raise the gear ratio. Up to 100 dB can be produced in the outlet duct at such speeds and attenuators may be required. The normal range of fan speeds is shown in Table 9.1.

Table 9.1 Fan speeds

Poles in motor	*Motor speed, Hz*	*Motor speed, RPM*
2	48	2880
4	24	1440
6	16	960
8	12	720

Fan data

Equations used to produce fan and duct system characteristic curves are:

Fan air flow $Q \frac{\text{l}}{\text{s}}$

Fan outlet area $A = \text{width } W \text{ mm} \times \text{height } H \text{ mm} \times \frac{1 \text{ m}^2}{10^6 \text{ mm}^2}$

Fan outlet area $A = \frac{WH}{10^6} \text{m}^2$

Fan outlet air velocity $v = \frac{\text{air flow } Q \frac{\text{l}}{\text{s}}}{A \text{ m}^2} \times \frac{1 \text{ m}^3}{10^3 \text{ l}}$

Fan outlet air velocity $v = \frac{Q}{A \times 10^3} \frac{\text{m}}{\text{s}}$

Fan velocity pressure $FVP = 0.5 \rho v^2$

$$FVP = 0.5 \times \rho \frac{\text{kg}}{\text{m}^3} \times \left(\frac{Q}{A \times 10^3} \frac{\text{m}}{\text{s}} \right)^2$$

$$FVP = 0.5 \times \rho \frac{\text{kg}}{\text{m}^3} \times \frac{Q^2}{A^2 \times 10^6} \frac{\text{m}^2}{\text{s}^2} \times \frac{1 \text{ Ns}^2}{1 \text{kg m}} \times \frac{1 \text{ Pa m}^2}{1 \text{ N}}$$

$$FVP = 0.5 \times \rho \times \frac{Q^2}{A^2 \times 10^6} \text{ Pa}$$

Fan velocity pressure is calculated from the fan air velocity and the density of the air flowing through the discharge duct. The density of the air at this location depends upon the dry bulb temperature and barometric pressure of the air within the duct. The air is subject to frictional heating by the fan and any heat output from a close coupled electric driving motor. Belt drive motors are remote from the air within the duct, but will be within a fan plenum. Duct air static pressure should be added to the barometric air pressure in order to ascertain the density.

Air density $\rho = 1.2 \times \frac{p_a + p_s}{101325} \times \frac{273 + 20}{273 + t} \frac{\text{kg}}{\text{m}^3}$

p_a = barometric presure, Pa

p_s = static presure within the duct Pa, it can be positive or negative

Fan static pressure FSP = fan total presure FTP – fan velocity pressure FVP

$FSP = FTP - FVP$

Where H = manometer head mm H_2O

$p = 9.807H$ Pa

Ductwork system resistance varies with the square of the air flow quantity; the general form for this is:

$\frac{\Delta p}{Q^2}$ = constant for the duct system

$$\frac{\Delta p_1}{Q_1^2} = \frac{\Delta p_2}{Q_2^2}$$

Suffixes 1 and 2 refer to different flow rates. Once the design flow rate and pressure drop due to ductwork resistance is known, that is at condition 1, the expected system pressure drop at any other rate of flow, condition 2, can be calculated from the ratio.

$$\Delta p_2 = \frac{\Delta p_1}{Q_1^2} Q_2^2$$

$$\Delta p_2 = \Delta p_1 \frac{Q_2^2}{Q_1^2}$$

Fan law is used to predict performance of a geometrically similar fan that is to be operated at a different speed, with air of a different density or with an impeller of a different diameter are:

Air flow $Q = K \times \rho \times N \times D^3$

Pressure rise $\Delta p = K \times \rho \times N^2 \times D^2$

Fan air power $= K \times \rho \times N^3 \times D^5$

K = constant of proportionality, dimensionless

N = rotational speed of the impeller, Hz

D = impeller diameter, m

Fan performance at a second condition is found by rearrangement of the laws of proportionality, where the known data are:

$$Q_1 = K \times \rho_1 \times N_1 \times D_1^3$$

$$K = \frac{Q_1}{\rho_1 \times N_1 \times D_1^3}$$

Required data are:

$$Q_2 = K \times \rho_2 \times N_2 \times D_2^3$$

The same fan is being used for 1 and 2, so K does not change.

$$K = \frac{Q_2}{\rho_2 \times N_2 \times D_2^3}$$

$$\frac{Q_1}{\rho_1 \times N_1 \times D_1^3} = \frac{Q_2}{\rho_2 \times N_2 \times D_2^3}$$

Volume flow provided by a geometrically similar fan at the second condition is found:

$$Q_2 = \frac{Q_1}{\rho_1 \times N_1 \times D_1^3} \times \rho_2 \times N_2 \times D_2^3$$

$$Q_2 = Q_1 \times \frac{\rho_2}{\rho_1} \times \frac{N_2}{N_1} \times \frac{D_2^3}{D_1^3}$$

Similarly, the fan pressure rise at a geometrically similar fan for a different air density, speed and impeller diameter is:

$$\Delta p_2 = \Delta p_1 \times \frac{\rho_2}{\rho_1} \times \frac{N_2^2}{N_1^2} \times \frac{D_2^2}{D_1^2}$$

And the fan air power required is:

$$P_2 = P_1 \times \frac{\rho_2}{\rho_1} \times \frac{N_2^3}{N_1^3} \times \frac{D_2^5}{D_1^5}$$

EXAMPLE 9.2

The Clune fan company has designed a mixed flow fan to exhaust humid air at atmospheric pressure and a temperature of 30°C d.b. from swimming pool buildings. The prototype fan passed 500 l/s, had an impeller of 300 mm diameter, developed a fan static pressure of 100 Pa when running at 1440 revolutions per minute with a two-pole motor. The input electrical power to the motor of the fan tested was 110 watts when connected to a single-phase 240 volt alternating current supply. Predict, for the manufacturer, the performance of a 1000 mm diameter fan that will be run at 12 hertz when it is fitted with an eight-pole 415 volt motor.

The larger fan is to be used in the same atmospheric pressure and air temperature as the prototype. The air densities:

$$\rho_1 = \rho_2$$

$$\frac{\rho_2}{\rho_1} = 1$$

$$N_1 = 1440 \frac{\text{rev}}{\text{min}} \times \frac{1\ \text{min}}{60\ \text{s}} \times \frac{1\ \text{Hz s}}{1\ \text{rev}}$$

$$N_1 = 24\ \text{Hz}$$

$$N_2 = 12\ \text{Hz}$$

$$Q_2 = Q_1 \times \frac{N_2}{N_1} \times \frac{D_2^3}{D_1^3}$$

$$Q_2 = 500 \times \frac{12}{24} \times \frac{1000^3}{300^3} \frac{\text{l}}{\text{s}}$$

$$Q_2 = 9260 \frac{\text{l}}{\text{s}}$$

$$\Delta p_2 = \Delta p_1 \times \frac{N_2^2}{N_1^2} \times \frac{D_2^2}{D_1^2}$$

$$\Delta p_2 = 100 \times \frac{12^2}{24^2} \times \frac{1000^2}{300^2}\ \text{Pa}$$

$$\Delta p_2 = 278 \text{ Pa}$$

$$P_2 = P_1 \times \frac{N_2^3}{N_1^3} \times \frac{D_2^5}{D_1^5}$$

$$P_2 = 110 \times \frac{12^3}{24^3} \times \frac{1000^5}{300^5} \text{W}$$

$$P_2 = 5658 \text{ W}$$

Fan testing

A fan's performance is specified from:

1. air volume flow rate;
2. inlet air static pressure;
3. outlet air static pressure;
4. electrical power consumed at the fan shaft;
5. fan rotational speed;
6. air temperature handled by the fan;
7. density of the air handled by the fan.

Duct system characteristics

The frictional resistance of the air duct system is found from the D'Arcy equation:

$$H = \frac{4flv^2}{2dg} \text{ m air}$$

H = frictional resistance of air duct, m air

f = duct surface friction factor, dimensionless

l = duct length, m

v = air velocity, m/s

d = duct internal diameter, m

Air pressure rise at the fan is,

$$\Delta p = \rho g H \text{ Pa}$$

$$\Delta p = \rho g \frac{4flv^2}{2dg} \text{ Pa}$$

Δp = air pressure drop in the duct, Pa

ρ = air density, kg/m^3

Velocity of air that flows through a circular duct is found from:

$$Q = \frac{\pi d^2}{4} \text{m}^2 \times v \frac{\text{m}}{\text{s}}$$

$Q =$ air volume flow rate, m^3/s

$$v = \frac{4Q}{\pi d^2} \frac{m}{s}$$

System pressure drop is found from:

$$\Delta p = \rho g \times \frac{4fl}{2dg} \times \left(\frac{4Q}{\pi d^2}\right)^2 \text{ Pa}$$

$$\Delta p = \rho g \times \frac{4fl}{2dg} \times \frac{16Q^2}{\pi^2 d^4} \text{ Pa}$$

$$\Delta p = \rho \frac{4fl}{2d} \times \frac{16Q^2}{\pi^2 d^4} \text{ Pa}$$

$$\Delta p = \frac{32\rho fl}{\pi^2 d^5} Q^2 \text{ Pa}$$

For a particular air ductwork system, the friction factor f is a fixed quantity as all the ducts, their material and shape and fittings are built. The length l m and the diameter d m, of the constructed duct system are fixed. Density of the air ρ flowing through the duct system will only vary as the result of the actions of the automatic control system and outdoor air variations. The basic design for the air density ρ is fixed. Air flow rate through the duct system Q will vary in response to the action of volume control dampers, temperature controls and the gradual increase in frictional resistance of the air filter. For a specific installation the pressure drop through the duct system is characterised by:

$$\Delta p = KQ^2 \text{ Pa}$$

$$\text{Where, } K = \frac{32\rho fl}{\pi^2 d^5}$$

Fans in series

The installation of a supply air fan and an extract air fan for a space is an example of fans that are connected in series. For comfort air conditioning and ventilation systems, the occupied room is almost always maintained at, or very close to, the ambient atmospheric pressure so as not to cause noticeable air movement at doorways and at openable windows. Buildings can be maintained at a small positive pressure above that of the atmosphere so that there will be no ingress of unconditioned outdoor air. Kitchens, chemical, biological and nuclear radiation areas and toilet accommodation are maintained at a static air pressure that is slightly below the surrounding rooms so that vitiated air only flows outward through the exhaust system that is dedicated to that area.

Normally constructed buildings are not airtight and outward leakages through the porosity of building materials, cracks around window and door frames, as well as the designed air flow through door openings, are used to exhaust some of the outdoor air intake to positively pressurised areas. The static air pressure maintained within the room or building, is a balance between the residual supply air fan pressure, air leakage from the space and the suction pressure that is generated by the extract air fan. The residual static pressure within the space, or room, is calculated as if it were a section of the duct system.

Two or more axial flow fans may be connected in series to increase the fan static pressure available and to provide step control of the fan static pressure. This method of connection does not increase the air flow rate Q as both fans pass the same air quantity. Guide vanes are fitted between the fan impellers to remove

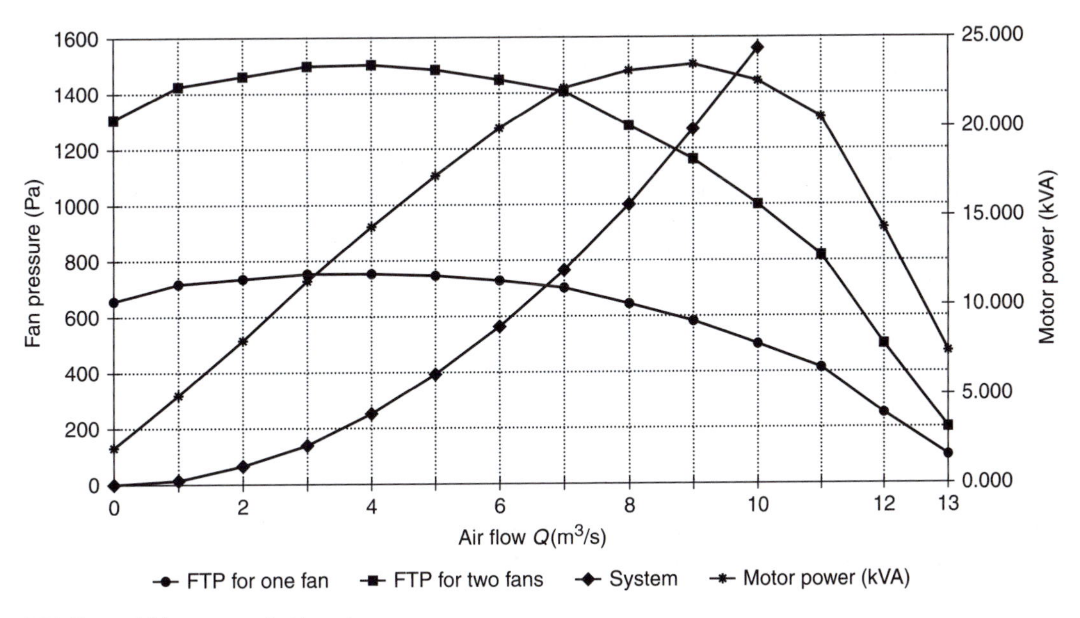

9.12 Two axial fans connected in series.

the swirl that is imparted to the air by each axial impeller. The overall performance characteristic for two axial flow fans that are connected in series is shown in Figure 9.12.

The cumulative curve is found by doubling the fan total pressure at each air flow rate. The fan velocity pressure curve remains the same as for one fan. The fan static pressure curve is found by subtracting the total and velocity pressures. The motor power consumed is twice that for the single fan. The sound power level of the combined identical fans is an increase of 3 dB over that for one fan.

Fan power

The input electrical power to be supplied to the motor is found from:

1. useful fluid power consumed;
2. hydraulic power loss at the impeller;
3. hydraulic power loss within the impeller casing;
4. power loss due to air leakage from the fan;
5. power loss in the fan shaft bearings;
6. power loss in the belt drive and pulleys;
7. loss of power within the electrical driving motor.

These losses of available power are classified into three groups; hydraulic, mechanical and electrical. Losses are usually expressed in percentage terms. The overall efficiency of the fan and drive system is found from the product of the hydraulic, mechanical and electrical efficiencies. The term for electrical efficiency is power factor.

$$\text{Power factor } PF = \frac{\text{motor useful power output kW}}{\text{apparent electrical input power kVA}}$$

The term kVA, kilo-volt-ampere, refers to the electrical input power to the motor. When the power factor is 1.0, kVA are equal to kW, but as with most motors, the useful output power of a fan motor, in watts, is less than the electrical input power, in kVA, that has to be paid for at the meter. Electrical motors typically have a power factor in the 0.5 to 0.7 range.

Such low power factors are corrected with capacitors that are connected in parallel with the motor. Capacitor power factor correctors store and release electrical charge during alternating currents and make it possible to achieve a power factor of 0.9 to 0.95. The original equipment designer and the subsequent energy auditor work to achieve the highest practical power factor, or overall electrical efficiency, at an affordable cost.

The electrical current per phase taken by the motor is:

$$\text{Current} = \frac{\text{volt ampere}}{\sqrt{3} \times \text{volt}}$$

The electrical services designer needs to know the full load phase running and starting currents. Input fan total power requirement of a fan and motor installation is found from:

$$\text{Input power } P = Q\frac{l}{s} \times FTP \text{ Pa} \times \frac{1}{PF} \times \frac{100}{Imp\%} \times \frac{100}{Motor\%} \times R$$

$$Imp\% = \text{fan impeller efficiency, } \%$$

$$Motor\% = \text{overall electrical motor efficiency, } \%$$

$$R = \text{overall motor drive power ratio}$$

$$\text{Power } P = Q\frac{\text{l}}{\text{s}} \times \frac{1\text{m}^3}{10^3\text{l}} \times FTP \text{ Pa} \times \frac{1\text{ N}}{\text{Pa m}^2} \times \frac{1}{\text{PF}} \times \frac{100}{Imp\%} \times \frac{100}{Motor\%} \times R$$

$$\text{Power } P = \frac{R \times Q \times FTP \times 100 \times 100}{10^3 \times PF \times Imp\% \times Motor\%} \frac{\text{Nm}}{\text{s}} \times \frac{1\text{Ws}}{1\text{ Nm}} \times \frac{1\text{ kW}}{10^3\text{ W}} \times \frac{1\text{ kVA}}{1\text{ kW}}$$

$$\text{Power } P = \frac{R \times Q \times FTP}{100 \times PF \times Imp\% \times Motor\%} \text{ kVA}$$

Motor input power does not drop to zero when the fan and motor are running against a closed duct damper. Although there is zero air flow through the fan, the motor is providing the energy to overcome the air friction of the moving components and the overhead losses in the drive system and the motor itself. Fan power load is found from tests that cover the range of air flows that the fan is to be used for. At zero air flow, the fan impeller efficiency will be low (60% or less), the drive system, motor efficiency and power factor will remain the same, or similar to those within the normal operating range. Motor power is consumed in generating the static pressure within the fan casing or plenum chamber, and in overcoming the aerodynamic forces on the rotating components. Duct systems are not always airtight and duct dampers do not seal completely. There may be a residual air flow of 5% or so of the design flow when all the dampers are closed. Leakage rate is tested during commissioning. Fan power at minimum air flow rate will vary from 20% to 70% of the peak power, depending upon the type of fan. The overall motor drive power ratio *R* allows for the overhead use of power at zero air flow by the fan.

EXAMPLE 9.3

A backward curved centrifugal fan delivers air at the rate of 5.6 m^3/s when the ductwork system resistance is 350 Pa. The 415 volt three-phase driving motor has a power factor of 0.7. The overall efficiency of the belt drive is 95%. The fan impeller has an efficiency of 85% at the design flow. The fan power at zero air flow, when it is only generating static pressure within its casing, is 40% of the design flow value. Calculate the motor power for the design and closed damper conditions.

At the design air flow:

Supply air flow *Q* 5600 l/s
Fan total pressure *FTP* 350 Pa
Power factor *PF* 0.7
Fan impeller efficiency *Imp*% 85%
Motor overall efficiency *Motor*% 95%
Motor power at zero air flow 40%, 0.4
Overall motor power ratio *R*,

$$R = 1.0 + 0.4$$

$$R = 1.4$$

$$\text{Power } P = \frac{R \times Q \times FTP}{100 \times PF \times Imp\% \times Motor\%} \text{kVA}$$

$$\text{Power } P = \frac{1.4 \times 5600 \times 350}{100 \times 0.7 \times 85 \times 95} \text{kVA}$$

$$\text{Power } P = 4.855 \text{kVA}$$

At the closed damper air flow:

$$\text{Power } P = 0.4 \times 4.855 \text{ kVA}$$

$$\text{Power } P = 1.942 \text{ kVA}$$

$$\text{Phase current} = \frac{\text{volt ampere}}{\sqrt{3} \times \text{volt}}$$

$$\text{Phase current} = \frac{1.942 \text{ kVA}}{\sqrt{3} \times 415 \text{ V}} \times \frac{10^3 \text{ VA}}{1 \text{ kVA}}$$

$$\text{Phase current} = 2.7 \text{ ampere}$$

Motor protection

Standard electric motors are designed to operate in ambient air temperatures of up to 50°C. Fan drive motors are protected with a bi-metallic thermal cut-out or a thermistor temperature sensor which is embedded in the motor windings. This thermal contact opens the motor contactor when an excessive temperature occurs in the windings. Such overheat protection can either be reset manually or through an auto-reset contactor once the windings have cooled. A manual reset is preferable so that the cause of the overheating can be investigated. A computer-based monitoring system will log such faults and generate an alarm signal.

Each fan drive motor is protected against excessive current by an overload circuit breaker or a high rupturing capacity fuse. In three-phase motors, unless all three fuses overheat and rupture simultaneously, the remaining windings can burn out during single phasing. A miniature circuit breaker (MCB) uses the heating effect of an excess current to open a bi-metallic strip and switch the current off. These can operate within a few seconds of the fault occurring.

Motors can be tropic-proofed by coating the windings with anti-fungal treatment. An explosion-proof motor is used when exhausting explosive mixtures of gases or dusts. The ingress protection standard IP54 is for electric motors in non-hazardous areas. IP54 provides complete protection against contact with live or moving parts inside the enclosure and water splashed against the machine will have no harmful effect. IP55 provides the additional protection against water projected by a nozzle from any direction. IP56 provides protection from sea water. IP65 provides protection against the ingress of dust and is for ignition proof electric motor installations.

Control

Methods of controlling the operation and performance of fans are:

Local isolator

This is a switch that is sited adjacent to the fan for disconnecting the electrical supply during servicing work.

Speed selector switch for multi-speed motors

An electric driving motor has pairs of poles that are wired to provide two to four possible fan speeds of 48, 24, 16 and 12 Hz – a low to moderate cost method of fan speed control.

Star/delta starter switch

Three-phase fan motors that can be manually started in star connection, which operates the motor at up to 70% of its full speed. While the motor is connected in star formation, the power consumption is half of its maximum. The motor can then be manually switched over to run in delta connected mode – low cost method of speed and power input control.

Star/delta start controller

This has the same operation and benefits as the star/delta switch but switching is achieved with an automatic controller.

Auto transformer control of the motor supply voltage

The motor is started at a low voltage through the transformer and then stepped up to the full 415 phase voltage in steps as the load on the fan increases; speed ratio reduction of up to 3:1 can be provided at moderate cost.

Triac semi-conductor motor speed controller

A triac soft starter is a solid state switch which alters the shape of the incoming alternating current sine wave to reduce the fan speed down to 10% of its full speed; power savings of up to 70% are achievable. A triac generates radio interference frequencies within the electrical system.

Frequency inverter, VFC, VSD

Called variable frequency control (VFC) or variable speed drive (VCD), this is the highest capital cost method of fan speed control. The fan driving motor is started at a low frequency and can be run up to full speed over any desired time period. Fan motor speed can be controlled at any time during the operation of the equipment. The control signal may come from an air pressure or temperature sensor in the supply air duct via the BMS or by manually programming the VSD controller. This is a very popular method of minimising energy use by fans and pumps in response to part load demands for heating and cooling in air conditioning systems. When retrofitting existing air conditioning systems, the financial payback from motor power, heating and cooling plant energy cost savings, can be within 2 years and in many cases a few months. It is the method used to provide variable air flows in response to temperature, economy, pressure or atmospheric contaminant sensors. It is used where static air pressure control is needed in critical facilities such as when dangerous substances are being contained within pressure controlled zones. The frequency inverter provides an output of 0–50 Hz as required. The inverter generates noise and has electrical losses of 4% or more of the input power. These losses generate heat output into the plant room. Each fan start takes place at the pre-set minimum frequency.

The performance of the ventilation or air conditioning system is controlled by either varying the temperature of the air at the heating and cooling coils, or by varying the volume flow rate of the air. Atmospheric contaminants, such as carbon monoxide or smoke density in a vehicle tunnel, can be the parameter that is used to generate the fan system control signal. Variable volume air conditioning systems use duct air static pressure to control the supply and recirculation fan speeds. Modulating air volume flow control dampers (VCD) are used to vary the proportion of outdoor air that is admitted into the occupied building, from say 10% to 100% of the supply air quantity, to provide an economy cycle of free cooling with outdoor air. Modulating air volume flow control dampers are located in the supply air ducts to variable air volume air conditioned rooms in response to comfort requirements.

Reducing downstream air flows has to be accompanied by the control of the fan output to avoid the generation of air noise at the dampers and unnecessary fan operating cost. Modulating dampers act to throttle the flow of air through the duct system. They increase the frictional resistance of the duct system and move the fan-system operating point along the fan performance curve to the controlled flow. The fan-system operating point will move away from the design efficiency and almost certainly result in lower fan efficiency. Increased noise generation at the fan is likely.

Methods of controlling the flow of air through a ventilation system are as follows:

1. Manually adjusted volume control dampers.
2. Automatically controlled volume control dampers.
3. Multi-step fan speeds.
4. Multiple fan installation.
5. Variable frequency control of fan speed.
6. Variable pitch angle axial flow fans.

Many comfort air conditioning systems and most ventilation systems operate with a constant fan speed throughout the year. This is the result of the designer's efforts to minimise the initial capital cost. The same applies to heating and cooling water circulation systems, with a constant pump speed through the year. The excess flow of water bypasses the terminal heating or cooling unit at a three-way diverting flow control valve and is harmlessly recycled. Exhaust ventilation systems from toilets and kitchens are operated from on/off switches at the design air flow irrespective of the imposed ventilation requirement. Running a fan or pump at its maximum design rotational speed and throttling the fluid flow in response to the comfort requirements, is the most expensive method of operating the plant. Rotodynamic machines consume less electrical power when they are run at reduced speed. Starting and operating fans and pumps at less than their maximum speed reduces wear on the shaft bearings and the belt drive, reduces noise and vibration, reduces the input electrical current, motor and cable temperatures, and generally prolongs the useful service period of the equipment. The largest air pressure drop in a ducted air conditioning system, 35 Pa to 150 Pa, often occurs through the air filter. Minimising the system air flow greatly reduces the filter resistance with a consequent saving in fan power and by extending the filter media use. Example 9.4 analyses the annual cost of the fans in an air conditioning system for the different methods of performance control.

EXAMPLE 9.4

A single duct air conditioning system has a design air flow of 8 m^3/s. The supply air duct system frictional resistance is 300 Pa. The supply air filter has a pressure drop of 50 Pa when clean and 175 Pa when it is ready to be changed. Use the average filter pressure drop for the analysis. The air conditioning system is operated for 260 hours per month. The return air ducted system has a design air flow of 8 m^3/s at a pressure drop of 150 Pa. The fan handles air at an average density of 1.2 kg/m^3 through the year. The fan impeller efficiency is 90%, motor efficiency is 60% and the motor power factor is 0.9. Assume that the overall efficiency remains constant at any air flow. The motor power used at zero supply air flow is 1.2 kW. The fan discharge duct is 1000 mm × 785 mm. Figure 9.13 shows the supply air fan performance curves. The building is in southern England where the load profile on the air conditioning system is indicated in Table 9.2. The supply air requirement for each month is the average for the month to match the thermal loads. A higher temperature differential between the supply and room air is maintained in winter and this reduces the supply air quantity. Free cooling is used in the mild weather. Electricity costs 9p/kVAh. Analyse the system running costs and potential cost savings from different methods of fan control. The file *vfc.xls* is provided for this analysis and other cases. Compare annual costs for:

No fan performance control;
Supply duct air volume control damper;
Two-speed fan motor producing 1400 and 700 RPM;
Variable frequency fan speed control.

Supply air duct system design pressure drop $\Delta p = (300 + 175)\,\text{Pa}$

Supply air duct system design pressure drop $\Delta p = 475$ Pa

Average system $\Delta p = (300 + 0.5 \times (50 + 175))\,\text{Pa}$

Average system $\Delta p = 413$ Pa

Design air flow $Q = 8 \dfrac{m^3}{s}$

$$\text{System characteristic } K = \frac{\Delta p_1}{Q_1^2} = \frac{413}{8^2} = \frac{\Delta p_2}{Q_2^2}$$

$$\Delta p_2 = 413 \times \frac{Q_2^2}{8^2}$$

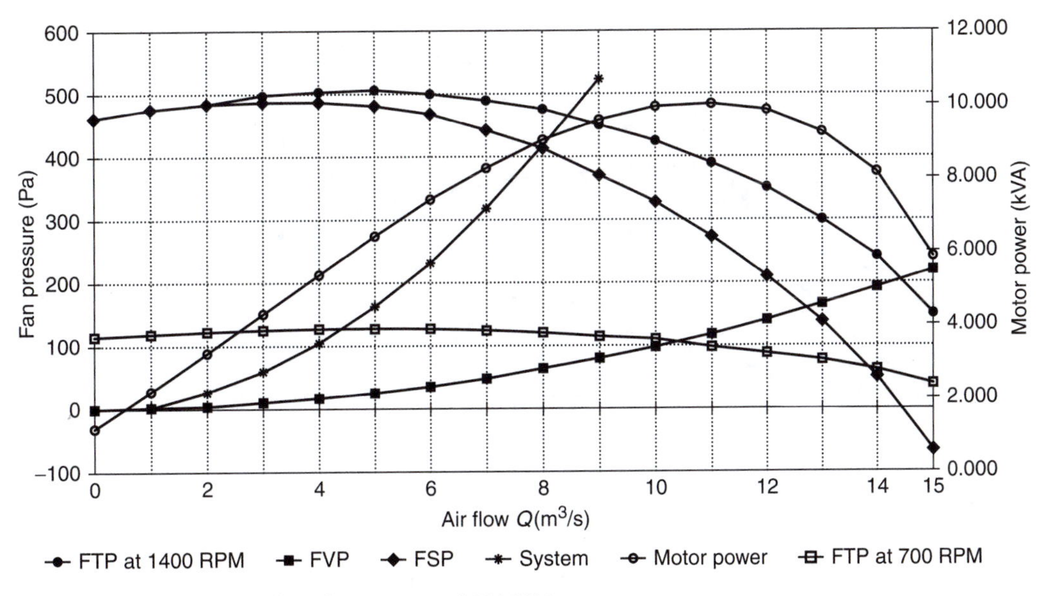

9.13 Fan example 9.4, centrifugal fan at 1400 and 700 RPM.

Table 9.2 Monthly supply air flow required

Month	*Heating/cooling load, %*	*Supply air required Q,* $\frac{m^3}{s}$
January	30	2.4
February	40	3.2
March	55	4.4
April	75	6.0
May	95	7.6
June	100	8.0
July	96	7.7
August	90	7.2
September	75	6.0
October	54	4.3
November	36	2.9
December	26	2.1

At an air flow of $4m^3/s$ the supply duct system average pressure drop is:

$$\Delta p_2 = 413 \times \frac{Q_2^2}{8^2}\,\text{Pa}$$

$$\Delta p_2 = 413 \times \frac{4^2}{8^2}\,\text{Pa}$$

$$\Delta p_2 = 103\ \text{Pa}$$

Data for the system resistance curve are shown in Table 9.3 and in Figures 9.13 and 9.14.

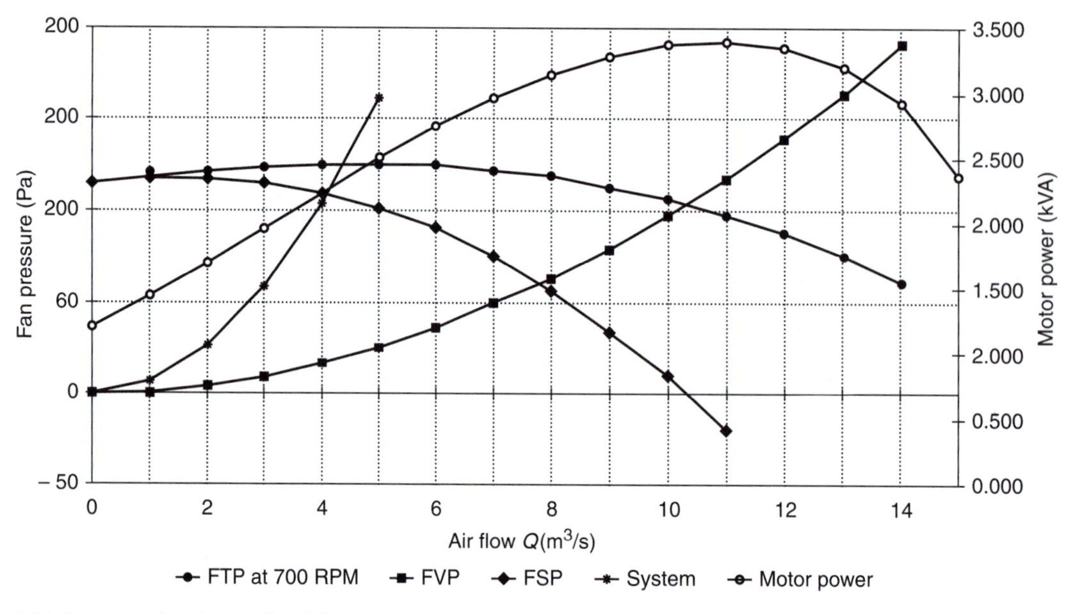

9.14 Fan example 9.4, centrifugal fan at 700 RPM.

Table 9.3 Duct system resistance data for example 9.4

Air flow Q $\frac{m^3}{s}$	*System resistance* Δp, *Pa*
0	0
1	6
2	26
3	58
4	103
5	161
6	232
7	316
8	413
9	523
10	645

System operating point, for the average filter pressure drop, is at a flow of 8.5 m^3/s and a fan total pressure of 465 Pa.

$$FTP = 465 \text{ Pa}$$

$$Q = 8.5 \frac{\text{m}^3}{\text{s}}$$

$$Q = 8500 \frac{\text{l}}{\text{s}}$$

$$\text{Fan discharge air velocity } v = Q\frac{\text{m}^3}{\text{s}} \times \frac{1}{\text{A m}^2}$$

$$\text{Fan discharge air velocity } v = 8.5\frac{\text{m}^3}{\text{s}} \times \frac{1}{(1\text{ m} \times 0.785\text{ m})}$$

$$\text{Fan discharge air velocity } v = 10.8\frac{\text{m}}{\text{s}}$$

$$FVP = 0.5 \times \rho \times \left(\frac{Q}{\text{area}}\right)^2$$

$$FVP = 0.5 \times 1.2 \times \left(\frac{8500}{10^3 \times 1 \times 0.785}\right)^2 \text{ Pa}$$

$$FVP = 70\text{ Pa}$$

$$PF = 0.9$$

$$Imp\% = 90\%$$

$$Motor\% = 60\%$$

$$FSP = FTP - FVP$$

$$FSP = (465 - 70)\text{ Pa}$$

$$FSP = 395\text{ Pa}$$

$$\text{Power } P = \left(\frac{Q \times FTP}{100 \times PF \times Imp\% \times Motor\%} + 1.2\right)\text{kVA}$$

$$\text{Power } P = \left(\frac{8500 \times 465}{100 \times 0.9 \times 90 \times 60} + 1.2\right)\text{kVA}$$

$$\text{Power } P = 9.333\text{ kVA}$$

The data and formulae for this example are in the downloadable file *vfc*. Each control method is analysed in the following sections.

No control of the supply fan output

The fan is operated at a constant speed of 1400 RPM. This is the constant volume, variable temperature air conditioning system that is frequently used. The system operating point, for the average filter pressure drop, is at a flow of 8.5 m^3/s and a fan total pressure of 465 Pa. Motor power input averages 9.333 kVA. Monthly fan operating energy use:

$$\text{Monthly fan energy} = 9.333\text{ kVA} \times 260\frac{\text{h}}{\text{month}}$$

$$\text{Monthly fan energy} = 2426.6\frac{\text{kVAh}}{\text{month}}$$

$$\text{Annual energy} = 2426.6\frac{\text{kVAh}}{\text{month}} \times 12\frac{\text{month}}{\text{yr}}$$

$$\text{Annual energy} = 29119\text{ kVAh}$$

Table 9.4 Monthly fan data for VAV system

Month	$Q\frac{l}{s}$	*FTP, Pa*	*FVP, Pa*	*FSP, Pa*	*kVA*	*kVAh*
January	2400	490	5	485	3.62	941.1
February	3200	500	10	490	4.49	1168.0
March	4400	510	18	492	5.82	1512.5
April	6000	510	34	476	7.5	1949.0
May	7600	485	54	431	8.78	2283.9
June	8000	475	60	415	9.0	2344.9
July	7700	485	56	429	8.88	2309.9
August	7200	480	49	431	8.31	2160.9
September	6000	510	34	476	7.5	1949.0
October	4300	510	17	493	5.71	1485.2
November	2900	495	8	487	4.15	1080.0
December	2100	485	4	481	3.3	856.9
						total: 20041.3

$$\text{Annual energy cost} = 29119 \text{ kVAh} \times \frac{9\text{p}}{\text{kVAh}} \times \frac{£1}{100\text{p}}$$

$$\text{Annual energy cost} = £2621$$

Supply duct air volume control damper

This is a variable air volume (VAV) air conditioning system. Modulating dampers are located in the terminal units. A supply duct damper controls the system supply air flow through the year. The fan runs at a constant 1400 RPM and the dampers absorb the excess fan static pressure to provide the building with the required air flow. The fan total pressures are read from Figure 9.14 and listed in Table 9.4. Fan velocity pressure, fan static pressure, motor power and the monthly electrical energy consumption are calculated as previously and are shown in Table 9.4.

$$\text{Annual electrical energy} = 20041.3 \text{ kVAh}$$

$$\text{Annual energy cost} = 20041.3 \text{ kVAh} \times \frac{9\text{ p}}{\text{kVAh}} \times \frac{£1}{100\text{ p}}$$

$$\text{Annual energy cost} = £1804$$

$$\text{Annual energy cost saving} = £2621 - £1804$$

$$\text{Annual energy cost saving} = £817$$

Two-speed fan motor

A two-speed fan motor is used to lower the electricity consumption when lower flow rates are needed. The fan total pressure that will be produced by the supply fan at the lower speed is:

$$\Delta p_2 = \Delta p_1 \times \frac{N_2^2}{N_1^2}$$

In the case, $\Delta p = FTP$ Pa

$$FTP_2 = FTP_1 \times \frac{N_2^2}{N_1^2}$$

From the fan characteristic curve, at an air flow rate of 4 m^3/s the FTP at 1400 RPM is 503 Pa. When the fan speed is reduced to 700 RPM, the FTP will be:

$$FTP_2 = FTP_1 \times \frac{N_2^2}{N_1^2}$$

$$FTP_2 = 503 \times \frac{700^2}{1400^2}\,\text{Pa}$$

$$FTP_2 = 126\ \text{Pa}$$

$$FVP_2 = 0.5 \times 1.2 \times \left(\frac{4000}{10^3 \times 1 \times 0.785}\right)^2\ \text{Pa}$$

$$FVP_2 = 16\ \text{Pa}$$

$$FSP_2 = (126 - 16)\,\text{Pa}$$

$$FSP_2 = 110\ \text{Pa}$$

$$\text{Power } P_2 = \left(\frac{4000 \times 126}{100 \times 0.9 \times 90 \times 60} + 1.2\right)\text{kVA}$$

$$\text{Power } P_2 = 2.237\ \text{kVA}$$

The calculated data for a fan speed of 700 RPM are shown in Table 9.5. Suffix 1 is for 1400 RPM and suffix 2 is for 700 RPM fan data. Intervals of 1 m^3/s have been selected. FTP_1 has been read from Figure 9.13. FTP_2, FVP_2 and P_2 have been calculated as shown above.

Fan and system curves for the two fan speeds of 700 RPM and 1400 RPM are shown in Table 9.6 and plotted in Figure 9.14. The fan total pressure curve for the 700 RPM is also shown on Figure 9.13. The lower fan speed is used up to the intersection of the duct system curve and the 700 RPM fan curve; this occurs at a flow of 4.5 m^3/s. When the air conditioning system requires an air flow that exceeds 4.5 m^3/s, the higher fan speed must be selected. FTP_2 is read from the characteristic curves. Motor power P_2 and the monthly energy consumed are calculated as previously.

Annual electrical energy 16106.4 kVAh.

Annual electrical energy cost:

$$\text{Annual energy cost} = 16106.4\ \text{kVAh} \times \frac{9\ \text{p}}{\text{kVAh}} \times \frac{£1}{100\ \text{p}}$$

$$\text{Annual energy cost} = £1449.58$$

$$\text{Annual energy cost saving} = £2621 - £1450$$

$$\text{Annual energy cost saving} = £1171$$

Table 9.5 Data for supply fan at 700 RPM

$Q \frac{l}{s}$	FTP_1, Pa	FTP_2, Pa	FVP_2, Pa	FSP_2, Pa	P_2, kVA
0	460	115	0	115	1.2
1000	475	119	1	118	1.445
2000	485	121	4	117	1.7
3000	495	124	9	115	1.964
4000	502	126	16	110	2.233
5000	505	126	24	102	2.5
6000	502	126	35	90	2.75
7000	490	123	48	75	2.96
8000	475	119	62	56	3.16
9000	450	113	79	34	3.28
10000	425	106	97	9	3.39
11000	390	98	118	−20	3.41
12000	350	88	140	—	3.36
13000	300	75	164	—	3.21
14000	240	60	191	—	2.93
15000	150	38	219	—	2.36

Table 9.6 Monthly fan data for two fan speeds

Month	$Q \frac{l}{s}$	Fan speed	FTP_2, Pa	P_2, kVA	kVAh
January	2400	700	123	1.807	470.0
February	3200	700	125	2.023	526.0
March	4400	700	126	2.341	608.6
April	6000	1400	502	7.4	1923.4
May	7600	1400	485	8.784	2284.0
June	8000	1400	475	9.02	2344.9
July	7700	1400	485	8.884	2309.9
August	7200	1400	480	8.31	2160.9
September	6000	1400	502	7.4	1923.4
October	4300	700	126	2.315	601.9
November	2900	700	124	1.94	504.4
December	2100	700	122	1.727	449.0
					total: 16106.4

Variable frequency control

A variable frequency control (VFC) generates the correct fan speed to supply the air flow that is required by the system. Download and use file *vfc.xls*. The fan total pressure is the same as the system air pressure drop, as dampers are not used to absorb excessive fan pressure as when fixed speeds are used. FTP_2 can be read from Figure 9.13, or more accurately, calculated from:

$$FTP_2 = 413 \times \frac{Q_2^2}{8000^2}$$

Fan total pressure at the system air flow of 8000 l/s is 413 Pa. Fan data are calculated as previously and shown in Table 9.7.

Table 9.7 Monthly fan data for variable fan speed

Month	$Q \frac{l}{s}$	FTP_2, *Pa*	P_2, *kVA*	*kVAh*
January	2400	37	1.38	359.5
February	3200	66	1.64	425.0
March	4400	125	2.33	606.2
April	6000	232	4.06	1056.7
May	7600	373	7.03	1828.6
June	8000	413	8.0	2079.6
July	7700	383	7.27	1889.7
August	7200	335	6.16	1602.4
September	6000	232	4.06	1057.7
October	4300	119	2.25	585.7
November	2900	54	1.52	395.8
December	2100	29	1.33	344.6
				total: 12231.5

Annual electrical energy 12231.5 kVAh.
Annual electrical energy cost:

$$\text{Annual energy cost} = 12231.5\text{kVAh} \times \frac{9\text{ p}}{\text{kVAh}} \times \frac{£1}{100\text{ p}}$$

$$\text{Annual energy cost} = £1100.84$$

$$\text{Annual energy cost saving} = £2621 - £1100$$

$$\text{Annual energy cost saving} = £1520$$

Summary

Annual cost saving for the different methods:

Damper volume control	£817
Two speed fan motor	£1171
Variable frequency fan speed control	£1520

A 10 kW fan motor represents a small power demand and the purchase cost of a 10 kW VFC unit is around £1000 producing a return of its capital cost in months. Benefits of the VFC include reduced stress on belt drive and bearings, continuing energy savings and controllability functions. VFCs are often recommended as an energy saving retrofit measure from energy audits.

EXAMPLE 9.5

A centrifugal supply air fan is being run at a speed of 1600 RPM to deliver the required system air flow of 1500 l/s. The fan total pressure rise at this air flow is 350 Pa. The inlet air to the fan does not come into contact with the heat emitted from the drive motor. The overall fan efficiency is 55% and this includes the losses in the impeller and the fan casing. Air enters the fan at 20°C d.b. Estimate the temperature of the supply air as it leaves the fan and comment on the result.

The motor and drive belt remain outside the air stream, so the only source of heating of the supply air is the turbulence and friction that are generated within the fan casing and as air passes through the impeller. Fan shaft bearings should have a high efficiency and not be generating noticeable heat output.

$$Q = 1500 \text{l/s}$$

$$FTP = 350 \text{ Pa}$$

$$\text{Impeller efficiency} = 55\%$$

The impeller is providing useful work, so the loss of power from the impeller and the fan casing is:

$$\text{Overall loss} = (100 - \text{efficiency})\,\%$$

$$\text{Overall loss} = (100 - 55)\,\%$$

$$\text{Overall loss} = 45\%$$

The overall loss applies to the input power to the fan shaft.

$$\text{Fan shaft input power loss} = Q\frac{\text{m}^3}{\text{s}} \times FTP \times \frac{100}{loss\%}$$

$$\text{Fan shaft input power loss} = 1.5\frac{\text{m}^3}{\text{s}} \times 350 \text{ Pa} \times \frac{100}{45}$$

$$\text{Fan shaft input power loss} = 1166 \text{ W}$$

This power loss appears as heat energy that is carried away by the air stream through the fan and into the supply air duct. Heat loss from the casing of the fan into the plant room may be significant; assuming that is negligible, the heat balance to be used is:

$$\text{Heat output by fan} = \text{heat absorbed by air stream}$$

$$\rho = 1.2\frac{\text{kg}}{\text{m}^3}$$

$$SHC = 1.012\frac{\text{kJ}}{\text{kg K}}$$

$$\text{Air inlet to fan} = t_1^{\circ}\text{C} = 20^{\circ}\text{C}$$

$$\text{Air outlet from fan} = t_2^{\circ}\text{C}$$

$$1166 \text{ W} = 1.5\frac{\text{m}^3}{\text{s}} \times 1.2\frac{\text{kg}}{\text{m}^3} \times 1.012\frac{\text{kJ}}{\text{kg K}} \times (t_2 - t_1)\text{K} \times \frac{\text{kWs}}{\text{kJ}} \times \frac{10^3 \text{ W}}{\text{kW}}$$

$$t_2 - t_1 = \frac{1166}{1.5 \times 1.2 \times 1.012 \times 10^3} \text{ K}$$

$$t_2 - t_1 = 0.64 \text{ K}$$

$$t_2 = 20.64°\text{C}$$

The loss of energy between the shaft input power and the useful power absorbed by the air produces an increase in the temperature of the air flowing through the fan. This will occur in most cases and also in fluid pumps. Should the air flow be reduced to 10% of the current value, the air temperature increase would rise tenfold, to 6.4 K. The designer needs to take account of the heat release rate into the fluid flowing. During heating processes, the fan and pump fluid temperature increase is beneficial to the system. For cooling applications, a temperature increase at the fan or pump is undesirable.

Commissioning and maintenance

Commissioning a new fan installation and its routine maintenance, include:

1. correct direction and free rotation of the fan impeller;
2. securely connected electrical wiring;
3. treat corrosion of metalwork and wiring system;
4. ensure that air ductwork and the fan intake are clear of debris and are clean;
5. correct installation of local isolator, fuses and circuit breakers;
6. correct operation of switches, fan interlock contactors, fuse and circuit breakers; test safety devices;
7. check that running motor current does not exceed name plate rating;
8. check that all bolted and riveted fastenings are secure;
9. check that anti-vibration mountings are in place and are operational;
10. check that protective guards are in place for all rotating components and drive belts;
11. check that drive belts are correctly tensioned and pulleys are aligned;
12. check that flexible duct connections are functioning properly;
13. where the fan shaft bearings are fitted with grease nipples, use the recommended lubricant at a maximum interval of 12 months or before the working hours limit has been reached; most modern fans have ball bearings that are 'sealed for life'; the life expectancy of sealed bearings is 40000 hours; standard ventilation fans are usually lubricated with lithium based grease that is suitable for continuous operation at up to 130°C; fans for spilling smoke require a silicon based grease that is suitable for continuous operation at up to 200°C; clean the grease nipples before use; purge the old grease through the relief port with the incoming grease pressure; only use compatible types of grease in a bearing;
14. isolate the fan drive motor from the electricity supply before any work is conducted on the rotating equipment.

Questions

Support answers with downloaded examples of current equipment and, where possible, sketches and photographs of installations known to you. Always explain the engineering and give reasons for what is there.

1. State the functions of fans within building services.
2. Explain how fans create movement in air.

3. State the limiting factors that are included in the design of fan and duct systems.
4. List the components of fan and drive systems. Comment upon how they are manufactured and their relative cost to the user.
5. List the types of fan and their principal applications.
6. Discuss the advantages and disadvantages of installing a large duty centrifugal fan within a builders' work plenum when compared with a metal plenum.
7. List the safety features that are to be considered when designing a fan installation.
8. Explain how the maintenance access to fans and their associated items of plant, such as air filters, drive system and heat exchange devices, is provided during the design of the overall system and building. State the importance of such access, the likely frequency of maintenance work and the safety precautions that should be provided.
9. Discuss the problems that are associated with manual entry into a positively or negatively pressurised plenum or ventilated space.
10. Sketch the characteristic curves for a backward curved centrifugal fan and ductwork system to show the relationship between fan total, static and velocity pressures, motor power and the ductwork system resistance. Do not look at any published graphics while attempting this question.
11. List the 'overhead' power requirements of a fan and electric motor drive plant. State how these minimum power requirements affect the shape of the fan power characteristic curve.
12. State the problems that are associated with running a fan against closed duct dampers.
13. What problems may be created within the ventilated rooms when starting a centrifugal supply or extract fan, or a combination of both?
14. State the designer's objectives when selecting the operating point on a fan characteristic graph.
15. Explain how fan manufacturers test fans and how they generate performance data for a range of fan sizes and rotational speeds.
16. State the electric motor speeds that are used and how these speeds are achieved.
17. Explain, with the aid of practical examples, how the rotational speed of a fan impeller is maintained at a different speed from that of the driving motor.
18. Explain what happens to the fan total pressure rise when the fan static pressure drops to zero.
19. Why can the duct system characteristic curve be calculated and drawn without reference to the duct system resistance data?
20. State the fan laws and show how they are used to generate predictions of fan performance.
21. State the data that are obtained from standard tests on fans.
22. Explain why and when fans would be connected in series with each other. Sketch the combined characteristics of two fans that are closely connected in series. Sketch the duct system resistance curve and identify the overall fan and system operational point.
23. Discuss how the equipment manufacturer and the building services designer obtain the maximum energy efficiency from a fan and drive system.
24. Explain why the fan motor power consumption does not diminish to zero when the duct system dampers are all closed.
25. Discuss the equipment that is needed to start, control and monitor the safe operation of the speed of electric drive motors for air conditioning system fans. State the economic and technical factors that are included in the decision as to which method is applied to an application.
26. Explain the characteristics of the different methods of controlling the flow of air through a ductwork system in relation to the energy cost of using the whole installation.
27. List the advantages that are gained by operating an air conditioning fan at the lowest possible rotational speed while meeting the required duty.
28. Design the layout for a commissioning task sheet for the fan and duct systems within an air conditioned building. The completed instructions will be implemented by a contract company as the result of

competitive tendering for the work. The document is to state the plant that is to be commissioned and precisely what work is to be conducted.

29. Design the layout for a maintenance manual for the air conditioned building in question 28. The completed instructions will be implemented by a contract company as a result of competitive tendering for the work. The document is to state the plant that is to be maintained, how it is to be started, stopped, shut down in an emergency and the timing for maintenance operations. State precisely what work is to be undertaken at the appropriate time intervals. The maintenance instructions must be easily understandable by the client's representative, who is not necessarily a building services engineer.
30. A forward curved centrifugal fan maintains a static air pressure of 75 Pa above that of the atmosphere within the fan acoustic plenum. An access door of 750 mm width and 2200 mm height allows entry for regular maintenance. The door is hinged along one vertical side and has a handle on the opposite side. Calculate the manual force that is necessary to open the door.
31. The supply air outlet duct from a packaged roof mounted air conditioning unit is 500 mm wide and 700 mm high. The supply air flow to the duct system is to be 2700 l/s of air at 33°C d.b. during winter use. Calculate the velocity pressure of the supply air as it leaves the packaged unit.
32. The Daylesford Impeller Company manufactures a range of centrifugal fans for air conditioning systems. A prototype fan was tested at an air flow rate of 1800 l/s with an impeller of 420 mm diameter producing a fan static pressure of 150 Pa when the impeller was running at 1200 RPM. The measured electrical input power to the motor of the fan tested was 800 watts. Predict the performance of a 700 mm diameter fan impeller that will be run at 16 hertz and is to be geometrically similar to the prototype.
33. An axial flow fan delivers air at the rate of 12 m^3/s when the ductwork system resistance is 120 Pa. The 415 volt three-phase driving motor has a power factor of 0.7. The overall efficiency of the belt drive is 96%. The fan impeller has an efficiency of 60% at the design flow. The fan power at zero air flow, when it is only generating static pressure within its casing, is 20% of the design flow value. Calculate the motor power for the design and closed damper conditions.
34. An axial flow fan runs at 940 RPM and has a blade angle of 28°. It passes air at a density of 1.21 kg/m^3 into a duct system that has a resistance of 100 Pa when the air flow is 5 m^3/s. The outlet air velocity from the fan is 11 l/s. The 415 volt three-phase drive motor has a power factor of 0.92. The motor and drive have an overall efficiency of 65% and the impeller has an efficiency of 80%. The minimum power consumption at zero air flow would be 0.25 kVA. The fan performance data are shown in Table 9.8. Calculate and plot the fan characteristic curves for fan velocity and static pressures, the duct system resistance and the motor input power. Use the file *axial.xls*. Find the fan and system operating conditions and the current that will be taken by the motor.

Table 9.8 Fan data for question 34

$Q\ \frac{m^3}{s}$	*FTP, Pa*
0	160
1	155
2	150
3	140
4	125
5	90
6	50
7	0

35. A mixed flow exhaust fan runs at 1450 RPM and handles air at a density of 1.2 kg/m^3. The fan outlet diameter is 500 mm. The duct system has a resistance of 600 Pa when the air flow is 1100 l/s. The outlet air velocity from the fan is 7.6 m/s. The 415 volt three-phase drive motor has a power factor of 0.7, the motor and drive have an overall efficiency of 70% and the impeller has an efficiency of 82%. The minimum power consumption at zero air flow would be 0.4 kVA. The fan performance data are shown in Table 9.9. Calculate and plot the fan characteristic curves for fan velocity and static pressures, the duct system resistance and the motor input power. Use file *axial.xls*. Find the fan and system operating conditions and the current that will be taken by the motor.

Table 9.9 Fan data for question 35

$Q \frac{l}{s}$	*FTP, Pa*
0	750
250	800
500	850
750	850
1000	780
1250	630
1500	450
1750	200

36. A belt driven centrifugal air conditioning supply fan runs at 700 RPM and handles air at a density of 1.18 kg/m^3. The fan outlet diameter is 750 mm. The duct system has a resistance of 500 Pa when the air flow is 4000 l/s. The outlet air velocity from the fan is 7.6 m/s. The 415 volt three-phase drive motor has a power factor of 0.92. The motor and drive have an overall efficiency of 75% and the impeller has an efficiency of 71%. The minimum power consumption at zero air flow would be 1.0 kVA. The fan performance data are shown in Table 9.10. Calculate and plot the fan characteristic curves for fan velocity and static pressures, the duct system resistance and the motor input power. Use file *axial.xls*. Find the fan and system operating conditions and the current that will be taken by the motor.

Table 9.10 Fan data for question 36

$Q \frac{l}{s}$	*FTP, Pa*
0	750
500	775
1000	780
1550	775
2000	765
2500	750
3000	700
3500	650
4000	576
4500	500
5000	360
5500	80

37. An air conditioning system has a design air flow of 3 m^3/s. The supply air duct system frictional resistance is 450 Pa. The supply air filter has a pressure drop of 40 Pa when clean and 180 Pa when

it is ready to be changed. Use the average filter pressure drop for the analysis. The air conditioning system is operated for 200 hours per month. The fan handles air at an average density of 1.2 kg/m^3 through the year. The fan impeller efficiency is 80%, motor efficiency is 66% and the motor power factor is 0.92. Assume that the overall efficiency remains constant at any air flow. The motor power used at zero supply air flow is 0.8 kW. The fan discharge duct is 700 mm × 600 mm. Table 9.11 shows the supply air fan performance data. The building is in southern England where the load profile on the air conditioning system is indicated in Table 9.12. The supply air requirement for each month is the average for the month to match the thermal loads. A higher temperature differential between the supply and room air is maintained in winter and this reduces the supply air quantity. Free cooling is used in the mild weather. Electricity costs 9 p/kVAh. The fan is operated at the constant speed of 960 RPM.

Table 9.11 Supply air fan performance data for question 37

Air flow Q $\frac{l}{s}$	*FTP, Pa*
0	800
500	810
1000	820
1500	820
2000	800
2500	750
3000	650
3500	500
4000	300
4500	50

Table 9.12 Monthly supply air flow required for question 37

Month	*Heating/cooling load %*	*Supply air required Q* $\frac{m^3}{s}$
January	35	1.05
February	42	1.26
March	56	1.68
April	76	2.28
May	97	2.91
June	100	3.0
July	97	2.91
August	90	2.7
September	77	2.31
October	57	1.71
November	40	1.2
December	31	0.93

Manual calculations and the file *vfc.xls* can be used to compare annual costs for the following:

(a) no fan performance control;
(b) supply duct air volume control damper;
(c) two-speed fan motor producing 960 and 550 RPM;
(d) variable frequency fan speed control.

10 Fluid flow

Learning objectives

Study of this chapter will enable the reader to:

1. calculate the flow rate of water, methane and air through pipes and ducts;
2. reproduce published pipe and duct size data;
3. find pipe and duct sizes to satisfy heating and cooling loads;
4. compare the heat carrying capacity of hot water pipe and air duct systems;
5. calculate the energy consumed by pump and fan heat distribution systems;
6. understand how thermal storage systems are used;
7. apply thermal storage to heating and cooling systems.

Key terms and concepts

Colebrook and White equation 305 ; fluid flow rate and pressure drop rate 305; fluid properties 310; pump and fan power 314; refrigeration 317; storage tank 316; thermal storage 316; turbulent noise 306.

Introduction

All air conditioning systems rely upon the flow of fluid through pipes and ducts. The derived equations used here for air flow in ducts, hot and chilled water and methane in pipes, have advantages over solely using charts and tables, in that the student's own calculator and computer can reproduce published data with acceptable accuracy.

Heat carrying capacities of low, medium and high temperature hot water pipe and air duct distributions are analysed and sizes compared. The performance of thermal storage arrangements is calculated. Pump and fan power consumption of distribution systems is calculated. All the stages of calculations are shown.

Pipe and duct equations

Sizes of pipes and ducts are found from charts and extensive tabulated data in the CIBSE Guide C (2007) and design engineers will have this within an arm's length at all times. Fluid flows are invariably fully developed

Table 10.1 Diameters of copper pipe

Nominal diameter, mm	*Internal diameter, m*
6.0	0.004 80
8.0	0.006 80
10.0	0.008 80
12.0	0.010 80
15.0	0.013 60
22.0	0.020 22
28.0	0.026 22
35.0	0.032 63
42.0	0.039 63
54.0	0.051 63

Data in Table 10.1 are approximate and for use within this book only

turbulent, as laminar flows would be at very low velocity, require large pipe diameters and produce no benefit. The formula describing such flows is the D'Arcy equation. The Colebrook and White equation is used to find the design friction factor for the D'Arcy formula. Insertion of the appropriate fluid properties and pipe material factors produces a useable equation that can acceptably replicate published data, and this is what we use here.

Water flow at 10°C

Mass flow can be found for water at 10°C in copper pipe:

$$m = -70.24 \Delta p^{0.5} d^{2.5} \log\left(\frac{4.05 \times 10^{-7}}{d} + \frac{7.33 \times 10^{-5}}{\Delta p^{0.5} d^{1.5}}\right) \frac{kg}{s}$$

$m =$ mass flow rate of water at 10°C, $\frac{\text{kg}}{\text{s}}$

$\Delta p =$ pressure drop per metre of pipe $\frac{\text{Pa}}{\text{m}}$

$d =$ pipe internal diameter, m

Internal diameters of pipes are given in Table 10.1.

The equation is solved by inserting known or estimated values for Δp and d, evaluating m and checking to see if the flow rate exceeds the design requirement. If not, a second estimate is tried. In the case of hot and cold water pipework for tap services, the pressure drop rate will be known from the available head and index circuit length, needing only diameter to be estimated.

EXAMPLE 10.1

Calculate the flow rate of water at 10°C that will flow through a 15.0 mm copper pipe when the frictional resistance is 1500 Pa/m.

For this pipe, d is 0.01360 m and Δp 1500 Pa/m

$$m = -70.24\Delta p^{0.5} d^{2.5} \log\left(\frac{4.05 \times 10^{-7}}{d} + \frac{7.33 \times 10^{-5}}{\Delta p^{0.5} d^{1.5}}\right) \frac{\text{kg}}{\text{s}}$$

First evaluate log bracket,

$$\log\left(\frac{4.05 \times 10^{-7}}{d} + \frac{7.33 \times 10^{-5}}{\Delta p^{0.5} d^{1.5}}\right) = \log\left(\frac{4.05 \times 10^{-7}}{0.0136} + \frac{7.33 \times 10^{-5}}{1500^{0.5} \times 0.0136^{1.5}}\right)$$

$$\log\left(\frac{4.05 \times 10^{-7}}{d} + \frac{7.33 \times 10^{-5}}{\Delta p^{0.5} d^{1.5}}\right) = -2.9125$$

$$m = -70.24\Delta p^{0.5} d^{2.5} \times (-2.9125) \frac{\text{kg}}{\text{s}}$$

$$m = -70.24 \times 1500^{0.5} \times 0.0136^{2.5} \times (-2.9125) \frac{\text{kg}}{\text{s}}$$

$$m = 0.171 \frac{\text{kg}}{\text{s}}$$

EXAMPLE 10.2

Find the pipe diameter required to carry 0.780 kg/s of water at 10°C if the available pressure loss rate is 1000 Pa/m in copper pipe.

Try a 28.0 mm pipe having d 0.02622 m, Δp 1000 Pa/m. From the flow equation, m is 0.803 kg/s. That is greater than the design requirement and is satisfactory.

Water flow at 75°C

For copper pipe carrying water at 75°C the flow equation becomes:

$$m = -69.36\Delta p^{0.5} d^{2.5} \log\left(\frac{4.05 \times 10^{-7}}{d} + \frac{2.15 \times 10^{-5}}{\Delta p^{0.5} d^{1.5}}\right) \frac{\text{kg}}{\text{s}}$$

EXAMPLE 10.3

Calculate the flow rate of water at 75.0°C that can be carried by a 42 mm copper pipe when the allowable pressure loss rate due to friction is 220 Pa/m.

Manual calculation shows d 0.03963m, Δp 220 Pa/m and flow rate capacity of m 1.194 kg/s.

For quiet operation of pipe work systems, water velocity should not exceed 1 m/s. Where turbulence generated noise can be tolerated, velocities of 3–6 m/s can be used. Such occasions will be where pipes are enclosed within concrete service ducts a long way from offices, or where the background noise level is high, such as in a factory. However, factory areas are increasingly high technology facilities and such a choice must be carefully considered against all the relevant facts.

High water velocity also means high pressure loss rate, large pump pressure rise and high pump running cost due to the power consumption of the losses of energy in the pipework. Conversely, high water velocity means that smaller pipe diameters can be used that reduce capital costs of the pipe, pipe fittings, valves, thermal insulation materials, service ducts and reduced heat loss through smaller pipe surface areas. The pipeline designer optimises all these factors in relation to the circumstances.

Water velocity can be calculated from the mass flow rate m kg/s, pipe internal diameter d m and water density ρ kg/m^3

$$\text{Density of water at } 10^\circ\text{C}, \rho = 999.73 \frac{\text{kg}}{\text{m}^3}$$

$$\text{Density of water at } 75^\circ\text{C}, \rho = 974.85 \frac{\text{kg}}{\text{m}^3}$$

$$Q = \text{fluid volume flow rate } \frac{\text{m}^3}{\text{s}}$$

$$v = \text{fluid velocity } \frac{\text{m}}{\text{s}}$$

$$\rho = \text{fluid density } \frac{\text{kg}}{\text{m}^3}$$

$$\text{Fluid velocity } v = Q \frac{\text{m}^3}{\text{s}} \times \frac{4}{\pi d^2 m^2}$$

$$\text{Fluid velocity } v = \frac{4Q}{\pi d^2} \frac{\text{m}}{\text{s}}$$

$$Q = m \frac{\text{kg}}{\text{s}} \times \frac{\text{m}^3}{\rho \text{kg}}$$

$$\text{Fluid velocity } v = \frac{4m}{\rho \pi d^2} \frac{\text{m}}{\text{s}}$$

EXAMPLE 10.4

Calculate the water velocity in a 28.0 mm copper pipe carrying 0.519 kg/s at 75°C.

For a 28.0 mm pipe, d is 0.02622 m.

$$\rho = 974.85 \frac{\text{kg}}{\text{m}^3}$$

$$\text{Velocity } v = \frac{4m}{\rho \pi d^2} \frac{\text{m}}{\text{s}}$$

$$\text{Velocity } v = \frac{4 \times 0.519}{974.85 \times \pi \times 0.02622^2} \frac{\text{m}}{\text{s}}$$

$$\text{Velocity } v = 0.986 \frac{\text{m}}{\text{s}}$$

Flow of methane in pipes

The equation for copper pipe carrying methane gas during turbulent flow is:

$$Q = -2.695 \Delta p^{0.5} d^{2.5} \log \left(\frac{4.05 \times 10^{-7}}{d} + \frac{2.3 \times 10^{-5}}{\Delta p^{0.5} d^{1.5}} \right) \frac{\text{m}^3}{\text{s}}$$

Q = volume flow rate of methane, $\frac{\text{m}^3}{\text{s}}$

Δp = pressure drop per metre of pipe, $\frac{\text{Pa}}{\text{m}}$

d = pipe internal diameter, m

EXAMPLE 10.5

Calculate the flow rate of methane in a 35 mm copper pipe when the pressure loss rate is 6 Pa/m, from Table 10.1, d is 0.03263 m.

$$Q = -2.695 \Delta p^{0.5} d^{2.5} \log \left(\frac{4.05 \times 10^{-7}}{d} + \frac{2.3 \times 10^{-5}}{\Delta p^{0.5} d^{1.5}} \right) \frac{\text{m}^3}{\text{s}}$$

$$Q = -2.695 \times 6^{0.5} \times 0.03263^{2.5} \times \log \left(\frac{4.05 \times 10^{-7}}{0.03263} + \frac{2.3 \times 10^{-5}}{6^{0.5} \times 0.03263^{1.5}} \right) \frac{\text{m}^3}{\text{s}}$$

$$Q = 0.00355 \frac{\text{m}^3}{\text{s}}$$

Air flow in ducts

The flow of air at 20°C d.b., 43% saturation and a barometric pressure of 1013.25 mb through clean galvanised sheet metal ducts having joints made in accordance with good practice, is given by:

$$Q = -2.0278 \Delta p^{0.5} d^{2.5} \log \left(\frac{4.05 \times 10^{-5}}{d} + \frac{2.933 \times 10^{-5}}{\Delta p^{0.5} d^{1.5}} \right) \frac{\text{m}^3}{\text{s}}$$

Q = volume flow rate of air, $\frac{\text{m}^3}{\text{s}}$

Δp = pressure drop per metre of duct, $\frac{\text{Pa}}{\text{m}}$

d = duct internal diameter, m

EXAMPLE 10.6

A 500 mm internal diameter galvanised sheet steel duct is carrying air at 20°C d.b. at a design pressure loss rate of 1 Pa/m. Calculate the air volume flow rate being passed and the air velocity.

$$Q = -2.0278 \Delta p^{0.5} d^{2.5} \log\left(\frac{4.05 \times 10^{-5}}{d} + \frac{2.933 \times 10^{-5}}{\Delta p^{0.5} d^{1.5}}\right) \frac{m^3}{s}$$

$$Q = -2.0278 \times 1^{0.5} \times 0.5^{2.5} \times \log\left(\frac{4.05 \times 10^{-5}}{0.5} + \frac{2.933 \times 10^{-5}}{1^{0.5} \times 0.5^{1.5}}\right) \frac{m^3}{s}$$

$$Q = 1.357 \frac{m^3}{s}$$

$$\text{Air velocity } v = \frac{Q \frac{m^3}{s}}{A m^2}$$

$$v = 1.357 \frac{m^3}{s} \times \frac{4}{\pi \times 0.5^2 m^2}$$

$$v = 6.91 \frac{m}{s}$$

EXAMPLE 10.7

Find a suitable diameter for a galvanised sheet steel duct that is to carry 3250 l/s of air at 20°C d.b. at a maximum velocity of 5 m/s when the pressure loss rate is not to exceed 1 Pa/m. Duct diameters are in increments of 50 mm.

An iterative procedure is needed, try d of 1 m:

$$Q = 3.25 \frac{m^3}{s}$$

$$\text{Air velocity } v = \frac{Q \frac{m^3}{s}}{A m^2}$$

$$v = 3.25 \frac{m^3}{s} \times \frac{4}{\pi \times 1^2 m^2}$$

$$v = 4.14 \frac{m}{s}$$

This is less than the limit. Try d of 950 mm:

$$v = 3.25 \frac{m^3}{s} \times \frac{4}{\pi \times 0.95^2 m^2}$$

$$v = 4.59 \frac{m}{s}$$

A 900 mm diameter duct produces an air velocity of 5.1m/s and that exceeds the stated limit. A 950 mm diameter duct will be used, provided that the frictional pressure loss rate does not exceed 1 Pa/m.

Insert $\Delta p = 1$Pa/m and $d = 0.95$m into the duct formula and calculate the maximum carrying capacity, irrespective of limiting velocity.

$$Q = -2.0278 \Delta p^{0.5} d^{2.5} \log\left(\frac{4.05 \times 10^{-5}}{d} + \frac{2.933 \times 10^{-5}}{\Delta p^{0.5} d^{1.5}}\right) \frac{m^3}{s}$$

$$Q = -2.0278 \times 1^{0.5} \times 0.95^{2.5} \times \log\left(\frac{4.05 \times 10^{-5}}{0.95} + \frac{2.933 \times 10^{-5}}{1^{0.5} \times 0.95^{1.5}}\right) \frac{m^3}{s}$$

$$Q = 7.365 \frac{m^3}{s}$$

$$Q = 7.365 \frac{m^3}{s} \times \frac{10^3 l}{1\ m^3}$$

$$Q = 7365 \frac{l}{s}$$

Carrying capacity of a 950 mm diameter duct is greater than required, so the actual pressure loss rate will be less than 1 Pa/m, around 0.22 Pa/m from calculation.

Heat carrying capacity

The engineering services designer uses the most cost effective fluid to transfer heating or cooling power to the occupied space. This may mean using high temperature hot water pipework distribution when there are long distances between the energy conversion source and the target location. It is more economical to transfer heated or cooled water than conditioned air, as smaller diameter pipes are used and pumping costs are less than those for fan and air duct networks. The circulation of air is limited to the smallest practical area with the minimum lengths and diameters of air ductwork.

Table 10.2 gives physical data on the heat transfer fluids used, to enable a comparison to be made between their relative transportation capacities.

To compare the heating or cooling carrying capacities of water and air, find the flow rates required for 1.0 kW of heat transfer, from:

Heat flow = mass flow rate × *SHC* × allowable temperature drop

$$m = \frac{\text{Heat flow } Q \text{ kW}}{SHC \frac{kJ}{kg\,K} \times \Delta t \text{ K}}$$

Table 10.2 Fluid physical data

Fluid	*Mean temperature t°C*	$\Delta t, K$	$SHC \frac{kJ}{kgK}$	$\rho \frac{kg}{m^3}$
Water	10	4	4.193	999.7
	75	12	4.194	974.9
	120	30	4.248	943.1
Air	10	4	1.012	1.242
	30	8	1.012	1.160
	40	12	1.012	1.123

EXAMPLE 10.8

Compare the mass flow rates of water needed to transfer 1.0 kW of heating power with low temperature hot water (LTHW) and medium temperature hot water (MTHW) heating systems.

From Table 10.2 mean water temperature for LTHW is 75°C and for HTHW is 120°C requiring the water to be pressurised to avoid boiling.

$$m = \frac{\text{Heat flow } Q \text{ kW}}{SHC \frac{\text{kJ}}{\text{kg K}} \times \Delta t \text{K}}$$

$$m = Q \text{ kW} \times \frac{\text{kg K}}{SHC \text{ kJ}} \times \frac{1}{\Delta t \text{ K}} \times \frac{1 \text{ kJ}}{1 \text{ kWs}}$$

$$m = \frac{Q}{SHC \times \Delta t} \frac{\text{kg}}{\text{s}}$$

(a) LTHW, Q 1.0 kW, SHC 4.194 kJ/kg K, Δt12 K:

$$m = \frac{Q}{SHC \times \Delta t} \frac{\text{kg}}{\text{s}}$$

$$m = \frac{1}{4.194 \times 12} \frac{\text{kg}}{\text{s}}$$

$$m = 0.02 \frac{\text{kg}}{\text{s}}$$

(b) MTHW, Q 1.0 kW, SHC 4.248 kJ/kg K, Δt30 K:

$$m = \frac{1}{4.248 \times 30} \frac{\text{kg}}{\text{s}}$$

$$m = 0.008 \frac{\text{kg}}{\text{s}}$$

The advantage of MTHW, and HTHW, is that larger temperature drops between flow and return can be used, reducing the water flow rate, while providing a high mean water temperature at the heat emitter, rapid heat transfer and small terminal units further away from the occupants, all to economise on the pipework costs.

EXAMPLE 10.9

Find suitable copper pipe diameters for LTHW and MTHW heating systems to transfer 12 kW from a boiler to a convector if the pressure loss rate is not to exceed 500 Pa/m and water velocity not to exceed 1 m/s.

(a) LTHW :

$$m = \frac{12}{4.194 \times 12} \frac{\text{kg}}{\text{s}}$$

$$m = 0.238 \frac{\text{kg}}{\text{s}}$$

Try a 22 mm pipe d 0.02022 m:

$$Q = m \frac{\text{kg}}{\text{s}} \times \frac{\text{m}^3}{\rho \text{kg}}$$

$$Q = 0.238 \frac{\text{kg}}{\text{s}} \times \frac{\text{m}^3}{974.9 \text{ kg}}$$

$$Q = \frac{2.441}{10^4} \frac{\text{m}^3}{\text{s}}$$

$$v = \frac{2.441}{10^4} \frac{\text{m}^3}{\text{s}} \times \frac{4}{\pi \times 0.02022^2 \text{ m}^2}$$

$$v = 0.76 \frac{\text{m}}{\text{s}}$$

Calculate the carrying capacity of the 22 mm pipe for Δp500 Pa/m:

$$m = -69.36 \Delta p^{0.5} d^{2.5} \log \left(\frac{4.05 \times 10^{-7}}{d} + \frac{2.15 \times 10^{-5}}{\Delta p^{0.5} d^{1.5}} \right) \frac{\text{kg}}{\text{s}}$$

$$m = -69.36 \times 500^{0.5} \times 0.02022^{2.5} \log \left(\frac{4.05 \times 10^{-7}}{0.02022} + \frac{2.15 \times 10^{-5}}{500^{0.5} \times 0.02022^{1.5}} \right) \frac{\text{kg}}{\text{s}}$$

$$m = 0.311 \frac{\text{kg}}{\text{s}}$$

This flow capacity is in excess of the design requirement. The actual pressure loss rate will be less than 500 Pa/m. The pipe size is satisfactory. The reader should try 15 mm.

(b) MTHW: Try a 15 mm pipe whose d is 0.0136 m:

$$m = \frac{12}{4.248 \times 30} \frac{\text{kg}}{\text{s}}$$

$$m = 0.094 \frac{\text{kg}}{\text{s}}$$

$$Q = 0.094 \frac{\text{kg}}{\text{s}} \times \frac{\text{m}^3}{943.1 \text{ kg}}$$

$$Q = \frac{9.984}{10^5} \frac{\text{m}^3}{\text{s}}$$

$$v = \frac{9.984}{10^5} \frac{\text{m}^3}{\text{s}} \times \frac{4}{\pi \times 0.0136^2 \text{ m}^2}$$

$$v = 0.69 \frac{\text{m}}{\text{s}}$$

This is within the design criteria. Carrying capacity:

$$m = -69.36 \times 500^{0.5} \times 0.0136^{2.5} \log\left(\frac{4.05 \times 10^{-7}}{0.0136} + \frac{2.15 \times 10^{-5}}{500^{0.5} \times 0.0136^{1.5}}\right) \frac{\text{kg}}{\text{s}}$$

$$m = 0.107 \frac{\text{kg}}{\text{s}}$$

This flow capacity is in excess of the design requirement. The actual pressure loss rate will be less than 500 Pa/m.

EXAMPLE 10.10

A mechanical ventilation system is to be used to heat a theatre having a total heat loss of 60 kW. The boiler plant room is 30 m from the theatre. Compare the diameters of LPHW pipes and warm air ducts that can be used, stating your preference and reasons.

(a) (*LPHW*) system:

$$m = \frac{60}{4.194 \times 12} \frac{\text{kg}}{\text{s}}$$

$$m = 1.192 \frac{\text{kg}}{\text{s}}$$

Try 42 mm copper pipe, *d* 0.03963 m:

$$Q = 1.192 \frac{\text{kg}}{\text{s}} \times \frac{\text{m}^3}{974.9.1 \text{ kg}}$$

$$Q = \frac{1.223}{10^3} \frac{\text{m}^3}{\text{s}}$$

$$v = \frac{1.223}{10^3} \frac{\text{m}^3}{\text{s}} \times \frac{4}{\pi \times 0.03963^2 \text{ m}^2}$$

$$v = 0.991 \frac{\text{m}}{\text{s}}$$

An estimate of the pressure loss rate is 220 Pa/m. Calculate the carrying capacity of the 42 mm pipe:

$$m = -69.36 \times 220^{0.5} \times 0.03963^{2.5} \times \log\left(\frac{4.05 \times 10^{-7}}{0.03963} + \frac{2.15 \times 10^{-5}}{220^{0.5} \times 0.03963^{1.5}}\right) \frac{\text{kg}}{\text{s}}$$

$$m = 1.194 \frac{\text{kg}}{\text{s}}$$

Carrying capacity is slightly in excess of the design requirement. The pipe size is satisfactory.

(b) Air duct system:

Warm supply air would leave the plant room at 40°C, travel through galvanised steel ducts, enter the theatre at 38°C, heat and ventilate the theatre, leave the theatre at the room air temperature of 22°C and arrive back in the plant room at 20°C, having lost some heat to the cold external air. A heating coil in the plant room raises air from 20°C to 40°C with LPHW from the boiler.

$$\Delta t \text{ of air in room} = (38 - 22)\text{K}$$

$$\Delta t \text{ of air in room} = 16\text{K}$$

$$\text{Mass flow rate of air through room } m = 60\text{kW} \times \frac{\text{kg K}}{1.012 \text{ kJ}} \times \frac{1}{16 \text{ K}} \times \frac{\text{kJ}}{\text{kWs}}$$

$$m = 3.706\frac{\text{kg}}{\text{s}}$$

Try an 800 mm diameter duct, d 0.8 m:

$$Q = 3.706\frac{\text{kg}}{\text{s}} \times \frac{\text{m}^3}{1.233 \text{ kg}}$$

$$Q = 3.3\frac{\text{m}^3}{\text{s}}$$

$$v = 3.3\frac{\text{m}^3}{\text{s}} \times \frac{4}{\pi \times 0.8^2\text{m}^2}$$

$$v = 6.6\frac{\text{m}}{\text{s}}$$

An estimate of the pressure loss rate through the air duct is 0.7 Pa/m. Calculate its carrying capacity:

$$Q = -2.0278\Delta p^{0.5}d^{2.5}\log\left(\frac{4.05 \times 10^{-5}}{d} + \frac{2.933 \times 10^{-5}}{\Delta p^{0.5}d^{1.5}}\right)\frac{\text{m}^3}{\text{s}}$$

$$Q = -2.0278 \times 0.7^{0.5} \times 0.8^{2.5} \times \log\left(\frac{4.05 \times 10^{-5}}{0.8} + \frac{2.933 \times 10^{-5}}{0.7^{0.5} \times 0.8^{1.5}}\right)\frac{\text{m}^3}{\text{s}}$$

$$Q = 3.89\frac{\text{m}^3}{\text{s}}$$

Air duct carrying capacity is slightly in excess of the design requirement. This size is satisfactory.

The choice facing the designer is between 42 mm LPHW flow and return pipes and 800 mm supply and recirculation air ducts. The more concentrated form of energy, water, is used for long distance heat transportation. A smaller underground service duct would be needed and less thermal insulation material used.

Pump and fan power consumption

The power expended in overcoming frictional resistance in a pipe or duct is:

$$\text{Power } P \text{ watts} = Q\frac{\text{m}^3}{\text{s}} \times H \text{ Pa}$$

H is the pressure overcome by the fan or pump in terms of total frictional resistance overcome or the vertical head increase for water systems if the discharge is open ended.

$$1\ \text{Pa} = 1\frac{\text{N}}{\text{m}^2}$$

$$1\frac{\text{Nm}}{\text{s}} = 1\ \text{W}$$

$$\text{Power } P = Q\frac{\text{m}^3}{\text{s}} \times H\ \text{Pa} \times \frac{1\ \text{N}}{1\ \text{Pa m}^2} \times \frac{1\ \text{Ws}}{1\ \text{Nm}}$$

$$\text{Power } P = QH\ \text{W}$$

H = total pressure drop overcome by the pump or fan, Pa

$$H = \Delta p\frac{\text{Pa}}{\text{m}} \times l\ \text{m}$$

l = pipe or duct length, m

$$\text{Input power} = \frac{\text{fluid power used, W}}{\text{overall efficiency of the pump, fan and motor}}$$

Overall efficiency will incorporate the electrical power losses in the motor, mechanical energy absorbed in bearings and drive systems plus the fluid energy losses due to casing friction and changes in flow direction. Belt drives lose some energy and tend to deteriorate during use and require regular maintenance. Typically overall efficiency will be 60%.

EXAMPLE 10.11

Calculate the pump and fan energy consumptions for the alternative systems in example 10.10. The overall electro-mechanical efficiency of each machine is 60%. The resistance of pipe fittings is 30% of that for straight pipes. The resistance of air duct fittings is 300% of that for straight ducts.

The theatre is 30 m from the plant room, so 60 m of pipe or duct would be used.

(a) $$\text{LPHW pipe system resistance } H = 1.3 \times 220\frac{\text{Pa}}{\text{m}} \times 60\ \text{m}$$

$$\text{LPHW pipe system resistance } H = 17160\ \text{Pa}$$

$$\text{Pump power } P = Q\frac{\text{m}^3}{\text{s}} \times H\ \text{Pa}$$

$$\text{Pump power } P = \frac{1.223}{10^3}\frac{\text{m}^3}{\text{s}} \times 17160\ \text{Pa}$$

$$\text{Pump power } P = 21\ \text{W}$$

$$\text{Pump input power} = \frac{21}{0.6}\ \text{W}$$

$$\text{Pump input power} = 35\ \text{W}$$

(b) Air duct $H = 3 \times \left(0.7\frac{\text{Pa}}{\text{m}} \times 60\ \text{m}\right)$

Air duct $H = 126$ Pa

Fan power, $P = Q\frac{m^3}{s} \times H$ Pa

Fan power $P = 3.3\frac{\text{m}^3}{\text{s}} \times 126$ Pa

Fan power $P = 415.8$ W

Fan input power $= \frac{415.8}{0.6}$ W

Fan input power = 693 W

This demonstrates that a fan and air duct system absorbs several times the electrical power of a hot water system to deliver the same amount of heating energy.

Thermal storage

Storage of heat in a water tank, rocks, bricks, underground concrete labyrinth, ground beneath our feet, geothermal source/sink, hollow concrete floor air passageways, concrete floors, walls and ceilings of the building being conditioned, ice bank or phase change material bank, nearby river, lake or ocean, is used to connect an intermittent demand for energy with an intermittently used source or sink. None of these methods is a new idea. Their principle of operation is well known and we have all used it during our lives. Just think about the difference between living in a tent with no thermal storage, and a cave 50 m or more below ground where the air and ground temperature never varies.

Demand for heat may come from the heating system of a building or the use of hot water by sanitary appliances. Thermal storage may be necessary as a buffer between a biomass wood pellet-burning water heater that needs to run continuously, and the highly variable demand for heat from an occupied building.

Low temperature heat storage with a chilled water tank is used between a refrigeration compressor and the air conditioning system it serves. It is necessary to limit the on/off operation of the refrigeration compressor to six starts per hour to minimise the temperature and wear of the starting equipment.

Typical demands for heating and cooling are cyclically variable during each 24 hour period and follow weather patterns. The economical supply of energy to meet such a demand has to be equally variable. Simple on/off and multi-step controls are designed for the convenience of the boiler or refrigeration compressor and not for the thermal comfort of the occupants of the building.

Intermittent supplies of energy come from night-time electrical use when the building is unoccupied, the on/off operation of boilers and refrigeration compressors, solar collectors and sources of waste heat. Some examples of thermal storage arrangements are as follows.

Hot water storage cylinder for sanitary services

(a) A low power output fossil-fuel-burning appliance heats the cylinder during long running periods at 100% output and maximum efficiency.
(b) An array of solar collector panels, or concentrators on the roof, is used to preheat the cold water supplied to the storage cylinder.
(c) Off-peak night-time electrical immersion heaters replace the hot water used during the day.

Table 10.3 Thermal storage data

Material	*ρ kg/m*^{3}	*SHC kJ/kg K*
Water at 75°C	974.9	4.194
Water at 10°C	999.7	4.193
Cast iron	7000	480
Concrete	2300	840
Granite	2650	1000

Space heating

(a) Off-peak electrical immersion heaters are used to raise the temperature of a storage tank or cylinder to 80°C overnight to meet the building heat demand for the following day.
(b) Off-peak electrical storage radiators that have brick, concrete or cast iron heat stores.
(c) Passive architecture utilises solar heating of exposed masonry walls to absorb heat energy that is released up to 10 hours later. The heat output can be partly controlled by increasing air circulation with fans.
(d) Passive architecture uses glazing to allow solar gains into the building. Shading is used to control heat input. A heat storage wall is often placed close to the glazing to avoid short-term overheating and allow control of heat release over a period of hours.
(e) Phase change salts and chemicals can absorb energy from waste heat sources or off-peak electricity in order to change phase. Heat release from the store is achieved by allowing reversal of the phase change when needed.

Refrigeration

A chilled water storage tank is cooled by on/off refrigeration compressors. Compressors can be run at full load for long periods to chill the water and then remain off for an hour or more. This is particularly advantageous when the cooling demand within the building is low during mild weather. Compressors are not to be allowed to start more than six times per hour as the starting current is around twice the running current and motor overheating is to be avoided.

Table 10.3 gives the physical data needed for thermal storage calculations.

To calculate the quantity of heat stored:

$$\text{Heat stored} = \text{m kg} \times SHC \frac{\text{kJ}}{\text{kg K}} \times (t_1 - t_2)\ \text{K}$$

$$\text{Mass of stored material} = \text{volume m}^3 \times \rho \frac{\text{kg}}{\text{m}^3}$$

$$t_1 = \text{mean storage temperature °C}$$

$$t_2 = \text{lowest usable storage temperature °C}$$

EXAMPLE 10.12

A house has an average rate of heat loss of 6.4 kW between 07.00 hours and 23.00 hours. A hot water storage tank is to be raised to 80°C between 23.00 hours and 07.00 hours by off-peak electrical immersion heaters. Calculate a suitable design for the system.

$$\text{Heat demand period} = (23.00 - 07.00)\ \text{h}$$

$$\text{Heat demand period} = 16\ \text{h}$$

$$\text{Electrical charge period} = 8\ \text{h}$$

The heat store will lose energy through its thermal insulation. The rate of heat loss can be accurately calculated when the tank dimensions are known. An estimate of 15% can be added initially.

$$\text{Heat storage required} = 1.15 \times 6.4\ \text{kW} \times 16\ \text{h}$$

$$\text{Heat storage required} = 117.76\ \text{kWh}$$

$$\text{Heat stored overnight} = \text{heater power kW} \times \text{charge period h}$$

$$\text{Heater power} = \frac{117.76\ \text{kWh}}{8\ \text{h}}$$

$$\text{Heater power} = 14.7\ \text{kW}$$

Five stages of 3 kW immersion heaters, 15 kW, would be suitable.
Calculate the water mass, volume and size of the water storage tank,

$$\text{Heat stored} = \text{water}\ m\ \text{kg} \times SHC \frac{\text{kJ}}{\text{kg K}} \times (t_1 - t_2)\ \text{K}$$

A suitable lower limit for stored hot water, t_2, could be 60°C.

$$t_1 = 80°\text{C}$$

$$\rho = 974.9 \frac{\text{kg}}{\text{m}^3}$$

$$SHC = 4.194 \frac{\text{kJ}}{\text{kg K}}$$

$$m = \frac{\text{heat stored kWh}}{(t_1 - t_2)\ \text{K}} \times \frac{\text{kg K}}{4.194\ \text{kJ}}$$

$$m = \frac{117.76\ \text{kWh}}{(80 - 60)\text{K}} \times \frac{\text{kg K}}{4.194\ \text{kJ}} \times \frac{1\ \text{kJ}}{1\ \text{kWs}} \times \frac{3600\ \text{s}}{1\ \text{h}}$$

$$m = 5054\ \text{kg}$$

$$\text{Water volume stored} = 5054\ \text{kg} \times \frac{\text{m}^3}{974.9\ \text{kg}}$$

$$\text{Water volume stored} = 5.18\ \text{m}^3$$

The storage tank will be no higher than 1 m, so its plan dimensions will be $\sqrt{5.18}$ m. One tank of 2.3 m × 2.3 m × 1 m, with the economic thickness of thermal insulation, would be needed.

EXAMPLE 10.13

A 150 kW refrigeration compressor system chills water from 10°C to 6°C in a building's air conditioning system. The compressor is not to cycle on/off more than six times per hour. During mild weather, the building cooling load is 25 kW. Heat gains to the stored chilled water amount to 10% of the cooling energy used. Calculate a suitable design for an intermediate chilled water storage tank.

$$\text{Building cooling hourly demand} = 25\ \text{kW} \times 1.1 \times 1\ \text{h}$$

$$\text{Building cooling hourly demand} = 27.5\ \text{kWh}$$

If the refrigeration system operates at 150 kW for 12 minutes:

$$\text{Cooling output} = 150\ \text{kW} \times 12\ \text{min} \times \frac{1\ \text{h}}{60\ \text{min}}$$

$$\text{Cooling output} = 30\ \text{kWh}$$

So, the compressors can run for 12 minutes each hour to lower the storage tank temperature to 6°C, satisfying the building cooling load. In order to store 30 kWh, water storage will be:

$$m = 30\ \text{kWh} \times \frac{\text{kg K}}{4.193\ \text{kJ}} \times \frac{1}{(10-6)\text{K}} \times \frac{1\ \text{kJ}}{1\ \text{kWs}} \times \frac{3600\ \text{s}}{1\ \text{h}}$$

$$m = 6440\ \text{kg}$$

$$\text{Water volume stored} = 6440\ \text{kg} \times \frac{\text{m}^3}{999.7\ \text{kg}}$$

$$\text{Water volume stored} = 6.44\ \text{m}^3$$

The storage tank will be no higher than 1 m, so its plan dimensions will be $\sqrt{6.44}$ m. One tank of 2.54 m × 2.54 m × 1 m, with the economic thickness of thermal insulation, would be needed.

Questions

1. State the engineering and economic objectives of heat distribution systems. Give examples of good practice. Outline the economic considerations needed.
2. Calculate the maximum carrying capacity of a 28 mm copper pipe at a pressure loss rate of 1200 Pa/m when the water temperature is 75°C.
3. Find the copper pipe size appropriate to a cold water flow rate of 0.123 kg/s and a pressure loss rate of 850 Pa/m.
4. A low temperature hot water heating system that serves air conditioning heater coils, has a two-pipe circuit that is to be sized on the basis of a constant pressure loss rate of 500 Pa/m. Calculate the maximum flow capacity of copper pipes from 15 mm to 54 mm nominal diameters to assist the designer.
5. Calculate the water velocity in a 42 mm copper pipe carrying 1.52 kg/s at 75°C.
6. Find the flow rate of natural gas, methane, in a 15 mm copper pipe at a pressure loss rate of 12.5 Pa/m.
7. The pressure loss rate available for a gas supply installation is 8 Pa/m. Copper pipe is to be used. Calculate the gas volume flow capacity of pipes from 22 mm to 54 mm nominal diameter.

8. A 300 mm internal diameter galvanised sheet steel duct is to carry air at 20°C d.b. with a pressure loss rate of 2 Pa/m. Calculate the air flow rate that can be carried.
9. Calculate the air velocity in a 450 mm internal diameter galvanised sheet steel duct when the air temperature is 20°C d.b. and the pressure loss rate is 0.6 Pa/m.
10. An air conditioning system is to have galvanised sheet metal ducts carrying air at 20°C d.b. at a maximum velocity of 7.5 m/s. The maximum allowable frictional resistance of straight duct is to be 3 Pa/m. The air flow rates to be carried in different sections of duct are 0.12, 0.3, 0.6, 1 and 2 m^3/s. Ducts are manufactured in sizes from 100 mm internal diameter upwards in 50 mm increments. Calculate the most economical size for each duct.
11. Calculate the water mass flow rates needed for LTHW and MTHW heating systems in order to transfer 3.5 kW and 420 kW. State that system would be used for each heat load.
12. Find a suitable copper pipe diameter for a LTHW heating system to transfer 25 kW from the boiler to a system of heat emitters if the pressure loss rate is not to exceed 750 Pa/m and the water velocity is not to be above 1 m/s. State the actual pressure loss rate and water velocity.
13. District heating is to have a total connected heat load of 550 kW using a MTHW system having a flow temperature of 150°C and a return of 120°C. Water velocity can be up to 3 m/s in the underground distribution mains. The pressure loss rate is not to exceed 750 Pa/m. If copper pipe is to be used, what would be the appropriate size, actual pressure loss rate and water velocity?
14. The air conditioning system in a 20-storey office block is to comprise of air handling plants and a basement boiler room. A total connected load of 100 kW was calculated. Compare the diameters of LTHW heating system pipes with air ducts that would be needed to transport the heating capacity vertically through the building. State how the design engineer would configure the heating service to minimise the spaces occupied by the distribution services. Maximum water velocity is to be 1 m/s. Air velocity is not to exceed 8 m/s.
15. Calculate the pump and fan energy consumptions to operate a LTHW system or a ducted air heating system where the boiler plant is 75 m from the heat load of 75 kW. Overall electro-mechanical efficiency of the pump and fan is 65%. Water velocity can be up to 2 m/s and air velocity up to 10 m/s. Frictional resistance of the pipeline fittings amounts to 25% of the length of straight pipe and that for the air duct is 75% of the duct length.
16. Discuss why thermal storage is needed in heating and cooling services installations and list the principal methods used.
17. A living room has an average rate of heat loss of 2.2 kW during a winter day while occupied between 07.00 hours and 23.00 hours. An electrically charged off-peak storage heater is to be installed to maintain comfort conditions. The charge period is 23.00 hours to 07.00 hours. Calculate the input power and the total energy storage required for the heater.
18. A building having an average rate of daily heat loss of 15 kW between 07.00 hours and 23.00 hours is to be heated from a hot water storage tank operating between 85°C and 60°C. The tank has electric immersion heaters operating between 23.00 hours and 07.00 hours. Calculate the storage tank size needed, heater power and quantity of heat stored in MJ.
19. A 230 kW refrigeration compressor system chills water from 12°C to 7°C in an air conditioning system. The compressor is not to cycle on/off more than four times per hour. During mild weather, the building cooling load is 100 kW. Heat gains to the stored chilled water amount to 15% of the cooling energy used. Calculate a suitable design for an intermediate chilled water storage tank.
20. Now that all users of my books are experts in using spreadsheet calculation files, create a calculation workbook for the pipe and duct equations provided in this chapter; each different equation to be entered onto its own sheet of the workbook. Hold a class competition to find the best workbook design and application. Easy now isn't it?

11 Air duct acoustics

Learning objectives

Study of this chapter will enable the reader to:

1. identify the need for sound attenuation in air duct systems;
2. calculate the sound pressure level in a room from the acoustic power input to the air duct system by a fan;
3. find the noise rating produced in a room;
4. use directivity, sound absorption coefficient, mean absorption coefficient and room absorption constant;
5. calculate and use noise rating data;
6. know the formulae and data used in the acoustic design of air duct systems;
7. be able to carry out sound pressure level and noise rating design calculations for air duct systems, try different solutions to attenuate the fan noise and be able to produce a practical design to meet a design brief.

Key terms and concepts

Introduction

This chapter uses the worksheet file *dbduct.xls* to find the noise rating that will be produced within an occupied room by direct transmission of sound through the air conditioning ductwork system from the fan. Sufficient data are provided in the worksheet for the examples within this chapter and for some design applications. The user can easily change to or add reference data from any source. This chapter facilitates the practical analysis of noise transmission through ductwork. Applications can be assessed quickly with *dbduct.xls* and without having to deal with the equations themselves. Data are provided for frequencies from 31.5 Hertz to 8000 Hertz as this range is likely to cover the important noise levels for comfort. The user can add to the range of frequencies should the need arise.

The purpose of this chapter is to provide an easily understandable method of analysing practical air duct acoustic applications. The user should find no difficulty in entering data and acquiring suitable results for educational reasons and in practical design office cases.

An introduction to acoustics is provided in Chadderton (2013), Chapter 15, Room acoustics.

The user should make use of reference data and formulae from CIBSE Guide B (2005), AIRAH Handbook. Approximate data are provided to use for the examples in this book.

Room acoustics

The sound pressure level, *SPL* dB, of a total sound field, direct plus reverberant pressure waves, in the target room, is generated by the sound power level, *SWL* dB, sound watts level, of the incoming air flow and moderated by the sound absorbing quality of the room. The sound power (watts) level of the incoming air flow must include that generated by the fan, air ductwork system, air terminal unit and the discharge air diffuser. These sound power levels are entered in the worksheet. When other equipment, such as compressors and boilers, are present at the fan end of the ductwork system, the combined sound power level of the various items is entered as the fan value. Normally, only the fan sound power will need to be entered as this will be the dominant level. If a damper, terminal heat exchanger, terminal volume controller, terminal fan or discharge air diffuser generate noise at, or near, the end of the ductwork system, their combined sound power level is entered as the diffuser value. Normally, there will not be any additional noise generated by this part of the ductwork system.

Sound power levels, *SWL* dB values, cannot be summed directly. Sound power, watts, values are added together. The resulting sound power level, *SWL* dB, is that produced by the total sound power. The greatest increase of sound power level is produced by two equal sound power sources combined. When a high sound power source is combined with a low power sound power, there may not be any increase in the sound power level above that of the larger source. The sound power level of two sources is found by making an addition to the larger number from the data in Table 11.1. The formulae used are explained in this chapter.

EXAMPLE 11.1

A fan which produces a sound power level of 88 dBA at 1000 Hertz is within the same plant room as a refrigeration compressor that produces a sound power level of 91 dBA at 1000 Hertz. It is expected that noise from the compressor will break into the air duct prior to it leaving the plant room. Calculate the overall sound power level that enters the air duct system.

Table 11.1 Addition of two sound power levels

Difference between two SWL, dB	*Add to higher SWL, dB*
0	3
1	2
2	2
3	1
4	1
5	1
6 and above	0

Data in this table are approximate and only for use with examples in this book

$$SWL_1 = 88 \text{ dB}$$

$$SWL_2 = 91 \text{ dB}$$

$$SWL_2 - SWL_1 = (91 - 88) \text{ dB}$$

$$SWL_2 - SWL_1 = 3 \text{ dB}$$

From Table 11.1, for a difference of 3 dB, add 1 dB to the higher, SWL_2dB.

$$\text{Combined } SWL_3 = (91 + 1) \text{ dB}$$

$$\text{Combined } SWL_3 = 92 \text{ dB}$$

An *SWL* of 92 dB at 1000 Hz enters the air ductwork system leaving the plant room. The sound pressure level produced in the target occupied room is found from:

$$SPL = SWL + 10\log\left(\frac{Q}{4\pi r^2} + \frac{4}{R}\right) \text{ dB}$$

SPL = sound pressure level produced in room, dB

SWL = sound power level of acoustic source, dB

log = logarithm to base 10, dimensionless

Q = geometric directivity factor, dimensionless

r = distance from sound source to the receiver, m

R = room sound absorption constant, m^2

An inlet air diffuser or grille that is on a plane surface radiates all its sound energy into a hemispherical sound field moving away from the surface. This has a directionality factor Q of 2. When the air inlet is at the junction of two adjacent surfaces that are at right angles to each other, such as the junction of a wall and ceiling, Q is 4. When there are three adjacent surfaces at the air inlet, such as two walls and a ceiling, Q is 8.

Distance r is that from the air inlet grille to the receiving person.

A return air or an extract air ductwork system can also be analysed by using this worksheet. There is a tendency for sound to travel with the general air flow direction. The transmission of sound from the extract air fan back along the duct against the air flow, may lead to an over estimate of the room noise rating. However, it is often necessary to install lining to the return air ductwork between the room and the fan.

Absorption of sound

The room sound absorption constant, R m^2, is found from the total surface area of the enclosing room, S m^2, and the mean sound absorption coefficient of the room surfaces $\bar{\alpha}$ at each of the relevant frequencies.

$$R = \frac{S\bar{\alpha}}{1 - \bar{\alpha}}$$

$\bar{\alpha}$ = mean absorption coefficient of room surfaces

S = total room surface area, m^2

Table 11.2 Absorption coefficients of common materials

Material	Absorption coefficient at					
	125 Hz	*250 Hz*	*500 Hz*	*1000 Hz*	*2000 Hz*	*4000 Hz*
25 mm plaster, 18 mm plasterboard, 75 mm cavity	0.3	0.3	0.6	0.8	0.75	0.75
18 mm board floor on timber joists	0.15	0.2	0.1	0.1	0.1	0.1
Brickwork	0.05	0.04	0.04	0.03	0.03	0.02
Concrete	0.02	0.02	0.02	0.04	0.05	0.05
12 mm fibreboard, 25 mm cavity	0.35	0.35	0.2	0.2	0.25	0.3
Plastered wall	0.01	0.01	0.02	0.03	0.04	0.05
Pile carpet on thick underfelt	0.07	0.25	0.5	0.5	0.6	0.65
Fabric curtain folds	0.05	0.15	0.35	0.55	0.65	0.65
15 mm acoustic ceiling tile, suspended 50 mm mineral fibre wool or glass fibre	0.5	0.6	0.65	0.75	0.8	0.75
50 mm polyester acoustic blanket, metalised film	0.25	0.55	0.75	1.05	0.8	0.7
50 mm glass fibre blanket, perforated surface finish	0.15	0.4	0.75	0.85	0.8	0.85
Glass, large areas	0.18	0.06	0.04	0.03	0.02	0.02
Glass, medium areas	0.35	0.25	0.18	0.12	0.07	0.04

Data in this table are approximate and only for use with examples in this book

Mean absorption coefficient $\bar{\alpha}$ is found from the area and absorption coefficient for each surface of the enclosing space.

$$\bar{\alpha} = \frac{A_1\alpha_1 + A_2\alpha_2 + A_3\alpha_3}{A_1 + A_2 + A_3}$$

A_1 = surface area of surface number one, m^2

α_1 = absorption coefficient of surface number one

Room absorption constant is calculated for each frequency in the range from 31.5 Hz to 8000 Hz. Typical sample absorption coefficients of some common surface materials are provided in Table 11.2 for use in the worksheet *dbduct.xls*.

Reverberation time

Reverberation time of a room is found from,

$$T = \frac{0.161V}{S\bar{\alpha}}$$

(CIBSE Guide B, 2005)

T = reverberation time, s

V = room volume, m^3

Fan sound power level

Fan manufacturers provide the results of acoustic test data for the building services design engineer. Current acoustic data, rather than catalogue information, are required and the manufacturer then becomes responsible for the numbers used. The designer needs the sound power level at each frequency that is to be analysed. These are normally 125 Hz to 4000 Hz. Often the critical frequency for design will be 1000 Hz and this corresponds to a sensitive band in the human ear response. Lower and higher frequencies tend not to be troublesome. The expected fan sound power level can be predicted from air flow rate and fan static pressure from:

$$SWL = 40 + 10\log(Q) + 20\log(FSP), \text{ dB}$$

$$Q = \text{fan air flow rate}, \frac{\text{m}^3}{\text{s}}$$

$$FSP = \text{fan static pressure, Pa}$$

$$SWL = \text{fan sound power (watts) level, dB}$$

EXAMPLE 11.2

Calculate the anticipated fan sound power level when a centrifugal fan is passing 2500 l/s at a fan static pressure of 250 Pa.

$$Q = 2.5\frac{\text{m}^3}{\text{s}}$$

$$FSP = 250 \text{ Pa}$$

$$SWL = 40 + 10 \log (Q) + 20 \log (FSP) \text{ dB}$$

$$SWL = 40 + 10 \log (2.5) + 20 \log (250) \text{ dB}$$

$$SWL = 91 \text{ dB}$$

For the worked examples and questions within this book, sound power levels are provided, either in the form of a discrete value for each frequency or a single value for all frequencies for the plant item. The worksheet *dbduct.xls* automatically calculates the spectral sound power level for the type of fan, by subtracting variances from the initial single value. These sample data are not to be used in real design work as they are provided for illustration purposes only. Sample data are listed in Table 11.3. The user can enter the manufacturer's fan sound power level data onto the worksheet. When real data are present, the worksheet automatically uses it in preference to the sample data.

Table 11.3 Illustrative fan sound power level variances

Fan type	*Sound power level dB variance at frequency, Hz*								
	31.5	63	125	250	500	1 *k*	2 *k*	4 *k*	8 *k*
Centrifugal fan	−2	−5	−8	−10	−14	−18	−23	−30	−40
Axial fan	−2	−3	−4	−6	−8	−10	−13	−16	−20

Data in this table are approximate and are only for use with examples in this book

Table 11.4 Static insertion loss of air duct attenuators

Type	*Attenuation SWL dB at frequency, Hz*							
	63	125	250	500	1 *k*	2 *k*	4 *k*	8 *k*
Circular open duct, diameter × length mm:								
250 × 900	0	2	8	22	37	34	18	16
600 × 600	3	4	7	13	14	9	8	6
600 × 1150	5	8	11	21	23	17	15	10
1150 × 1150	3	4	9	14	12	8	7	6
1200 × 2400	7	9	15	21	18	11	11	10
Circular duct, central pod, diameter × length mm:								
600 × 600	4	6	10	18	22	19	15	11
1150 × 1150	5	7	12	19	18	14	12	9
1200 × 2400	9	12	20	26	28	26	19	16
Rectangular duct, straight splitters, length × width × height mm:								
900 × 550 × 550	6	12	23	36	45	43	36	28
1500 × 1375 × 1375	9	19	35	49	50	50	49	42
2400 × 1375 × 1375	13	28	48	50	50	50	50	50

Data in this table are approximate and only for use with examples in this book

Transmission of sound through air ducts

Acoustic output power of the fan and the turbulence of the air passing through the fan generate noise within the air ductwork system and radiate sound into the plant space. Metal ducts are not usually rigid. Their flexibility provides some attenuation. Internal lining of the ducts with acoustically absorbent material provides useful attenuation. A packaged attenuator may be installed on the discharge side of the fan. A similar attenuator may be fitted to the fan inlet when this is necessary. The inlet attenuator reduces the sound transmitted into the return air duct system. It also reduces the noise that could be projected to the outside environment, as the outside air inlet duct is connected to the return air system. Sample data for manufactured duct attenuators are given in Table 11.4 and repeated in the worksheet. Static insertion loss in sound power level, *SWL* dB, is given for a range of attenuators. The user can add other attenuator data. The air pressure drop Δp through a splitter type attenuator will be up to 100 Pa. The design engineer checks each possible noise path between the noise source and the recipient locations. This chapter assesses the noise transmission through the duct routes.

The attenuation provided by unlined galvanised sheet metal ducts, 0.6 mm thickness, and bends with turning vanes, for air ducts 600 mm wide and 600 mm height, is given in Table 11.5. These attenuation values are typical for various duct sizes in normal systems. Unlined air ducts are downstream of the packaged attenuator or the lengths of lined ductwork, so they are usually in the smaller sizes. The data can be used for other duct sizes when high accuracy is not required. Ducts larger than 1 m × 1 m have much reduced attenuation. The user can enter new attenuation data from any source when necessary.

Attenuation provided by an acoustically lined duct is given by:

$$\text{Attenuation} = 12.56 \times \left(\frac{P\bar{\alpha}^{1.4}}{S}\right), \frac{\text{dB}}{\text{m}}$$

P = perimeter of duct, m

$P = 2 \times (\text{width} + \text{height}),\ \text{m}$

S = cross-sectional area of duct, m^2

S = width × height, m^2

$\bar{\alpha}$ = mean absorption of internal lining material, dimensionless

Lined ducts are installed adjacent to the fan and air handling unit. Both inlet and outlet air ducts are usually lined so that the sound power output from the fan is contained and absorbed close to the source. Exhaust air that is discharged into the external atmosphere and outside air intakes may also require attenuation. Those ducts close to the air handling plant are larger in size when compared to the downstream distribution ducts. The attenuation data for lined ducts and bends given in Table 11.4 are for a duct of 900 mm wide and 600 mm high. Such a duct would be connected to an air handling unit passing up to 4000 l/s. For this duct:

$$\text{Attenuation} = 12.56 \times \left(\frac{P\bar{\alpha}^{1.4}}{S}\right)\frac{\text{dB}}{\text{m}}$$

$$\text{Attenuation} = 12.56 \times \left(\frac{2 \times (0.9 + 0.6) \times \bar{\alpha}^{1.4}}{0.9 \times 0.6}\right)\frac{\text{dB}}{\text{m}}$$

$$\text{Attenuation} = 70\bar{\alpha}^{1.4}\frac{\text{dB}}{\text{m}}$$

Typical absorption values for 25 mm and 50 mm thick acoustic lining material, rigidly fixed inside sheet metal air ducts and having a perforated surface facing the air flow, are given in Table 11.6. When the absorption values are used with the attenuation equation for a 900 mm × 600 mm duct, the attenuation rates shown in Table 11.5 are found.

An upper limit of 45 dB per meter of duct and 50 dB for an attenuator is imposed. This is because noise break-in to the air stream can occur if very low sound levels are produced. Noise break-in to the duct will come from other services systems, outdoor traffic, voices and other noise sources within the building. Using 50 mm thick duct lining improves the attenuation at the lower frequencies.

The attenuation data for a typical 400 mm diameter flexible duct connection between the rectangular distribution duct and the terminal unit are given in Table 11.5. The attenuation of this flexible duct is assumed to be equivalent to that of a circular metal duct. A terminal attenuator may be fitted prior to the

Table 11.5 Attenuation data for air ducts

Item	*Attenuation at frequency, Hz*						
	63	125	250	500	1 *k*	2 *k*	4 *k*
Unlined rectangular duct $\frac{\text{dB}}{\text{m}}$	0.77	0.31	0.15	0.15	0.15	0.15	0.15
Unlined bend with turning vanes dB	0	3	4	3	3	3	3
25 mm lined straight duct $\frac{\text{dB}}{\text{m}}$	0.5	1.7	5	13	38	45	45
Lined bend with turning vanes dB	1	3	6	8	9	10	10
50 mm lined straight duct $\frac{\text{dB}}{\text{m}}$	0.7	6.8	19	45	45	45	
400 mm diameter flexible duct $\frac{\text{dB}}{\text{m}}$	0.02	0.07	0.09	0.12	0.17	0.22	0.22
600 mm duct end reflection dB	6	3	1	0	0	0	0

Data in this table are approximate and only for use with examples in this book

Table 11.6 Attenuation of lined air ducts

Item	*Attenuation at octave band frequency, Hz*						
	63	*125*	*250*	*500*	*1 k*	*2 k*	*4 k*
25 mm thick alpha	0.03	0.07	0.15	0.3	0.65	0.85	0.85
Attenuation $\frac{dB}{m}$	0.5	1.7	5	13	38	45	45
50 mm thick alpha	0.04	0.19	0.4	0.8	0.85	0.85	0.85
Attenuation $\frac{dB}{m}$	0.7	6.8	19	45	45	45	45

Data in this table are approximate and only for use with examples in this book

discharge of air into the occupied room. This attenuator may be a circular duct type or part of the terminal mixing or volume control box. The attenuation of the reflection of sound waves from the end of a run of 600 mm width duct is shown in Table 11.5.

By the careful selection of terminal air outlet diffusers and grilles, additional sound power should not be generated. Where the terminal air handling unit has a flow variable control such as a variable air volume controller, fan coil unit or a fan powered variable air volume controller, additional acoustic power may be generated in the air system.

Sound pressure level in the target room

The sound source is normally the mechanical services plant. The reverberant sound pressure level in a room can be taken as:

$$SPL_1 = SWL + 10\log T_1 - 10\log V_1 + 14,\ \text{dB}$$

SPL_1 = reverberant sound pressure level, dB

SWL = sound power level of source, dB

T_1 = reverberation time of room, s

V_1 = volume of room, m^3

Reverberant sound pressure level is independent of measurement location within the room. When a sound pressure level is required at a known location, the earlier equation is used with the radius from the source r m:

$$SPL = SWL + 10\log\left(\frac{Q}{4\pi r^2} + \frac{4}{R}\right)\ \text{dB}$$

This equation is used for the target room as it takes into account the distance between the air inlet diffuser and the recipient.

EXAMPLE 11.3

An axial flow fan has an overall sound power level of 92 dB on the A scale. The plant room has a reverberation time of 2.4 s and a volume of 100 m^3. Calculate the plant room reverberant sound pressure level.

$SWL = 92$ dBA

$T_1 = 2.4.$ s

$V_1 = 100\text{m}^3$

$SPL_1 = SWL + 10\log T_1 - 10\log V_1 + 14$ dB

$SPL_1 = 92 + 10\log 2.4 - 10\log 100 + 14$ dB

$SPL_1 = 89$ dB

Noise rating

The human ear has a different response to each frequency within the audible range of 20 Hz to 20000 Hz. It has been found that a low-frequency noise can be tolerated at a greater sound pressure level than a high-frequency noise. Noise rating *NR* curves are used to specify the loudness of sounds. Each curve is a representation of the response of the human ear in the range of audible frequencies.

The design engineer makes a comparison between the sound pressure level produced in the room at each frequency and the noise rating curve data at the same frequency. When all the noise levels within the room fall on or below a noise rating curve, that noise rating is attributed to the room. Noise rating curves for *NR* 20 to *NR* 85 are shown in Figure 11.1. The values are plotted from:

$SPL = NR_f \times B_f + A_f$ dB

SPL = sound pressure level at frequency f and noise rating NR, dB

NR_f = noise rating at frequency f Hz, dimensionless

B_f and A_f = physical constants, dB

f = frequency, Hz

The values of the physical constants to calculate noise rating are shown in Table 11.7 (Australian Standard AS 1460-1983).

EXAMPLE 11.4

A fan and a refrigeration compressor are located within the same plant room. The manufacturers state that the sound power levels are 89 dB for the fan and 92 dB for the compressor. Calculate the combined sound power level in the plant room.

$SWL_1 = 89$ dB

$SWL_2 = 92$ dB

Only acoustic powers, watts, can be added, not sound power levels, *SWL* dB. Two or more acoustic powers can be summed and then the overall sound power levels found, or refer to Table 11.7. Calculate

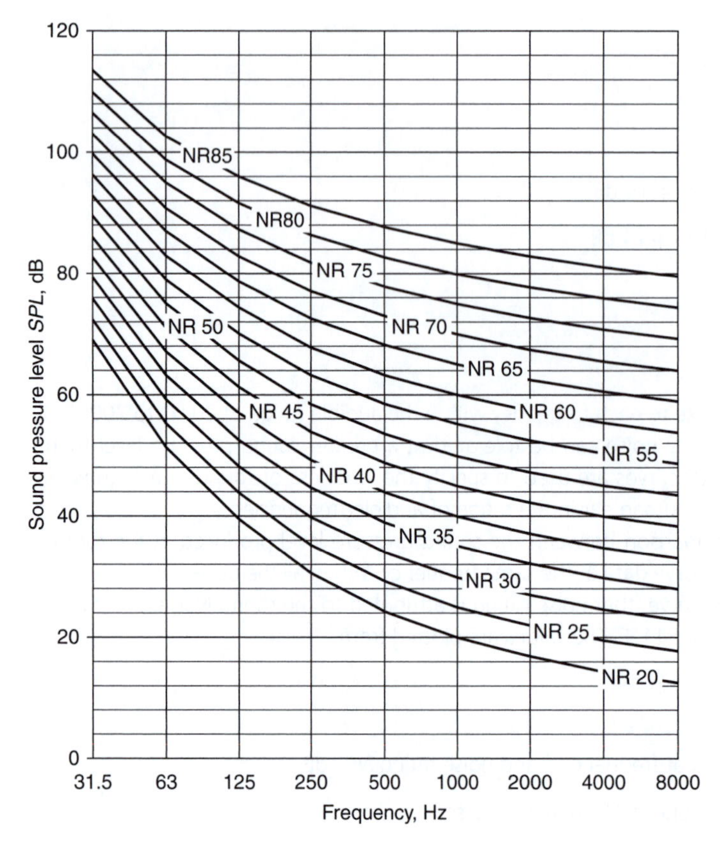

11.1 Noise rating curves.

Table 11.7 Physical constants for noise rating calculation

Frequency f, Hz	A_f, *dB*	B_f, *dB*
31.5	55.4	0.681
63	35.5	0.79
125	22.0	0.87
250	12.0	0.93
500	4.8	0.974
1000	0	1.0
2000	−3.5	1.015
4000	−6.1	1.025
8000	−8.0	1.03

The normal applications of noise rating are shown in Table 11.8

the acoustic output power of sound source number one and sound source number two from the known *SWL* dB values by:

$$SWL_1 = 10\log\left(\frac{W_1}{10^{-12}}\right)\ \text{dB}$$

SWL_1 = sound power level, dB

Table 11.8 Noise rating applications

Application	*Noise rating*	*Comment*
TV, radio studio, concert hall	$\leq NR$ 20	Excellent listening
Large conference room	*NR* 25	Very good listening
Hospital, home, hotel	*NR* 30	Sleeping, relaxing
Library, private office	*NR* 35	Good listening
Office, restaurant, retail	*NR* 40	Fair listening
Cafeteria, corridor, workshop	*NR* 45	Moderate listening
Commercial garage, factory	*NR* 50	Minimum speech interference
Manufacturing	*NR* 55	Speech interference
Heavy engineering to industrial	*NR* 60+	Risk of hearing damage

W_1 = acoustic power, watt

10^{-12} = reference acoustic power, pico watt

The pico watt is the threshold of sound intensity, a zero power value. Any acoustic power level is a multiple of this value on a logarithmic scale. Logarithm to base 10 is used, $\log_{10}$. Arrange the equation to find the power:

$$SWL_1 = 10\log\left(\frac{W_1}{10^{-12}}\right)$$

$$\frac{SWL_1}{10} = \log\left(\frac{W_1}{10^{-12}}\right)$$

This is of the form:

$$X = \log_{10} \mathrm{N}$$

This can be written in its anti-logarithm form:

$$10^X = \mathrm{N}$$

$$10^{\left(\frac{SWL_1}{10}\right)} = \frac{W_1}{10^{-12}}$$

$$W_1 = 10^{-12} \times 10^{\left(\frac{SWL_1}{10}\right)}$$

This is of the form:

$$W_1 = 10^a \times 10^b$$

This is written as:

$$W_1 = 10^{a+b}$$

$$W_1 = 10^{\frac{SWL_1}{10} - 12}$$

Acoustic power output of the fan:

$$SWL_1 = 89 \text{ dB}$$

$$W_1 = 10^{\frac{SWL_1}{10} - 12} \text{ watts}$$

$$W_1 = 10^{\frac{89}{10} - 12} \text{ watts}$$

$$W_1 = 10^{-3.1} \text{ watts}$$

$$W_1 = 7.943 \times 10^{-3} \text{ watts}$$

Acoustic power output of the compressor:

$$SWL_2 = 92 \text{ dB}$$

$$W_2 = 10^{\frac{92}{10} - 12} \text{ watts}$$

$$W_2 = 1.5849 \times 10^{-3} \text{ watts}$$

$$W_3 = W_1 + W_2$$

$$W_3 = \left(7.943 \times 10^{-3} + 1.5849 \times 10^{-3}\right) \text{ watts}$$

$$W_3 = 2.379 \times 10^{-3} \text{ watts}$$

$$SWL_3 = 10 \log\left(\frac{W_3}{10^{-12}}\right) \text{ dB}$$

$$SWL_3 = 10 \log\left(\frac{2.379 \times 10^{-3}}{10^{-12}}\right) \text{ dB}$$

$$SWL_3 = 93 \text{ dB}$$

The difference between SWL_1 and SWL_2 is (92 – 89) dB, 3 dB. From Table 11.1, add 1 dB to 92 dB to find the combined *SWL* of 93 dB.

EXAMPLE 11.5

A model FC123 supply air centrifugal fan passes 3000 l/s at a fan total pressure of 400 Pa. The fan impeller runs at 16 Hz. The supply air ductwork system comprises of 8 m of lined 750 mm × 700 mm duct, two lined bends with turning vanes, no branches, two unlined bends, 12 m of unlined duct, one duct end reflection. The diffuser does not generate any noticeable sound power. The target room is an open plan office. The office is 15 m × 20 m and 3 m high. The nearest recipient's head is 1 m from a diffuser. the office floor is carpeted α_1 0.07, the ceiling has 15 mm suspended acoustic tiles and 50 mm glass fibre mat α_2 0.5, and the walls are plastered brick α_3 0.01 and 20 m^2 of single glazed windows α_4 0.35, all at 125 Hz. Find the noise rating that is not exceeded in the office.

The solution of this example is shown in the original copy of file *dbduct.xls*.

Ensure that the duct branch air flows are all set at 100% as all the supply air passes through these, non-existent, branches. A zero branch air flow proportion is an error. The worksheet shows that a frequency of 125 Hz produces the highest noise rating. The solutions shown here are only for 125 Hz.

Fan $SWL_1 = 40 + 10 \log Q + 20 \log FSP$ dB

$Q = 3000 \frac{l}{s}$

$Q = 3 \frac{m^3}{s}$

$FSP = 400$ Pa

$SWL_1 = 40 + 10 \log 3 + 20 \log 400$ dB

$SWL_1 = 96$ dB

Fan speed = 16 Hz

Fan speed = 960 RPM

Centrifugal fan SWL variance (from Table 11.2) = −8 dB

Calculated fan $SWL_1 = (96 - 8)$ dB

Calculated fan $SWL_1 = 88$ dB

SWL entering duct = 88 dB

Attenuation of lined duct $= 8 \text{ m} \times 1.7 \frac{\text{dB}}{\text{m}}$

Attenuation of lined duct = 13 dB

(There are no decimal places of decibels.)

Attenuation of lined duct bends = 2 × 3 dB

Attenuation of lined duct bends = 6 dB

Attenuation of unlined duct $= 12 \text{ m} \times 0.31 \frac{\text{dB}}{\text{m}}$

Attenuation of unlined duct = 3 dB

Attenuation of unlined duct bends = 2 × 3 dB

Attenuation of unlined duct bends = 6 dB

Attenuation of end reflection = 1 × 3 dB

Attenuation of end reflection = 3 dB

Diffuser $SWL_2 = 0$ dB

Office $SWL_3 = (88 - 13 - 6 - 6 - 3 - 3)$ dB

Office $SWL_3 = 57$ dB

This is the acoustic power entering the office. Its effect on the aural environment is modified by the physical properties of the room, that is, reflection and absorption by the surfaces, furniture and occupants. We are ignoring any sound absorbing beneficial effects from furniture and occupants in this example.

Office floor $A_1 = 15\ \text{m} \times 20\ \text{m}$

Office floor $A_1 = 300\ \text{m}^2$

Office ceiling $A_2 = 300\ \text{m}^2$

Office volume $V_1 = 300\ \text{m}^2 \times 3\ \text{m}$

Office volume $V_1 = 900\ \text{m}^3$

Area of windows $A_3 = 20\ \text{m}^2$

Glazing tends to vibrate from variations in *SPL* in the room and from outdoors and so has some sound reduction properties.

Area of walls $A_4 = (2 \times (15 + 20) \times 3 - 20)\text{m}^2$

Area of walls $A_4 = 190\ \text{m}^2$

Surface area of office $S_1 = (2 \times 300 + 190 + 20)\ \text{m}^2$

Surface area of office $S_1 = 810\ \text{m}^2$

Absorption of floor $\alpha_1 = 0.07$

Absorption of ceiling $\alpha_2 = 0.5$

Absorption of walls $\alpha_3 = 0.01$

Absorption of windows $\alpha_4 = 0.35$

$$\bar{\alpha} = \frac{300 \times 0.07 + 300 \times 0.5 + 190 \times 0.01 + 20 \times 0.35}{810}$$

$$\bar{\alpha} = 0.222$$

$$R = \frac{S_1 \times \bar{\alpha}}{1 - \bar{\alpha}}\ \text{m}^2$$

$$R = \frac{810 \times 0.222}{1 - 0.222}\ \text{m}^2$$

$$R = 231.1\ \text{m}^2$$

$$T_1 = \frac{0.161 \times V_1}{S_1 \times \bar{\alpha}}\ \text{s}$$

$$T_1 = \frac{0.161 \times 900}{810 \times 0.222}\ \text{s}$$

$$T_1 = 0.81\ \text{s}$$

Directivity $Q = 2$

Distance from diffuser to recipient $r = 1\ \text{m}$

Barrier attenuation $B = 0$ dB

Office $SWL_3 = 57$ dB, the acoustic power entering from the supply diffuser

$$\text{Final } SPL_2 = SWL_3 + 10\log\left(\frac{Q}{4\pi r^2} + \frac{4}{R}\right)\text{dB}$$

$$\text{Room effect } SPL_1 = 10\log\left(\frac{Q}{4\pi r^2} + \frac{4}{R}\right)\text{dB}$$

$$SPL_1 = 10\log\left(\frac{2}{4 \times \pi \times 1^2} + \frac{4}{231.1}\right)\text{dB}$$

$$SPL_1 = -7 \text{ dB}$$

$$\text{Final room } SPL_2 = SWL_3 + SPL_1 \text{ dB}$$

$$\text{Final room } SPL_2 = (57 - 7)\text{dB}$$

Final room $SPL_2 = 50$ dB, the sound pressure level in the office at 125 Hz.

For *NR* 35:

$$SPL = (NR\ 35 \times 0.87 + 22) \text{ dB}$$

$$SPL = 52 \text{ dB}$$

As room SPL_2 created by the fan and system = 50 dB, NR 35 criteria are not exceeded in the office. All other frequencies can be manually calculated as shown. File *dbduct.xls* shows the range of sound levels created in the room and that NR 35 is not exceeded. In fact, the critical frequency turns out to be 63 Hz. Any negative values calculated for SPL_2 are irrelevant and are automatically replaced with zero in the worksheet. Calculated sound levels are displayed as integer numbers only, as there are no decimal places for decibels.

The alternative method of calculating the room reverberant sound pressure level is:

$$SPL_2 = SWL_3 + 10\log T_1 - 10\log V_1 + 14 \text{ dB}$$

$$SPL_2 = 57 + 10\log 0.83 - 10\log 900 + 14 \text{ dB}$$

$$SPL_2 = 40 \text{ dB}$$

This corresponds to the recipient being at the opposite end of the 20 m long room from the inlet air sound source. It is preferable to include the direct sound radiation from the source, as has been done in the worksheet.

EXAMPLE 11.6

A model AX43 supply air axial flow fan passes 5000 l/s at a fan static pressure of 150 Pa into a swimming pool hall. The fan impeller runs at 20 Hz. A circular attenuator is located downstream of the fan. The supply air ductwork system comprises of 20 m of unlined straight 1100 mm diameter spirally wound circular duct from the fan to the first grille. The duct has a blank end and it has no bends. Air leaves the duct through grilles in the side of the duct to discharge air into a swimming pool building at high level. There is no noise generation in the duct system or at the terminal grille. The pool building is 30 m × 20 m and 7 m high. The nearest recipient's head to the diffuser is at a distance of 5 m. There is a tiled concrete floor around the pool, the ceiling is insulated rippled steel sheeting and the walls are concrete block with single glazing. There is 200 m^2 of single glazing in the walls. Directivity *Q* is 2. Ensure that the duct branch air flows are all set at 100% as all the supply air passes through these, non-existent, branches. When the design data have been entered, the worksheet shows that a frequency of 250 Hz produces the highest noise rating, so calculate sample values for this frequency.

(a) Find the noise rating that should be provided for the building.
(b) Enter the data onto a working copy of the worksheet and state how the ventilation system is to be attenuated to meet the design requirement.

(a) Appropriate noise rating for a swimming pool building to create reasonable listening is NR 45.
(b) For a fan supply air flow *Q* of 5 $\frac{m^3}{s}$:

$$\text{Fan } SWL_1 = 40 + 10 \log Q + 20 \log FSP \text{ dB}$$

$$\text{Fan } SWL_1 = 40 + 10 \log 5 + 20 \log 150 \text{ dB}$$

$$\text{Fan } SWL_1 = 90 \text{ dB}$$

$$\text{Fan speed } N = 1200 \text{ RPM}$$

Fan *SWL* variance −6 dB from Table 11.2 at 250 Hz

$$\text{Calculated fan } SWL_1 = 84 \text{ dB}$$

Before inclusion of a duct attenuator:

$$SWL_2 \text{ entering duct} = 84 \text{ dB}$$

$$\text{Attenuation of unlined duct} = 20 \text{ m} \times 0.15 \frac{\text{dB}}{\text{m}}$$

$$\text{Attenuation of unlined duct} = 3 \text{ dB}$$

$$\text{Attenuation of end reflection} = 1 \text{ dB}$$

$$\text{Diffuser } SWL_2 = 0 \text{ dB}$$

$$SWL_3 = (84 - 3 - 1) \text{ dB}$$

$$SWL_3 = 80 \text{ dB}$$

Floor $A_1 = 600 \text{ m}^2$ including pool water, taken as concrete absorption.

$$\text{Ceiling } A_2 = 600 \text{ m}^2$$

$$\text{Volume } V_1 = 4200 \text{ m}^3$$

$$\text{Walls } A_3 = 500 \text{ m}^2$$

Glazing $A_4 = 200\ \text{m}^2$

Surfaces $S_1 = 1900\ \text{m}^2$

Surface absorption coefficients, ducts and attenuator are taken from tabulated data provided. Sample values at 250 Hz are used for these calculations:

Absorption of floor $\alpha_1 = 0.02$

Absorption of ceiling $\alpha_2 = 0.4$

Absorption of walls $\alpha_3 = 0.02$

Absorption of windows $\alpha_4 = 0.06$

$$\bar{\alpha} = \frac{600 \times 0.02 + 600 \times 0.4 + 500 \times 0.02 + 200 \times 0.06}{1900}$$

$$\bar{\alpha} = 0.144$$

$$R = \frac{1900 \times 0.144}{1 - 0.144}\ \text{m}^2$$

$$R = 319.6\ \text{m}^2$$

$$T_1 = \frac{0.161 \times 4200}{1900 \times 0.144}\ \text{s}$$

$$T_1 = 2.47\ \text{s}$$

Directivity Q is 2. Distance from diffuser to recipient r is 5 m. Barrier attenuation B is 0 dB. SWL_3 is 80 dB

$$SPL_1 = 10 \log \left(\frac{2}{4 \times \pi \times 5^2} + \frac{4}{319.6} \right)\ \text{dB}$$

$$SPL_1 = -17\ \text{dB}$$

Final pool hall $SPL_2 = SWL_3 + SPL_1$ dB

Final pool hall $SPL_2 = (80 - 17)$ dB

Final pool hall $SPL_2 = 63$ dB

For NR 45:

$SPL = (NR\ 45 \times 0.93 + 12)$ dB

$SPL = 53$ dB. This is the maximum allowed sound pressure level at 250 Hz to achieve NR 45.

$SPL_2 > 53$ dB so NR 45 is exceeded and attenuation is needed.

The recommended noise rating for a swimming pool building is NR 45. Enter data for a lined section of duct and test whether an attenuator would be suitable. An attenuator shifts the problem frequency to 2000 Hz and 4000 Hz. Also, the critical frequency turns out to be 125 Hz. The simplest solution seems to be 3 m of acoustically lined duct after the fan, and this is better than a circular attenuator at reducing high frequency noise. This reduces the swimming pool building to NR 45 as required. Note that NR 45 only applies to the swimming pool area and not the other rooms that are associated with a sports facility.

Questions

1. Explain how the sound field in an occupied room is perceived.
2. List the sources of a sound field in normally occupied rooms and spaces within air conditioned buildings.
3. State all the plant, equipment and systems that provide an acoustic environment in rooms.
4. Explain how two or more sources of acoustic energy are combined mathematically.
5. A fan which produces a sound power level of 83 dBA at 1000 Hz is within the same plant room as a gas-fired boiler that produces a sound power level of 85 dBA at 1000 Hz. Calculate the combined sound power level that could enter the air duct system.
6. A centrifugal fan which produces a sound power level of 88 dBA at 500 Hz is within the same plant room as a refrigeration compressor that produces a sound power level of 88 dBA at 500 Hz. Calculate the overall sound power level within the plant room.
7. Explain the difference between direct and reverberant sound fields.
8. Sketch and describe the ways in which acoustic energy is absorbed in an air ductwork system serving an office.
9. State what is meant by reverberation time.
10. Find the anticipated fan sound power level when a centrifugal fan is passing 1750 l/s at a fan static pressure of 190 Pa for the range of frequencies from 31.5 Hz to 8000 Hz.
11. Find the anticipated fan sound power level when an axial flow fan is passing 3500 l/s at a fan static pressure of 100 Pa for the range of frequencies from 31.5 Hz to 8000 Hz.
12. Explain how sound enters, travels through, is attenuated by and escapes from, a ducted air duct system.
13. Write a technical report on the equipment and materials that are used to reduce noise transmission between the plant and the occupants of a building. Include sketches and data on typical products that are available to the designer.
14. Explain what is meant by the static insertion loss of an air duct silencer. State the types and performance data of duct silencers that are available.
15. Explain how air duct systems attenuate fan noise.
16. State how lined and unlined air ducts reduce the transmission of sound.
17. Obtain sound absorption data for a variety of building materials and acoustic linings for air ducts from different sources. These can be used to update and expand the worksheet data bank.
18. Sketch and describe the most effective locations for absorbent duct lining materials and duct attenuators.
19. State why the whole of the available attenuation is not taken into account during the design process.
20. Create a combined table of data to compare the attenuation provided by circular and rectangular attenuators, a 25 mm thick duct lining and a 50 mm thick duct lining for a 900 mm × 600 mm air duct. A 2 m length of the duct is to be lined and this is to include one lined bend. Discuss the relative merits of each of these methods of providing attenuation.
21. Explain why there are practical limits to the attenuation of air flow in ducts.
22. An axial flow fan has an overall sound power level of 78 dB. The plant room has a reverberation time of 2.5 s and a volume of 240 m^3. Calculate the plant room reverberant sound pressure level.
23. A fan and a refrigeration compressor are located within the same plant room. The manufacturers state that the sound power levels are 79 dB for the fan and 80 dB for the compressor. Calculate the combined sound power level in the plant room.
24. A model C44 supply air centrifugal fan passes 6500 l/s at a fan total pressure of 550 Pa. The fan impeller runs at 14 Hz. The supply air ductwork system comprises of 8 m of lined 850 mm × 750 mm duct, one lined bend with turning vanes, no branches, two unlined bends, 20 m unlined duct, one duct end reflection. The diffuser does not generate any noticeable sound power but is near the wall and ceiling junction, so directivity Q is 4. The target room is a lecture theatre for 250 people. The theatre is

30 m × 25 m and 6 m high. The nearest recipient's head is 3 m from a diffuser. The floor is carpeted. The ceiling has 15 mm suspended acoustic tiles and 50 mm glass fibre mat. The walls are plastered brick. There are no windows. Find the noise rating that is not exceeded in the theatre when empty and whether it is suitable.

25. A model AX66 600 mm diameter supply air axial flow fan passes 2000 l/s at a fan static pressure of 95 Pa. The fan impeller runs at 18 Hz. A circular attenuator is located downstream of the fan. The supply air ductwork system comprises of 10 m of unlined straight 900 mm × 700 mm duct from the fan to the first grille. The duct has a blank end and there are two unlined bends in the duct. Air leaves the duct through a grille in the side of the duct. There is no noise generation in the duct system or at the terminal grille. The ventilated office is 12 m × 8 m and 3 m high. The nearest recipient's head to the diffuser is at a distance of 1 m. The floor is concrete, the ceiling is exposed 18 mm floorboards on timber joists and the walls are plastered brick with 12 m^2 of single glazing. Find the noise rating that is not exceeded at any frequency for the occupied room and whether it is suitable.
26. An air conditioned food supermarket is 30 m × 30 m and 4 m high. The conditioned air is supplied by a model CE45 centrifugal fan that passes 7500 l/s at a fan static pressure of 390 Pa. The fan impeller runs at 21 Hz. The fan sound power levels are: 87 dB overall, 60 dB at 31.5 Hz, 55 dB at 63 Hz, 65 dB at 125 Hz, 72 dB at 250 Hz, 76 dB at 500 Hz, 83 dB at 1 kHz, 84 dB at 2 kHz, 74 dB at 4 kHz and 64 dB at 8k Hz. The supply air ductwork system comprises of 4 m of lined straight 900 mm × 900 mm duct, including two lined bends, and 12 m of the same size of unlined duct between the fan and the first air outlet grille. The duct has a blank end. There is no noise generation in the duct system or at the terminal grille. The nearest recipient's head to the diffuser is at a distance of 2 m. Directivity is 2. The floor has plastic tiles on concrete, the ceiling is 15 mm acoustic tile with 50 mm mineral fibre insulation and the walls are plastered brick. The single glazed shop window is 25 m long and 3 m high. Find the noise rating that is not exceeded at any frequency in the supermarket and whether it is suitable.
27. An air conditioned library is 25 m × 35 m and 5 m high. The conditioned air is supplied by a model CE86 centrifugal fan that passes 4500 l/s at a fan static pressure of 275 Pa. The fan impeller runs at 15 Hz. The fan sound power levels are: 77 dB overall, 40 dB at 31.5 Hz, 45 dB at 63 Hz, 55 dB at 125 Hz, 62 dB at 250 Hz, 66 dB at 500 Hz, 73 dB at 1 kHz, 74 dB at 2 kHz, 64 dB at 4 kHz and 60 dB at 8K Hz. The supply air ductwork system comprises of 10 m of 800 mm ×700 mm duct, three bends, a blank duct end and supply air diffusers. There is no noise generation at the diffusers. Directivity Q is 2. The nearest recipient's head to a diffuser is at a distance of 2 m. The floor has carpet tiles with underfelt on concrete, the ceiling is 15 mm suspended acoustic tiles with roof insulation and the walls are plastered brick. There are single glazed windows 8 m long and 2 m high. Find the noise rating that is to be provided in the library. Enter the data onto a working copy of the worksheet and calculate how to achieve the required acoustic design criteria.
28. A model CE14 extract centrifugal fan removes six air changes per hour from a hotel lounge room. The lounge is 8 m × 6 m and 3 m high. The fan static pressure is 130 Pa and the fan impeller runs at 17 Hz. The fan sound power levels are: 67 dB overall, 30 dB at 31.5 Hz, 35 dB at 63 Hz, 45 dB at 125 Hz, 52 dB at 250 Hz, 56 dB at 500 Hz, 63 dB at 1 kHz, 64 dB at 2 kHz, 54 dB at 4 kHz and 50 dB at 8kHz. The extract air ductwork system comprises of 3 m of 600 mm × 400 mm duct, one bend, a blank duct end and an extract air grille. The nearest recipient's head to the grille is at a distance of 1 m. Directivity Q is 8. The floor is carpeted, the ceiling is hard plastered to the equivalent of a plastered wall and the walls are plastered brick. There are single glazed windows 3 m long and 2 m high. Fabric curtains and soft furniture have a surface area of 10 m^2. Find the noise rating that is to be provided in the lounge. Enter the data into a working copy of the worksheet and calculate how to achieve the required acoustic design criteria.
29. An 1860 heritage building is typical of its era. No permanent structural alterations are allowed by the Heritage Council. The owner asks a consultant to advise on how to lower the noise level in

the restaurant. The room is 35 m × 20 m and 5 m high. Two long walls are bare granite. The end walls are all single glazing. The floor is bare polished timber on foundation walls. The roof structure is bare varnished timber planks on timber frame with no thermal insulation and corrugated iron sheet; take its absorption data as for the floor. 150 patrons can be seated and served by eight staff. There is no noise entering from outdoors or the kitchen. Chairs scraping on the floor are annoying. Background white noise, no distinct frequencies, appears loud.

A model D45 supply air centrifugal fan passes 10000 l/s at a fan total pressure of 700 Pa. The fan impeller runs at 12 Hz. The air conditioning supply air duct system comprises of 30 m of 950 mm × 950 mm duct, four bends and four duct end reflections. The final diffuser does not generate any noticeable sound power but is near the wall and ceiling junction, so directivity Q is 4. The nearest recipient's head is 3 m from a diffuser. Use file *dbduct.xls* to calculate the aural environment when the restaurant is full and decide on the consultant's advice.

12 Air conditioning system cost

Learning objectives

Study of this chapter will enable the reader to:

1. understand and use an industry standard method of costing a building services engineering air conditioning project;
2. become familiar with parts of *Spon's Mechanical and Electrical Services Price Book*;
3. know how to calculate the cost of materials and labour to install building services systems;
4. calculate the project selling price for an air conditioning installation.

Key terms and concepts

Air ductwork 345; all-in labour cost 342; commissioning 343; cost of material 344; cost variations 342; labour total 344; measured work price 342; overhead costs 342; project price 344.

Introduction

The capital cost and the contractor's selling price of an air conditioning system can be found from the use of published price data, trade prices and the file *accost.xls*. The user enters the quantities of materials measured from plans and elevations of the ducted air conditioning system. Current prices, labour hours, labour rates, user-defined costs, discounts and price updates are easily entered in the place of the examples shown in the worksheet.

The user of the worksheet is less likely to forget to include cost items as they are all listed. The total cost of the installation is automatically calculated. Spaces are allocated for user-defined cost items which have not been shown in the normally expected categories. The worksheet can easily be adapted for other mechanical and electrical services cost calculations such as for heating, plumbing and electrical systems, by replacing cost items with others from the source of reference, *Spon's Mechanical and Electrical Services Price Book*. This tailoring facility enables services designs to be costed rapidly.

Pricing method

Overall cost of an air conditioning system is generated from the data and method published in *Spon's Mechanical and Electrical Services Price Book*. Update the worksheet data as required. The standard price book method is to evaluate the price of each item that has been measured from the design drawings. There are measured work prices for the items to be installed into the building. Measured work price includes material cost, the labour used, discounts, overheads and profit for normal working conditions in the outer London area. Regional variation in the costs, additional items and changes to the overall price can be made in the worksheet.

Regional variation *A*% is applied to the total project material and labour costs at the end of the worksheet. The starting point for material cost is the net price; this price represents that paid by the installing contractor including delivery to site in the outer London area after the deduction of all trade discounts; it excludes any charges in respect of value added tax (VAT).

B% is an update to all the sample net prices entered onto the worksheet; enter 0% to use listed net price data as provided; entering ±% calculates a revised material cost.

The variable *C*% allows for wastage, contribution to overhead costs and for the contractor to realise a profit. This allows for the normally expected wastage of the material from unuseable offcuts of such materials as pipe, sheet metal, insulation and cable, the overhead administrative costs of finding, ordering and administering the handling of the item, and an amount of profit which is attributable to it.

Material cost is thus the cost to the purchasing contractor associated with that item.

Labour hours are the number of labour hours needed to handle and install the item. All-in hourly labour rates are inclusive of all factors in operating a gang of tradespeople in that industry for a typical large project or company. Use the sample hours and rates provided or look up current data. Labour total is the labour cost needed to install one item and found from the labour hours multiplied by the hourly labour rate.

The measured work price for each unit and the total price for all the items is intended for new installation work in the outer London area. It includes allowances for all the charges, preliminary items and profit. Adjustments can be made to each item, or to the overall total costs, where quantity discounts apply, site conditions are unfavourable or if commercial considerations indicate that prices should be raised or lowered. If large numbers of items are to be installed in easily repetitive situations with easy access, suitable variations can be applied to the discount, waste and updated percentages. Percentages can be +% or −%.

Air duct sizes are listed as the sum of two sides, so a 400 mm × 400 mm duct is shown as 800 mm size in the reference source.

Real tendered prices by suppliers can be used as net prices.

Enter user-defined items, net prices and labour hours, extending the range of cells as needed. A large range of user-defined items are listed before the end of the worksheet; these might not be exhaustive and cover such things as:

1. Prime cost sums, pre-defined fixed costs.
2. Provisional sums, declared fixed costs for unmeasured work, specified plant or planned payments.
3. Drawings and documentation.
4. Commissioning.
5. Pre-commission pipe cleaning and chemical treatment.
6. Builders' work.
7. Scaffolding and cranes.
8. Site accommodation.
9. Building information modelling.
10. Physical modelling.
11. Performance monitoring.

12. Fees for certificates and approvals.
13. Utility provisions and consumption.
14. Travel and accommodation.
15. Insurance and bonds.
16. Provision of spare parts.

Extra cost items, additional overhead *D*%, a safety or difficult risk item and previously unspecified commissioning can be entered into the totalling area of the worksheet.

The air conditioning installation contractor is required to take out a third party insurance policy which may cost 0.25% of the measured cost; this will be to provide multi-million pound insurance for any claims against the company for accidents to third parties, such as other subcontractors who may be injured by falling air conditioning ducts. This is *E*%.

The air conditioning company may be prepared to cut or raise their profit in order to win the contract against competitors, or to put themselves out of the running for the job; in which case, a negative or positive percentage which is entered as *F*%.

A difference in time between the tender price being quoted to the client and the dates when payment is to be made for the work completed can lead to changes in costs. These cost variations will be partly forecast, but some will occur at random, such as currency fluctuations on imported material prices, wage awards and general inflation. A contract which is carried out during a six month period, may suffer inflationary cost increases. An allowance for such fluctuations is made with a percentage being entered as *G*% unless such fluctuations can be charged within the contract terms.

The main contractor for the building project will require a cash discount of maybe 2.5%. The mechanical and electrical contractor's selling price is increased to provide a sum that gives the main contractor a 2.5% cash income. This is to pay for the time and expense incurred by the main contractor in attending to the subcontractor during the work.

The end result of the tender preparation is the project selling price to the main contractor, or customer.

Formulae

Representative samples of the formulae are given here. Each formula can be read on the spreadsheet by moving the cursor to a cell.

$$\text{Material price} = \text{net price} + \text{update } B\% + \text{Overheads waste profit preliminaries } C\%$$

$$\text{Material price} = \text{£ Net price} \times \frac{(100 + B\%)}{100} \times \frac{(100 + C\%)}{100}$$

EXAMPLE 12.1

A straight galvanised steel air duct of size 1400 mm, 1000 mm × 400 mm, can be purchased from an off-site manufacturer and delivered into site storage for £24.00 per metre length. By the time of delivery, a price update *B*% of 5% takes place. The mechanical contractor's overheads attributable to ductwork, plus an allowance for wastage due to handling damage and profit margin, amount to a *C*% of 20% to be added. Calculate the contractor's material cost.

$$\text{Material price} = \text{£ net price} \times \frac{(100 + B\%)}{100} \times \frac{(100 + C\%)}{100}$$

$$\text{Material price} = £24.00 \times \frac{(100 + 5\%)}{100} \times \frac{(100 + 20\%)}{100}$$

$$\text{Material price} = £30.24 \text{ per metre}$$

EXAMPLE 12.2

A galvanised steel straight air duct of size 1800 mm, 1000 mm × 800 mm costs the mechanical services subcontractor £45.00 per metre length and 36 m are ordered. By the time of delivery a price update $B\%$ of 2.5% takes place. The mechanical contractor's overheads attributable to ductwork, plus an allowance for wastage due to handling damage and profit margin, amount to a $C\%$ of 22.5% to be added. The all-in labour rate for the ductwork erectors is £29.50 per hour and the work takes 1.75 hours per metre run of duct. Calculate the contractor's measured unit and total cost for this part of the work.

$$\text{Material price} = £ \text{ Net price} \times \frac{(100 + B\%)}{100} \times \frac{(100 + C\%)}{100}$$

$$\text{Material price} = £45.00 \times \frac{(100 + 2.5\%)}{100} \times \frac{(100 + 22.5\%)}{100}$$

$$\text{Material price} = £56.50 \text{ per metre}$$

$$\text{Labour cost per measured unit} = \frac{£29.50}{\text{h}} \times 1.75 \text{ h}$$

$$\text{Labour cost per measured unit} = £51.63$$

$$\text{Total cost per measured unit} = £56.50 + £51.63$$

$$\text{Total cost per measured unit} = £108.13 \text{ per m}$$

$$\text{Total cost to mechanical contractor} = 36 \text{ m} \times \frac{£108.13}{\text{m}}$$

$$\text{Total cost to mechanical contractor} = £3892.61$$

There may be some fluctuation in the contract price due to the effect of unplanned changes in the costs. Examples of these fluctuations are specific inflation in the price of materials, unexpected wage awards or the result of fluctuation in the relative values of the currencies which are used in the acquisition of imported materials and plant. Such fluctuation can be upwards or downwards. An allowance for general inflation, or other fluctuations, which is anticipated to take place during the length of the contract period can be made. An increase or decrease percentage can be entered as $G\%$ for the entire project. Specific items of plant or materials, such as a large imported water chiller or copper pipe, may have price fluctuations entered on their costing lines.

The main contractor requires to be paid a cash discount by each subcontractor for attendance to their work, site meetings and work carried out for the subcontractor, such as forming holes in concrete, cutting, installing pipe and duct brackets into the structure, and other matters as they arise. This may be 2.5% of the final subcontract price. An amount has to be added to the total project cost which will generate a 2.5% discount when it is taken from the gross project price. A 2.5% discount equates to adding one thirty-ninth to the project price to arrive at the final selling price.

EXAMPLE 12.3

An air conditioning installation has a finalised price of £198000. Calculate the project selling price when the main contractor's 2.5% discount is included.

$$\text{Cash discount to main contractor} = \frac{£198000}{(100-2.5)} \times 100 - £198000$$

$$\text{Cash discount to main contractor} = £5076.92$$

$$\text{Project selling price to main contractor} = £198000 + £5076.92$$

$$\text{Project selling price to main contractor} = £203076.92$$

Check result by deducting 2.5% from project selling price,

$$\text{Subcontractor receives} = £203076.92 - \frac{2.5}{100} \times £203076.92$$

$$\text{Subcontractor receives} = £198000$$

Alternatively,

$$\text{Project selling price to main contractor} = £198000 + \frac{1}{39} \times £198000$$

$$\text{Project selling price to main contractor} = £203076.92$$

Questions

Use the file *accost.xls* for all questions. Model solutions are not provided as tenders can be based on current prices and labour hours and be in competition with others in a class situation.

1. A toilet extract system of air ductwork is part of a new construction project and is to have the following components: 12 m of 400 mm × 400 mm air ductwork, one stopped end, six 90° curved bends, one 45° bend, two branches, two transforms downwards from the duct size, one fire damper 400 mm × 400 mm, two access doors 235 mm × 90 mm, two test holes and seals, one 600 mm × 600 mm extract grille and damper, two axial flow fans, 0.47 m^3/s each, two sets of anti-vibration mountings for 25 kg fans, two push button starters for 0.5 kW fans, two contactor relays for fan interlocking, six 13 amp electrical socket outlets with PVC cable in conduit, commercial. Make your own decisions about what else is to be included in the contractor's selling price and calculate your tender figure.
2. An air handling unit supplies heated air to a manufacturing area of a single-storey building. The extract air has become contaminated and is discharged to the atmosphere. The factory is in the north of England where the regional price variation is +18%. The ductwork installation will require the use of scaffolding. Use the following information to calculate the contractor's tender price to the main contractor.

   Supply ductwork: Air intake weatherproof grille 2 m × 2 m and transition to the air handling plant; one air handling unit 1 m × 1 m in cross-section with air heater, filter and centrifugal fan which passes

$3\ m^3/s$; one motorised damper of 1 m × 1 m; one transition from air handling unit to a 800 mm × 600 mm duct; 50 m of 800 mm × 600 mm duct; five 800 mm × 600 mm 90° curved bends; six branch ducts for grilles; six supply grilles with damper 600 mm × 300 mm; one 800 mm × 600 mm blank duct end; four access doors 235 mm × 140 mm; three test holes and seals.

Extract ductwork: air discharge weatherproof grille 2 m × 2 m; one transition to the centrifugal extract fan which carries $3\ m^3/s$; one set of anti-vibration mountings for a 115 kg fan; one transition from the extract fan to the ductwork; one motorised damper 800 mm × 600 mm; 20 m of 800 mm × 600 mm duct; two 800 mm × 600 mm 90° square bends with turning vanes; one transition to a 1 m × 1 m grille; one extract grille and damper 1 m × 1 m; one 800 mm × 600 mm blank duct end; two access doors 235 mm × 140 mm; one test hole and seal.

Control equipment: three air temperature detectors; 1 Pa XY36 control unit, list price £300 and 3 hours labour; two 2 kW fan push button starters; one contactor relay; two damper drive motors; one 32 mm three-port hot water diverting valve.

Electrical supply: one four-way triple pole and neutral 415 volt distribution board; 120 m of $4\ mm^2$ 415 volt cable in conduit; six 240 volt 13 ampere socket outlets, PVC cable in conduit for a commercial application.

Hot water pipework: 220 m of 40 mm black medium steel low pressure hot water heating flow and return pipework, brackets and fittings; 220 m of 40 mm pipe thermal insulation; four 40 mm gate valves; one 40 mm commissioning double regulating valve; two air vents; two water pipe test points.

Concluding items: regional variation 18%; hire of scaffolding prime cost sum £2500; additional overhead 5%; safety risk item £750 for use of scaffolding; commissioning cost £275; decrease profit by 3% due to competition; fluctuation during contract 5%. There is only the mechanical and electrical contractor on this existing site.

3. An air conditioning system is to be installed in a new information technology room for 200 personal computer work stations in a university located in Scotland, regional price variation −15%. There is no space for air ductwork, so packaged cooling units are to be installed within the room. Refrigeration condensing units will be located on the flat roof of the room.

   The room cooling load for each computer work station is calculated to be 200 W from the computer, 110 W from the user, 25 W from the lighting, 100 W from solar heat gains and 20 W from the outdoor air ventilation. There are to be two refrigeration condensing units on the roof in order to provide 100% standby capacity. Each condensing unit is to have a 5.5 kW push button starter. A contactor relay is to be installed to avoid the possibility of simultaneous operation of the two condensing units.

   Each room cooling unit will have 12 m of 20 mm diameter refrigeration pipework from the condensing units. Assume that this will be in black medium weight steel tube and fittings for the purpose of this exercise only. Both pipes will be insulated. There are no control items or valves in the pipelines as all the controls are part of the condensing unit and room unit packages.

   Each room cooling unit has a 13 ampere 240 volt socket outlet with PVC insulated cable in conduit for a commercial application. Each condensing unit has 20 m of $6\ mm^2$ 415 volt three-phase cable in conduit. A four-way triple pole and neutral 415 volt distribution board is installed for the refrigeration system.

An extra cost item of £2500 is to be allowed for repairs to the structure and decoration which will be necessary as the result of the installation work. No additional overhead costs are expected. There are no additional safety costs. Commissioning the installed system will take 2 days work.

A highly competitive price is to be tendered and the contractor will reduce the allowance for profit by 8%. The contract price will be fixed for the duration of the agreed period and there is no main contractor or discount. Calculate the cooling load in kW and the number and capacity of the split package room cooling units which are to be used. Note that there are several possible solutions and these will be reflected in the project price. Evaluate a competitive project selling price.

13 Question bank

Learning objectives

Study of this chapter will enable the reader to:

1. practice answering short questions from a multiple choice of answers;
2. prepare for tests, assignments and written examinations where this form of questioning is provided;
3. evaluate why some answers are not entirely correct for the question;
4. have discussions with peers and instructors over the meaning of the incorrect answers;
5. lead into further study and investigation;
6. answer a wide range of questions from this book;
7. enjoy the stimulation of learning as much as the author has.

Key terms and concepts

Acronyms 349; air conditioning systems 349; air quality 354; Building Management System 349; *Carbon Plan* 349; cooling towers 351; data transmission 349; ducts 352; emissions 349; energy 349; fans 352; green building 351; humidity 351; inverter 350; Kelvin 351; *Legionella* 357; low energy building 353; mass cooling 360; noise 358; NR 359; odours 351; refrigeration 358; reverberant 359; sustainability 352; temperature 351; thermal comfort 354; zero carbon 349.

Introduction

This a general knowledge section of questions covering topics within this book. Some questions may require the reader to look up answers in additional resources or search the internet. There is only one correct answer to each question unless it is specified as having more. Incorrect answers may be partially true but are not considered by the author to be the entirely correct response for the purpose of this book; these may stimulate additional study, discussion, questioning with peers or instructors.

Question bank

1. Which of these acronyms is correct?

    1. ASHRAE means Australian Society for Heating, Refrigerating and Air Engineering.
    2. AIRAH means American Institute for Refrigeration and Air Heating.
    3. CIBSE stands for The Chartered Institution of Building Services Engineers.
    4. BSRIA stands for British Services Refrigeration Institute for Air Conditioning.
    5. CIC is the Council for Industry and Construction.

2. Which does the HM Government Carbon Plan (2011) do for air conditioning in buildings?

    1. Has little effect on primary energy consumption.
    2. Diminishes energy use in some buildings.
    3. Prescribes how buildings are to be air conditioned.
    4. Is an alternative to the EU ETS.
    5. Encourages reducing CO_2 emissions.

3. Do air conditioning and lift systems lead to obesity?

    1. Yes, absolutely.
    2. Obesity is due to overeating only.
    3. Obesity is due to working at computers too long each day.
    4. Metabolism slows and we eat more in air conditioned environments.
    5. Modern facilities tend to reduce our physical activity and we eat less natural food.

4. Which of these is not a common standard for data transmission in a building management system?

    1. Ethernet.
    2. RS484.
    3. RS232.
    4. RS124.
    5. C-bus.

5. Which is correct?

    1. Building designers accurately calculate future energy use.
    2. Architects do feedback studies of all their buildings.
    3. New buildings always work perfectly as design predictions.
    4. PROBE stands for probable recycling of built environment.
    5. Only careful analytical review establishes how a new building achieves its objectives.

6. What does sustainability mean for low energy buildings?

    1. There is no such thing as a modern sustainable building.
    2. Everything used in the building's service life comes from globally sustainable resources.
    3. The building is an example of good modern design practice.
    4. All waste output from the building is recycled.
    5. The building has been constructed from organically grown materials.

7. Which describes a zero carbon building?

    1. Supplied from solar power systems.
    2. Forested timber, no glass, naturally heated and ventilated.

3. Consumes a minimum amount of energy for all uses.
4. Net exporter of electricity.
5. Probably no such thing.

8. Which correctly describes air conditioning heat transfer?

   1. Sensible heat transfer comprises all types.
   2. Latent heat transfer raises temperature.
   3. Sensible heat transfer is logged by a thermocouple and thermistor.
   4. Latent heat transfer is hidden from view.
   5. A CHW cooling coil lowers air temperature and creates condensation.

9. How is air leakage by a building measured?

   1. Anemometer readings at every opening to outdoors.
   2. Fill with smoke and measure time taken to fully disperse.
   3. Cannot be done at all.
   4. A large fan sucks air out of whole building at −50 Pa and flow is measured.
   5. A large fan pressurises the building at 50 Pa and flow is measured.

10. AR, BER, SER, DEC and iSBEM all relate to which?

    1. Electrical engineering services and systems.
    2. They are meaningless acronyms.
    3. Types of zero carbon buildings.
    4. Emission rating of a building.
    5. Types of energy audit.

11. When designing the shape of a building:

    1. Maximise exposure to solar warming;
    2. Ignore location, make it impressive;
    3. Minimise solar overheating;
    4. Square in plan is always better for energy saving;
    5. Minimise external surface area.

12. We sense odours within air conditioned buildings by:

    1. Identifying smells;
    2. Breathing onto others;
    3. A measuring instrument;
    4. Tasting them in our mouth;
    5. Olfactory response.

13. What does inverter mean?

    1. Alternating current phases are reversed.
    2. It is an electronic soft starter for a three-phase motor.
    3. Incoming 50 Hz alternating current is digitally reformed into an output frequency to a motor in the range from 0 to 20000 Hz.
    4. Incoming 50 Hz alternating current is digitally reformed into an output frequency to a motor in the range from 0 to 50 Hz.
    5. Alternating current is electronically converted into direct current to drive a motor.

14. What are fluorinated hydrocarbons used for?

    1. Swimming pool water treatment.
    2. Biocide decontamination of cooling towers.
    3. Ozone-depleting refrigerants.
    4. Non-CFC foam insulation and furnishings.
    5. Combustible gaseous fuel.

15. Which is correct about the density of humid air?

    1. Decreases with increasing pressure.
    2. Increases with increasing air temperature.
    3. Varies with air temperature and pressure.
    4. Not affected by humidity.
    5. Increases as air velocity increases.

16. Satisfactory air conditioning may be deemed when:

    1. 100% of the full-time occupants are satisfied.
    2. 85% of the full-time occupants are satisfied.
    3. 50% of the full-time occupants are satisfied.
    4. Complaints cease.
    5. Odours have been eliminated.

17. What does Greenhouse Rating of a building stand for?

    1. The higher the greenhouse gas production due to the building, the higher the greenhouse rating.
    2. A 10 star building produces no greenhouse gases.
    3. Assessed greenhouse gas emission standard of a building.
    4. Any new building even if not painted green.
    5. An emission standard applied to all types of buildings.

18. Which is correct about Kelvin?

    1. Name of engineer who designed the first steam engine.
    2. Unit of heat.
    3. Measured in kJ/kg s.
    4. Temperature scale.
    5. Absolute temperature.

19. Which is correct about low energy buildings?

    1. A low energy building is one that requires the minimum amount of primary resource energy to build it.
    2. A low energy building may consume more energy to construct.
    3. A low energy building consumes less energy during its 100+ years of use than an equivalent building.
    4. We have no idea what an equivalent building is for a specific site.
    5. All buildings consume uncontrolled amounts of energy.

20. Which might be a means of reducing refrigeration system energy usage?

    1. Install smallest capacity compressors possible.
    2. Carry out frequent maintenance checks and parts replacement.

3. Switch reciprocating compressors off for as long as possible and maintain wide temperature differentials.
4. Use outdoor air cooled condensers.
5. Variable refrigerant volume scroll compressor with software controlled digital operation programmable for all variations in year round duties.

21. Which is correct about low energy building designs?

1. Are always modern and look impressive.
2. Are always found to be ideally comfortable by users.
3. Must have large windows and glazed walling.
4. Must have small windows and high levels of thermal insulation.
5. Should consume a minimum of primary energy when compared with similar types and sizes of buildings.

22. What does sustainability mean for low energy air conditioned buildings?

1. The mechanical and electrical services within this building all have a low maintenance requirement.
2. All the water, sewerage, paper and plastic waste output from this building go to recycling.
3. All the light bulbs and tubes from this building are recyclable.
4. Somebody has found a good argument why this design of building is less harmful to the global environment than competitive designs.
5. This building has consumed, and will continue to consume, more of the Earth's physical resources than it can ever put back.

23. How is the air pressure sealing, or leaking, ability of a building found?

1. Close all doors and windows. Close spill air and exhaust air dampers. Run supply and return air fans. Measure internal static air pressure for one hour to see if it can be maintained at a set value.
2. Close all doors, windows, air vents, exhaust air outlet ducts and spill air dampers. Run supply and return air fans. Raise building air static internal pressure to 50 Pa above outdoor atmosphere barometric pressure. Switch fans off. Measure rate of decay of indoor air pressure. Use a formula to calculate air leakage rate from building.
3. Switch off all fans. Seal all mechanical ventilation openings into building with polythene sheet. Fit false main entrance door having a pressurising fan, duct and air flow meter. Run pressurising fan to maintain a specified internal air static pressure. Measure steady inflow rate, this is building air tightness measurement.

24. Air dry bulb temperature is measured by:

1. Suspending a sensor about 1.0 m below the ceiling and waiting for it to stabilise.
2. Reading the Building Management System computer screen data from a fixed sensor in the room.
3. Leaving a sling psychrometer on a desk for an hour.
4. Shielding a mercury in glass thermometer from room air draughts.
5. Rotating a sling psychrometer at head height in room air for one minute and taking an immediate reading.

25. Which is not one atmospheric pressure?

1. 300 inches of mercury.
2. 1.0 bar.

3. 14.7 pounds per square inch, psi.
4. 1013.25 millibars, mb.
5. 101325 Pa.

26. Where could carbon monoxide, benzene and toluene gases have come from if detected within an air conditioned low energy building?

    1. Water chiller plant room refrigerant leakage.
    2. Hydrocarbon natural gas combustion water heating plant.
    3. Drains and sewers.
    4. Cleaning fluids and off-gassing from furnishings.
    5. Outside air intake to AHU or people smoking tobacco or cannabis.

27. Air dry bulb temperature is dependent upon:

    1. People and furniture.
    2. Size of room.
    3. Radiation sources within the room.
    4. Solar heat gain through the windows.
    5. Air velocity in the room.

28. Which is correct about peak summertime temperature in buildings?

    1. A low energy building is one that always overheats.
    2. Victorian era houses and large buildings never overheat.
    3. Any building anywhere in the world, even an igloo, can become overheated.
    4. We have no idea why buildings become too hot, they are all ventilated.
    5. Get used to it, HM Government Carbon Plan (2011) will stop the widespread use of air conditioning in the UK.

29. Atmospheric vapour pressure is:

    1. The total pressure of the atmosphere at the time.
    2. The pressure exerted on the ground by the dry gases of the atmosphere above sea level.
    3. The sum of the clouds, wind and static air forces on the ground.
    4. That part of the barometric pressure produced by the water vapour in humid air.
    5. None of these.

30. Where does legionaires' disease originate?

    1. French Foreign Legion.
    2. Drains and sewers.
    3. Cold water storage tanks.
    4. Hot water storage cylinders.
    5. Aerosols from cooling towers, shower heads, spray taps, spa baths and humidifiers.

31. Air conditioning engineers consider atmospheric air pressure to consist of which?

    1. Polluted air.
    2. Vapours, gases and water vapour.
    3. Around 800.0 mb from dry gases plus 213.0 mb due to water vapour.
    4. Around 700.0 mb from dry clean gases, 200 mb from polluting vapours and dusts plus 113.0 mb due to water vapour.
    5. Around 990.0 mb from dry gases plus 20.0 mb due to water vapour.

32. Air quality within a building depends upon:

    1. Number of people indoors.
    2. How much and where air pollutants are found.
    3. Relative humidity of room air.
    4. Dry bulb air temperature.
    5. Plants, animals and furnishings in the building.

33. Which is the likely outcome from inadequate outdoor air ventilation?

    1. Comfortably warm houses and offices.
    2. Less draught.
    3. Suppression of house dust mites, condensation and mould growth due to warmer environment.
    4. Inadequate removal of house dust mites, condensation and potential mould growth.
    5. Lower energy costs.

34. Which is correct about air humidity?

    1. Spraying water into room air heats up the room.
    2. Evaporating water consumes sensible heat energy.
    3. Evaporating water consumes latent heat energy.
    4. A room with a relative humidity of 25% feels humid.
    5. Every air conditioning system must have a humidifier system.

35. What is air percentage saturation?

    1. Water suspended in air relative to the same quantity of liquid water.
    2. Amount of moisture in air above a base of zero.
    3. Absolute moisture content of humid air.
    4. Same as relative humidity.
    5. A ratio.

36. Which of these is not a contaminant of indoor air quality (IAQ)?

    1. Cigar smoke.
    2. Carbon dioxide.
    3. Benzene, toluene, formaldehyde and ethylene glycol.
    4. Carbon tetrachloride.
    5. NO_X.

37. Which is correct about air humidity?

    1. Moisture in room air finds its own way out of the building.
    2. Moisture gained by room air will always condense somewhere and drain away.
    3. Moisture within building air will always condense into liquid at the lowest surface temperature location.
    4. Natural ventilation does not remove moist air from a building.
    5. Only mechanical exhaust systems remove moist air from a building.

38. An anemometer is:

    1. For measuring fan vane angles.
    2. For assessing animosity towards the room conditions.
    3. A calibrated device to measure air speed in a room, outdoors or an air duct.

4. A rotating vane with thermistor or heated wire sensor.
5. Only to be used by qualified personnel.

39. In well-insulated non air conditioned buildings having modest glazing areas and little air movement, which will operative temperature be closest to?

    1. Globe temperature.
    2. Mean radiant temperature.
    3. Wet bulb temperature.
    4. Dry bulb air temperature.
    5. Environmental temperature.

40. Which does not apply to heat stroke in a hot climate?

    1. To avoid it, get into a swimming pool.
    2. Occurs at a body temperature of 40.6°C.
    3. Sweating ceases.
    4. Body becomes involuntarily hyperactive.
    5. Body becomes comatose, brain damage from reduced blood supply and death is imminent.

41. Which is correct about a VAV air conditioning system?

    1. Stands for valve authority value.
    2. Only used in hotels and conference centres.
    3. Reducing room demand for cooling opens the VAV damper.
    4. Rise in zone air temperature causes the VAV damper to throttle the cool supply air flow further.
    5. Single duct all-air system with a throttling damper at each room supply air outlet.

42. Which is appropriate for an FCU system?

    1. Does not require distribution air ducts.
    2. Self-contained air conditioning unit.
    3. Only requires an outside air duct and electrical power connection.
    4. A cooling only terminal unit.
    5. Usually have air ducts, chilled water, hot water flow and return plus a condensate drain to sewer pipework.

43. Which of these applies to packaged room air conditioning units?

    1. Always connected to a ducted air system.
    2. Always connected to a central chilled water plant system.
    3. Each unit has a refrigeration compressor.
    4. Always very quiet operation.
    5. Power demand not exceeding 250 W.

44. Wet bulb thermometer:

    1. No such thing.
    2. Dry bulb mercury in glass thermometer immersed in a water tank.
    3. Does not work in humid air.
    4. Used inside a 38 mm diameter black copper globe.
    5. Mercury in glass thermometer having a wetted cotton sock covering the sensing bulb.

45. How is sick building syndrome defined?

    1. Many people consider working in the building makes them sick.
    2. Publicised condemnation of the building.
    3. That combination of health malfunctions that noticeably affect more than 5% of the building's population.
    4. Accumulation of health malfunctions noticeably affecting 25% of the building's users.
    5. User formalised surveys finding overall dislike for an inadequately comfortable environment.

46. Which is not correct about air ductwork?

    1. Spiral wound flexible fabric ducts make final connections to terminal units and diffuser boxes.
    2. Air ducts have taped joints for air tightness.
    3. Air duct leakage is unimportant.
    4. Galvanised sheet steel ductwork has riveted or flanged joints.
    5. Air ducts can be cleaned internally.

47. Difference between dry and wet bulb thermometer readings:

    1. Is called the wet bulb depression.
    2. Measures room atmosphere depression.
    3. Is used to find the vapour pressure of the room air.
    4. Wet bulb temperature is always higher than the dry bulb temperature due to evaporative heat transfer.
    5. Dry bulb temperature is always higher than the wet bulb temperature due to evaporative heat transfer.

48. Ventilation rates:

    1. Are never measured.
    2. Designed rates are never achieved due to duct losses.
    3. Vary from 4.0 to 25.0 air changes per hour for air conditioned office spaces.
    4. Cause drafts.
    5. Mean that supply air grilles in office ceilings must direct air away from sedentary personnel.

49. Which is a correct description of the dual duct air conditioning system?

    1. Duplicated supply and return air ducts.
    2. A reduced cost design.
    3. Simultaneous heating and cooling to adjacent rooms.
    4. Not used in commercial office buildings.
    5. Appropriate for low energy new buildings.

50. What is the meaning of chilled beam?

    1. Structural steel beam that is kept cool by the air conditioning system.
    2. Structural steel beams supporting the weight of the air conditioning system water chiller compressors.
    3. Air conditioning surface operating at below the occupied room air dew point temperature.
    4. A chilled water surface providing only radiant cooling.
    5. Finned pipes or flat panels at high level in offices providing convective cooling surfaces.

51. Which is correct about commissioning air duct systems?

    1. Air duct systems do not need to be inspected during commissioning.
    2. All air ducts must be internally cleaned prior to commissioning.
    3. All air ducts must be internally inspected with remote controlled lamps and cameras before use.
    4. Rough internal projections, rivets and metal cuttings are removed by the commissioning technician.
    5. Air duct systems are sealed in sections and pressure tested for an airtightness standard compliance.

52. Single duct air conditioning systems are used:

    1. In multi-roomed office and hotel bedroom applications.
    2. With hot and chilled water pipe distributions to each fan coil unit.
    3. To condition a single large volume occupied space such as a lecture theatre.
    4. To service several hospital wards and departments from one air handling unit.
    5. To minimise the size and cost of the refrigeration plant.

53. How could a chilled water cooling coil distribute bacteria into occupied air conditioned rooms?

    1. It cannot, as air temperature remains too cool.
    2. It will not under normal operation.
    3. The condensate water trap between the drain tray and sewer always maintains a water seal.
    4. The water seal in the P-trap between the drain tray and sewer may become dehydrated and allow sewer gases to pass into the air handling unit and supply duct.
    5. It will not when adequately maintained in accordance with codes and standards.

54. Which is a primary characteristic of a cooling tower?

    1. Quiet operation.
    2. Uses almost no water.
    3. Potential source of water-based *Legionella* bacteria for outdoor air.
    4. Compact unit usually installed within a chiller plant room.
    5. Functions equally well in any outdoor climate.

55. Which of these does a cooling tower not do?

    1. Rejects heat from the building.
    2. Cools condenser cooling water at 35°C when outdoor air is at 40°C d.b.
    3. Cools the evaporator circuit.
    4. Evaporates condenser water.
    5. Only functions when outdoor air wet bulb temperature remains below incoming condenser cooling water temperature.

56. What drives the fan in a large air handling unit?

    1. Gas engine prime mover.
    2. Six-pole, phase electric motor.
    3. Single-phase synchronous alternating current motor.
    4. 415 volt AC motor.
    5. 240 volt AC motor.

57. Sound waves repeat at a frequency due to:

    1. Absorption by porous surfaces.
    2. Wind forces.

3. Multiple sources of sound.
4. Passage of blades in a rotary machine such as a compressor, pump or turbine.
5. Variations in air pressure.

58. How is noise transmission from plant reduced?

    1. Cannot be reduced, only contained within plant room.
    2. Select quieter plant.
    3. Turn the plant room into an anechoic chamber.
    4. Locate plant room away from occupied rooms.
    5. Flexible rubber and spring mountings.

59. How is noise transmission from plant reduced?

    1. Porous sound absorbing materials.
    2. Thicker concrete walls and floors.
    3. Bolt refrigeration compressor to concrete slab.
    4. Line plant room with lead.
    5. Sit further away from it.

60. Which item of plant does not normally create noise?

    1. Fans and pumps.
    2. Refrigeration compressors.
    3. Fired water heaters.
    4. Air compressors.
    5. Piped systems.

61. Which frequency range is audible?

    1. 2 kHz to 200 kHz.
    2. Infinite range.
    3. 200 kHz to 1 MHz.
    4. 0 Hz to 10000 Hz.
    5. 20 Hz to 20 kHz.

62. A noisy machine on a plant room base:

    1. Radiates direct sound in straight lines only.
    2. Fills the plant room with noise.
    3. Sounds equally noisy from all directions.
    4. Produces a spherical field of sound waves.
    5. Produces a hemispherical sound field.

63. A noisy machine on a plant room concrete floor:

    1. Has no sound directivity.
    2. May direct sound more strongly in a particular direction.
    3. Sends a direct sound field through the floor.
    4. Only creates a reverberant sound field.
    5. Gains a benefit from sound energy absorbed by the floor.

64. Which does NR stand for?

    1. Noise resonance.
    2. Normal rating.
    3. No resonance.
    4. Noise ratification.
    5. Noise rating.

65. How are noises related to human ear response?

    1. Humans respond to sound power level within a range of audible frequencies.
    2. Humans respond to loudness produced over a range of audible frequencies.
    3. Sound pressure levels are added to create an overall relationship to ear response.
    4. Sound power levels are added to create an overall relationship to ear response.
    5. The loudest sound at any frequency is taken as ear response.

66. How is noise related to human ear response?

    1. Noise rating curves specify equal sound power levels for all frequencies.
    2. Noise rating curves specify equal sound pressure levels for any frequency.
    3. Noise rating curves specify equal loudness for a range of frequencies.
    4. Noise ratings are subjectively assessed.
    5. Machines are given a noise rating value.

67. How is noise transmission from plant reduced?

    1. Cannot be reduced, only contained within plant room.
    2. Select quieter plant.
    3. Seal plant room doors.
    4. Locate plant room away from occupied rooms.
    5. Flexible rubber and spring mountings.

68. What is a reverberant sound field?

    1. Sound transmitted over a large distance.
    2. Sound passing through a structure.
    3. What remains within an enclosure after source energy is absorbed by the building structure.
    4. Reflected sound.
    5. Sound pressure level measured in an anechoic laboratory chamber.

69. Which is correct for sound reverberation time?

    1. Time between echoes.
    2. How long a sound level continues.
    3. Time lag between directly received sound and reverberated sound.
    4. Time for a sound to decrease to zero within a room after the source is switched off.
    5. Time taken for a sound to decrease by 60 dB.

70. Which is not correct about an anechoic chamber?

    1. It has no reverberant sound field.
    2. Its walls, floors and ceiling are perfect sound absorbers.
    3. It allows a spherical sound field from a centrally placed source.
    4. It is used for measuring reverberation time from a test item.
    5. It is a laboratory to measure sound power level of an item.

71. Sound waves repeat at a frequency due to:

    1. Absorption by porous surfaces.
    2. Wind forces.
    3. Multiple sources of sound.
    4. Passage of blades in a rotary machine such as a compressor, pump or turbine.
    5. Variations in air pressure.

72. Sound transmission occurs:

    1. From high SWL to low SPL.
    2. Through dense concrete.
    3. In the direction of airflow.
    4. From a location of higher sound pressure level to a location of lower sound pressure level.
    5. From a location of higher sound power level to a location of lower sound power level.

73. Sound pressure level, SPL dB, within a room is:

    1. Reverberant sound field.
    2. Direct sound field.
    3. Reflected sound.
    4. Greatest noise source.
    5. Summation of direct and reverberant sound fields.

74. Which is correct about fabric energy storage?

    1. Inconsequential in sizing air conditioning plant.
    2. Thermal insulation greatly reduces heat flow into walls and roofs.
    3. Fast response by the building to temperature controllers is preferred.
    4. Warmth and coolness stored in concrete is useful.
    5. Labyrinths do not work as they are sources of dust and infestation contamination in the outside air stream.

75. What use are hollow concrete floors?

    1. None.
    2. They do not add or extract heat from air supplied through them.
    3. Only used as pipe and cable ducts.
    4. Sound hollow to foot traffic and are an annoyance to the lower floor occupants.
    5. Tempers supply air as part of a low energy air conditioning system.

76. And finally, what have I learnt from this study?

    1. Nothing, it is all a fog to me!
    2. Mechanical and electrical services within a building are not very important to the overall concept of the design and construction.
    3. I can design or construct buildings; someone else must worry about the fiddly bits.
    4. The building will work without the mechanical and electrical services anyway.
    5. I now appreciate the importance and main features of air conditioning!

14 Understanding units

Learning objectives

Study of this chapter will enable the reader to:

1. practice answering short questions from a multiple choice of answers;
2. test own understanding of using units of measurement;
3. prepare for tests, assignments and written examinations where this form of questioning is provided;
4. evaluate why some answers are not entirely correct for the question;
5. have discussions with peers and instructors over the meaning of the incorrect answers;
6. lead into further study and investigation;
7. answer a range of questions relevant to this book and air conditioning generally.

Key terms and concepts

Acceleration due to gravity 362; Atmospheric pressure 362; Celsius 367; Density of water 364; Electrical units 365; Exponential 364; Frequency 366; Humid air 363; Joule 365; Kelvin 367; Newton 365; Pressure 366; Pressure drop rate 369; Specific heat capacity 363; Standard atmosphere 362; Stefan–Boltzmann 363; Time 366; Volume 368; Watt 365.

Introduction

This a general knowledge section of questions covering measurement units. All questions should be understood by students of this topic area. There is only one absolutely correct answer to each question. Tackling these may stimulate additional study, discussion, questioning with peers or instructors.

Questions

1. Which of these equals one standard atmosphere at sea level?

   1. 1.013 tonne/m^2
   2. 1 bar
   3. 10000 N/m^2
   4. 1013.25 mb
   5. 10^6 N/m^2

2. Which of these equals one standard atmosphere at sea level?

   1. 1013 tonne/m^2
   2. 10^5 bar
   3. 10^9 N/m^2
   4. 14.7 lb/in^2
   5. 10^6 N/m^2

3. Which of these equals one standard atmosphere at sea level?

   1. 1×10^5 pascals, Pa
   2. 1.01325×10^5 N/m^2
   3. 1×10^4 N/m^2
   4. 30 m H_2O
   5. 1013.25 mm Hg

4. Which of these equals one standard atmosphere at sea level?

   1. 9.807 m H_2O
   2. 29.35 m H_2O
   3. 10.3 m H_2O
   4. 101325 kJ/m^2
   5. 1.205 kg/m^2

5. Which of these is the acceleration due to gravity, g?

   1. 10 m/s^2
   2. 30 ft/s^2
   3. 186000 miles per hour
   4. gravity is static
   5. 9.807 m/s^2

6. Which of these describes the acceleration due to gravity, *g*?

   1. Calculated from a 1 kg weight free falling from a height.
   2. Relative to distance from the Moon.
   3. Constantly 9.807 m/s^2.
   4. Varies with height above sea level.
   5. Inversely proportional to depth below sea level.

7. Which is the Specific Heat Capacity of air?

   1. Sensible heat content, kJ/kg.
   2. Total heat content, kJ/kg.
   3. 1.205 kJ/kg

4. 1.012 kJ/kg K
5. 4.186 kJ/kg K

8. Which is the Specific Heat Capacity of air?

   1. Ratio of C_p/C_v
   2. Cannot be defined.
   3. Varies with atmospheric pressure.
   4. 1.012 kg K/W
   5. 1.012 kJ/kg K

9. Which is the Specific Heat Capacity of water?

   1. 1.013 kW/kg K
   2. 1.012 MJ/kg K
   3. 4.186 kg K/kW
   4. 4.186 kJ/kg K
   5. 4.2 kg K/kJ

10. Which is correct about the Specific Heat Capacity of water?

    1. Varies with water pressure.
    2. 4.19 kW s/kg K
    3. A ratio.
    4. 1.102 kJ/kg K
    5. Used to calculate the flow rate of heating and cooling system water.

11. Which is correct about the Stefan–Boltzmann constant?

    1. Used to calculate convective heat transfer.
    2. 4.186 kJ/kg K
    3. 1.012 kJ/kg K
    4. 5.67×10^{-8} W/m^2 K^4
    5. Combines convective and radiant heat transfer.

12. Which is correct about the density of humid air?

    1. Decreases with increasing pressure.
    2. Increases with increasing air temperature.
    3. Varies with air temperature and pressure.
    4. Not affected by humidity.
    5. Increases as air velocity increases.

13. Which is correct about the density of humid air?

    1. 4.186 kg/m^3 at 21°C, 60% relative humidity.
    2. 1.013 kg/m^3 at 20°C, sea level.
    3. 0.802 m^3/kg
    4. 1.205 kg/m^3 at 20°C, 1013.25 mb
    5. 5.67 kg/m^3

14. Which is correct about the density of water?

    1. Always 10^3 kg/m^3
    2. 1013.25 m^3/kg

3. 101325 kg/m^3
4. 1.205 MJ/m^3 K
5. 1000 kg/m^3 at 4°C

15. Which is correct about the density of water?

1. Cannot be measured.
2. Cannot be measured accurately.
3. Always relative to the specific gravity number.
4. 1000 times that of air.
5. Specific gravity is 1.0.

16. Which is correct about the density of water?

1. 1.205×10^5 kg/m^3
2. 1 $tonne/m^3$ at 4°C.
3. 1.012×10^3 kJ/m^3.
4. 1.27 kJ/kg K.
5. 100 g/cm^3.

17. Which is correct about the density of water?

1. 1 g/cm^3.
2. 1.2×10^3 kg/m^3.
3. Specific gravity is 4.186.
4. 1000 $tonne/m^3$ at 10°C.
5. 1 kg per litre.

18. What does the exponential, e, mean?

1. A logarithm.
2. A variable number.
3. Always 10^x, ten to the power x.
4. 2.718.
5. Has no meaning.

19. What does the exponential, e, mean?

1. Something which is raised to a power.
2. $e = 10^x$.
3. The sum of an infinite series.
4. $e = \sqrt{-1}$.
5. $e^1 = 2.718$.

20. Which of these has the correct units?

1. Mass is measured in kilos.
2. 1 tonne $= 10^6$ kg.
3. 1 kg $= 10^9$ mg.
4. 10^3 kg/m^3 $=$ 1 kg/l.
5. 10^6 m $=$ 1 km.

21. Which of these are not the correct units?

    1. 1 hour = 3600 seconds.
    2. 60 hours = 3600 minutes.
    3. 3.6×10^3 seconds = 1 hour.
    4. 1 year = 8760 hours.
    5. 1 hour = 360 seconds.

22. Which of these has the correct units?

    1. 1 newton = 1 kg $\times$ 1 m/s^2.
    2. 1 joule = 1 kg $\times$ 1 m.
    3. 1 watt = 1 kg $\times$ g m/s^2.
    4. 10^3 joules = 3600 kN/m^2.
    5. 1 joule = 1 N/m^2.

23. Which of these has the correct units?

    1. 1 joule = 1 newton $\times$ 1 metre.
    2. 1 joule = 1 watt/s.
    3. 10^3 J = 1 kW/s.
    4. 1 watt = 10^3 joule/s.
    5. 1 MJ = 10^3 kW/s.

24. Which of these has the correct units?

    1. 1 W = 1 Nm s.
    2. 1 W = 1 Js.
    3. 1 W/s = 10^3 J.
    4. 1 W = 1 Nm/s.
    5. 1 kW/h = 10^3 J/h.

25. Which of these has the correct electrical units?

    1. 1 MW = 10^3 W.
    2. 10^3 kJ = 10^3 kW/s.
    3. 1 watt = 1 volt $\times$ 1 ampere.
    4. Electrical energy meters accumulate kW/h.
    5. 10^3 W = 10^3 V $\times$ 10^3 A.

26. Which of these has the correct electrical units?

    1. 10^3 kW/h = $10^3 \times$ 3600 W/s.
    2. kWh = energy.
    3. 1 GJ = 10^6 V $\times$ 1 A.
    4. 1 MJ = 10^6 V $\times$ 1 A.
    5. 1 kJ/s = 10^3 V $\times$ 1 A.

27. Which of these has the correct electrical units?

    1. 10^3 W = 10^3 V $\times$ 1 A.
    2. 1 MWh = 10^3 Ws.
    3. 1 kWh = 10^3 Ws.

4. 1 kWh = 1000 W/s.
5. 1 kWh = 1000 W/h.

28. Which of these has the correct units?

1. 1 atmosphere = 10^3 b.
2. 1 pascal = 1 N/m^2.
3. Pascal is a unit of radiation measurement.
4. 1 kN/m^2 = 1 b.
5. 1 mb = 10^3 N/m^2.

29. Which of these has the correct pressure units?

1. 1.01325 mb = 1 atmosphere.
2. 1 MN = 10^3 kN/m^2.
3. 1 b = 1 kN/m^2.
4. 13.6 mb = 13.6 N/m^2.
5. 1 b = 10^5 N/m^2.

30. Which of these has the correct pressure units?

1. 1 mb = 1 N/m^2.
2. 1 b = 10^3 mb.
3. 1 mb = 10^3 N/m^2.
4. 10^3 kN/m^2 = 1 b.
5. 1 mb = 10^6 b.

31. Which of these has the correct pressure units?

1. 1 Nm = 1 Pa.
2. 1000 Pa = 1 atmosphere.
3. 1 kPa = 1 kN/m^2.
4. 1 Pa = 1 mb.
5. 1 Pa = 1 N/m^2.

32. Which has the correct meaning for frequency?

1. Number of times an event is repeated.
2. Cyclic repetition of an event.
3. Number of complete rotations per unit time.
4. Statistical correlation.
5. Occasional reoccurrence.

33. Which has the correct meaning for frequency?

1. Alternating current rate of increase.
2. Electrical single or three phase.
3. Torque of a motor.
4. Air changes per hour.
5. Revolutions per minute.

34. Which is not correct in relation to frequency?

1. 3000 RPM = 50 Hz.
2. 1 Hz = 1 Nm/s.

3. High frequency fluorescent lamps work at 20000 Hz.
4. VFD means variable frequency drive.
5. 60 Hz = 3600 RPM.

35. Which is correct about Kelvin?

1. Name of engineer who designed the first steam engine.
2. Unit of heat.
3. Measured in kJ/kg s.
4. Temperature scale.
5. Absolute temperature.

36. Which is correct about Kelvin?

1. Where absolute zero gravity starts.
2. Something to do with temperature.
3. First name of Dr. K. Celsius.
4. °C + 273.
5. Engineered the first closed circuit piped heating system.

37. Which is correct about Kelvin degrees?

1. Celsius scale plus 180.
2. Are always negative values of Celsius degrees.
3. Symbol K.
4. Awarded by Kelvin University, Peebles, Scotland.
5. $K = C \times \frac{9}{5} + 32$.

38. Which is correct about Kelvin?

1. Name of a famous Scottish scientist.
2. Invented first bicycle in Scotland.
3. K = °C + 273.
4. Kelvin McAdam invented tarmacadam for road surfacing.
5. Degrees measured above absolute zero at −180°F.

39. Which is correct about Kelvin degrees?

1. Measurement of room air temperature.
2. Always used in heat transfer units.
3. Used to specify absolute temperature and temperature difference.
4. Fahrenheit plus 180.
5. Zero scale commences at −40°C.

40. Which is correct about Celsius?

1. Latin name of inventor of Roman hypocaust under floor heating system in 200 BC.
2. Fahrenheit minus 32.
3. $C = F \times \frac{5}{9} + 32$.
4. $C = 32 - F \times \frac{5}{9}$.
5. $C = (F - 32) \times \frac{5}{9}$.

41. Which is correct about Celsius?

    1. Called °C units.
    2. Kelvin degrees plus 273.13.
    3. $C = (F + 32) \times \frac{5}{9}$.
    4. Commonly used for cryogenic applications.
    5. $F = (C - 32) \times \frac{9}{5}$.

42. Which is correct about Celsius?

    1. Temperature scale in the centimetre, gram, second (CGS) metric system.
    2. Name of the Roman Senator in 35 AD who stabbed Caesar.
    3. $C = (F - 180) \times \frac{5}{9}$.
    4. Defines normal human body temperature of 98.4 degrees.
    5. Temperature scale in the metre, kilogramme, second (MKS) metric system.

43. Which is correct about volume?

    1. 1 cubic centimetre water occupies 1 litre.
    2. 1 tonne water occupies 1000 m^3.
    3. 1 m^3 = 1000 litre.
    4. 1 litre water weighs 100 kg.
    5. 1 litre water weighs 10 kg.

44. Which is correct about volume?

    1. 1 m^3 air weighs around 100 kg.
    2. 1 m^3 air weighs around 10 kg.
    3. 1 m^3 air weighs around 1 kg.
    4. 1 litre occupies 1 m^2 area and 100 mm height.
    5. 1 litre occupies 1 m^2 area and 10 mm height.

45. Which is correct about volume?

    1. 1 litre water is contained in a cube of 100 mm sides.
    2. 1 litre air is contained in a cube of 1000 mm sides.
    3. There is such a thing as a volume sensor for a control system.
    4. 100 concrete blocks of 300 mm × 200 mm × 100 mm occupy a volume of 6 m^3.
    5. 1 tonne water occupies 10 m^3.

46. A room 12 m long, 8 m wide and having an average height of 4 m, has a volume of?

    1. 400 m^3
    2. 62 m^3
    3. 462 m^3
    4. 384 m^3
    5. 192 m^3

47. Which is the correct length of a 1200 m^3 sports hall of average height 4 m and width 12 m?

    1. 25 m
    2. 10 m

3. 250 m
4. 120 m
5. 12.5 m

48. What are the units for pressure drop rate in a pipeline?

    1. m head H_2O/m run
    2. N/m^2
    3. mb/m
    4. N/m^3
    5. kN/m^3

49. What does N/m^3 stand for?

    1. Nanometres per m^2 pressure drop per metre run of pipe.
    2. Neurons per cubic metre of room volume.
    3. Newtons per square metre pressure drop per metre run of pipe or duct.
    4. Newton per cubic metre is a density.
    5. Nano-particles of radon gas per cubic metre of air in a building.

50. What does N/m^3 stand for?

    1. Normalised volumetric air change rate for a room.
    2. Number of people in a building divided by building volume.
    3. Volumetric coefficient.
    4. Noise rating divided by room volume.
    5. Pressure drop rate in a pipe or duct.

15 Answers to questions

1 Uncooled low energy design

2. Because the internal air temperature is the unknown. Heat gain into the room and heat carried away by the ventilation air flow depends on the difference between outdoor air and indoor air temperatures. An iterative solution might solve the equations but is not easy. Also, time lag is involved with heat flows into and out of thermal mass of the structure and that adds further difficulty.
3. 3
4. 5
5. 2
6. The viewing deck is uninhabitable without air conditioning from 06.30 to –20.30 h as t_{ei} remains above 26°C, peaking at 43°C at 15.00 h, as we expected.
7. Our assumption of a rectangular building, not the rounded shape, north and south entrance glazing that is very well shaded from solar gain by verandas, and full occupancy, have led to our overestimation of the heat gains. We have no published information on the natural ventilation rate. For leaky standard, empty velodrome, t_{ei} of 28°C is exceeded from 12.00 to 17.00 h but falls away rapidly in the evening, as is expected for the UK. When fully occupied, 28°C is exceeded from 10.00 to 19.00 h even with 8.3 air changes per hour, and then falls away. Indoor air temperature could be 1–2°C below environmental, as discussed earlier. We have predicted a peak t_{ei} of 32.4°C when there are 8.3 air changes/h and 42.1°C when there are 3 air changes/h, meaning a measurable range for t_{ai} to be 30–40°C d.b. This can be considered to be a very satisfactory outcome from such a simplified assessment, as the designers' published modelling predictions show track air to be 30°C and 36°C around the spectators and ceiling.
8. Tight air leakage standard selected. The occupants did not make use of the manually controlled ventilators, windows or openable roof lights; summer overheating led to the installation of room air conditioners. No air leakage test was conducted. 26°C is exceeded during working hours of 10.30–19.30 h.
9. Little point in making the calculation. Internal heat gains from people, lighting and computer systems may exceed heat gains from the external environment. Internal heat gains are a constant and may exceed heat loss from the building during cold weather. Mechanical ventilation and air conditioning is a necessity. There is no way any form of uncooled ventilation system could maintain comfort.
10. Using an average air leakage standard, a consistently warm internal environment during summer is maintained. Thermal storage time lags are so long, days rather than hours, once warmed that the

inside remains warm; rather like in deep caves. On a typical London summer day, the interior does not overheat and peaks at 26.2°C environmental temperature, not that the builders would have understood that at all. Note the evenly warm conditions only range between 22.3°C and 26.2°C. We have no reason to think that weather was significantly different in 1078 to today, apart from being told that today is warmer due to our use of hydrocarbon fuels. What has modern design learned? Obviously we design and build much faster in modern times as compared to ancient builders, but our glass-ridden, lightweight steel and concrete towers are uninhabitable without mechanical cooling and mechanical ventilation. Have we improved on the ancients' work? Have a look and visit buildings of 1000 years of age. All buildings require maintenance or they fall down, so there is no change there. We might conclude that modern life has seen a retrograde step in the design of buildings and led us to burn more hydrocarbon fuels that emit carbon dioxide into the atmosphere, which leads to climate change, global warming and a need to reduce our dependence on such fuels. What do you think?

11. Notice how the internal environmental temperature tracks the outside air temperature. With so little thermal lag in the structure of the lightweight caravan and a necessarily high ventilation rate, internal conditions are outside air plus solar heat gains for the ventilated box. 26°C environmental temperature is exceeded during 10.00–18.30 h. The occupants are on holiday, they might switch on a fan to improve air flow and adjust clothing to be more comfortable. If this was the office or factory of their employment they would complain, but here they accept discomfort, and almost certainly choose to sit in the shade outdoors.
12. 26°C environmental temperature is exceeded during 09.30–18.45 h with a peak of 32.5°C without insulation. Opening up the doors and windows to maximise outdoor air ventilation makes little difference. The lack of thermal insulation and storage in the structure allows indoor temperature to match that of outdoors overnight. Insulating the walls and roof reduces the peak to 29.6°C and maintains a warmer interior overnight. Increasing the ventilation lowers interior temperatures and makes the building more manageable. The principle advantage of using corrugated steel is in creating shade; it also heats up rapidly to a high temperature and convects heat away. In addition, it is a cheap and weatherproof building material, easily transported and erected. In temperate climates, it needs a lot of added thermal and sound insulation.
13. 26°C environmental temperature is exceeded during 09.00–19.00 h with a peak of 34.1°C using a tight leakage standard. Opening up the doors and windows to maximise outdoor air ventilation makes little difference. The lack of thermal insulation and storage in the structure allows indoor temperature to match that of outdoors overnight. When we live outdoors, we accept discomfort as part of the experience and adjust our activities to suit the conditions. We might complain, but there is no point, as being there is our own fault. However, we might appreciate our centrally heated, draught proof, cooled and electrically lit homes and workplaces more as a result.
14. Housing faces east and west in streets like Hurst. All occupants were at work all day. Large windows admitted daylight and morning sunshine into east facing bedrooms and the living room to wake people; late evening sunshine warmed the rear elevation and yard. The opposite side of the street did not enjoy these advantages. 26°C environmental temperature is exceeded during 08.00–20.00 h with a peak of 28.2°C during 16.00–18.00 h. Overnight temperature only dropped to 23.3°C so the houses remained warm in summer.
15. Unfortunately, a bad outcome is found. 26°C environmental temperature is continuously exceeded and has a peak of 29.1°C. The occupants will find indoor home conditions too hot and will not accept them. Opening windows to create through flow from the street to the back garden would help, but without a prevailing wind the house would be very uncomfortable. The occupants have air conditioning in their cars, in shopping malls, hotels, air travel, restaurants and in the offices where they work. They might be satisfied by purchasing several 100 W portable fans to cool them, but will be severely tempted to

install a reverse cycle packaged air conditioning unit for their home, adding another 1 kW of electrical demand.

Copy the *Oldham improved* file and save it as *Oldham unpowered*. Delete the occupants, lighting and the entire electrical load. The uninhabited house remains comfortable with an average t_{ei} of 22.9°C and a peak of 24.3°C. The problem is not the insulated house. It is the electrical demand that occupants create which causes overheating and a need for a mechanical cooling system. Will our craving for modern living overcome the objectives of The Kyoto Protocol (1990), and HM Government Carbon Plan (2011)? It is possible. What do you think?

16. 3

2 Air conditioning systems

11.

Side	*Glass, m²*	*Peak,* $\frac{W}{m^2}$	*Peak, kW*	*Date*	*Time, h*
S	80	510	40.8	Sept	13.00
E	80	477	38.16	June	09.00
W	80	477	38.16	June	17.00
N	80	161	12.88	July	13.00

Plant refrigeration load peak is not found from the sum of the peaks of each side.

27. Rotating drum, compressed air jet, spinning disc, steam injection, capillary cells, sprayed coil, pan humidifier, ultrasonic humidifier and infrared evaporator.
29. Ambient air, ground water, rivers or lakes, sea water, air cooled condenser, dry air cooler, open circuit cooling tower and evaporative condenser.

3 Heating and cooling loads

5. 44.8 m.
6. total height 46.15 m + 15.593 m = 61.744 m, 3.25 m each.
7. vertical height 20.139 m, sloping face length 20.59 m and area 617.7 m^2.
8. 52°, 69.3°.
9. D 69°, F 87°, 44.3 $\frac{W}{m^2}$.
10. D 0°, F 33°, 686.9 $\frac{W}{m^2}$.
11. From CIBSE the maximum may occur on 22 September at 12.00 h, I_{THd} 625 W/m², I_{TVd} 710 W/m², I_{dHd} 190 W/m², I_{TSd} 878.4 W/m². Check other dates and times.
12. I_{DVd} 303 W/m², Q_D 1569.5 W, Q_d 1738.8 W, total 3308.3 W.
13. I_{DSd} 632 W/m², Q_D 872 W, Q_d 256 W, total 1128 W.
14. t_g 39.8°C, Q 324 W/m², total T 0.52.
15. t_g 4.8°C, Q – 59 W/m², total T – 0.34, heat flow outwards from room.
16. F_u 0.54, F_v 0.92, F_y 0.96, Q_u 264 W, $\tilde{Q}_u$ 2402 W, net Q_u 2666 W.
17. F_u 0.98, F_v 0.5, F_y 0.88, Q_u 466 W, $\tilde{Q}_u$ –12 W, net Q_u 454 W.
18. Solar gain through the glazing 4304 W, 24 hour mean conduction through the structure –151 W, swing in the conduction gain –329 W, ventilation air infiltration –820 W, occupancy gain 180 W, electrical equipment emission 500 W, total heat gain 3684 W.

19. F_u 0.94, F_v 0.93, F_y 0.89, F_2 1.17, solar gain through the glazing 14208 W, 24 hour mean conduction through the structure −1554 W, net swing in the conduction gain 2324 W, outdoor ventilation air infiltration −912 W, indoor air infiltration 1216 W, occupancy gain 3600 W, electrical equipment emission 5050 W, mean conduction gain from below 6545 W, total net gain 30477 W, 30.477 kW, 14.5 W/m^2 of the office volume.
20. 485 W
22. Total mean gain Q 1586 W, mean t_{ei} 24.7°C, total swing in gains $\tilde{Q}_t$ 4095 W, $\tilde{t}_{ei}$ 3.2°C, peak t_{ei} 27.9°C.
23. Mean internal gain 48 kW, mean t_{ei} 21.2°C, total swing in gains $\tilde{Q}_t$ 399.375 kW, $\tilde{t}_{ei}$ 3.3°C, peak environmental temperature t_{ei} 24.5°C.
25. 73 mm.
26. 5.473 m^2.
28. Net gain of 531 W.
32. (c) approximately 10.6°C.

4 Psychrometric design

1. 46%, 0.846 m^3/kg, 10.1°C, 41.7 kJ/kg, 0.0077 kg/kg.
2. Entry 0.0038 kg/kg, 9.45 kJ/kg, 0.778 m^3/kg, 0°C dew point. Exit 30°C d.b., 14.5°C w.b., 14% saturation, 0.864 m^3/kg 40 kJ/kg, 0°C dew point.
3. Entry 12°C d.b., 4.1°C w.b., 20%, 0.00175 kg/kg, 16.5 kJ/kg, 0.81 m^3/kg, −9°C dew point. Heated to 37°C d.b., 15.3°C w.b., 4%, 0.00175 kg/kg, 41.5 kJ/kg, 0.881 m^3/kg, −9°C dew point. Humidified 15.8°C d.b., 14.8°C w.b., 90%, 0.0102 kg/kg, 41.5 kJ/kg, 0.832 m^3/kg, 14.4°C dew point.
4. Outdoor 43%, 0.00118 kg/kg, 60.3 kJ/kg, 0.875 m^3/kg, 16.3°C dew point. Cooled 11.1°C w.b., 0.0079 kg/kg, 32 kJ/kg, 10.5°C dew point.
5. Outdoor 58%, 0.0141 kg/kg, 64 kJ/kg, 0.872 m^3/kg, 19.2°C dew point. Cooled 5°C d.b., 5°C w.b., 100%, 0.0054 kg/kg, 18.6 kJ/kg, 0.794 m^3/kg, 5°C dew point. Reductions of 45.4 kJ/kg and 0.0087 kg/kg.
6. 26°C d.b., 18.6°C w.b., 48%, 0.0103 kg/kg, 52.4 kJ/kg, 0.861 m^3/kg, 14.4°C dew point.
7. Mixed 17.4°C d.b., 13°C w.b., 61%, 0.0076 kg/kg, 36.5 kJ/kg, 0.833 m^3/kg, 9.8°C dew point. Heated 35°C d.b., 19.5°C w.b., 21%, 0.0076 kg/kg, 54.8 kJ/kg, 0.883 m^3/kg, 9.8°C, dew point. Humidified 24°C d.b., 19.3°C w.b., 64%, 0.0121 kg/kg, 54.8 kJ/kg, 0.858 m^3/kg, 16.9°C, dew point. Duty 54.922 kW.
8. 113.779 kW
9. Inlet 52%, 0.0134 kg/kg, 63.3 kJ/kg, 0.874 m^3/kg, 18.5°C, dew point. Cooled 12.1°C w.b., 0.0084 kg/kg, 34.4 kJ/kg, 0.821 m^3/kg, 11.4°C dew point. Duty 132.265 kW.
10. 163.5 kW
11. Mixed 23.7°C d.b., 0.010 kg/kg, 55%, 17.5°C w.b., 49.5 kJ/kg. Cooled 7.8°C d.b., 0.0063 kg/kg, 95%, 7.4°C w.b., 23.7 kJ/kg. Duty 145 kW, 41.232 tonne refrigeration, condensate 74.867 kg/h.
12. −5.9°C w.b., 0.7615 m^3/kg, 0.00198 kg/kg, 10%, 76.428 kW
13. 14.2°C w.b., 91% saturation, 48.055 kW
14. 19.6°C d.b., 0.0075 kg/kg, 38.7 kJ/kg, 0.839 m^3/kg, 13.9°C w.b.
15. No, 21.2°C w.b., 0.877 m^3/kg, 6.681 kW
16. p_s 2.9808 kPa, g_s 0.01885 kg/kg, for t_{sl} 16°C p_s 1.8159 kPa, p_v 1.276 kPa, g 0.00793 kg/kg, PS 42 %, h 44.3, kJ/kg p_v 1276 Pa, v 0.8523 m^3 per kg dry air, density 1.173 kg/m^3, t_{dp} 11.03°C. Accuracy is acceptable for most purposes.
17. p_s 0.6564 kPa, g_s 0.00406 kg/kg, for t_{sl} 0.5°C p_s 0.633 kPa, p_v 0.6 kPa, g 0.0037 kg/kg, PS 91%, h 10.26 kJ/kg, v 0.781 m^3 per kg dry air, density 1.28 kg/m^3, t_{dp} −0.01°C

18. p_s 4.751 kPa, g_s 0.0306 kg/kg, for t_{sl} 22°C p_s 2.641 kPa, p_v 1.966 kPa, g 0.0123 kg/kg, PS 40%, h 63.65 kJ/kg, v 0.881 m^3 per kg dry air, density 1.135 kg/m^3, t_{dp} 17.3°C.

5 System design

1. 1.999 m^3/s.
2. Q 1.2 m^3/s, SH 7.287 kW.
3. Q 2.25 m^3/s, t_s 15.1°C.
4. Q 3.891 m^3/s.
5. Q 2 m^3/s, SH 16.49 kW.
6. Q 1.68 m^3/s, t_s 33°C.
7. Summer Q 5.315 m^3/s, N 12.8 $\frac{\text{air changes}}{\text{h}}$, v 0.822 m^3/s, M 6.466 $\frac{\text{kg}}{\text{s}}$. Winter t_s 28.3°C d.b., v 0.863 m^3/s, Q 5.58 $\frac{m^3}{s}$, recalculated t_s 28°C d.b.
8. Q 1.454 m^3/s, g_s 0.008 825 kg/kg.
9. $\frac{S}{T}$ ratio 0.96, Q 6.722 m^3/s, g_s 0.007709 kg/kg.
10. All air flows l/s, fresh air inlet 960, total supply 12821, office supply 10630, office extract duct 9942, corridor supply and extract 1962, toilet supply duct 229, transfer grille into corridor and toilet 688, toilet separate exhaust 917, plant exhaust 43, plant recirculation duct 11861.
11. 9.846 m^3/s, 0.0073 kg/kg.
12. Q 0.4 m^3/s, t_s 7°C d.b., 0.007 kg/kg.
13. SH gain 13.75 kW, Q 2.83 m^3/s, 0.00725 kg/kg.
14. Summer SH gain 12.974 kW, 14.4 W/m^3, Q 1.331 m^3/s, N 5.3 air changes/h, g_r 0.008 905 kg/kg, g_s 0.008 590 kg/kg, fresh air proportion 22.5%, t_m 24.4°C d.b., t_m 17.8°C w.b., h_m 50 kJ/kg, 0.856 m^3/kg, supply 15°C d.b., 13°C w.b., cooling coil 20.665 kW. Winter SH loss 9.57 kW, 10.6 W/m^3, Q 1.331 m^3/s, mixed air t_m 14.5°C d.b., g_m 0.006 kg/kg, room 45% saturation produced, h_m 29.75 kJ/kg, 0.822 m^3/kg, supply 25.1°C d.b., 14.7°C w.b., h_s 40.55 kJ/kg, heating coil 17.488 kW.
15. Summer SH gain 2.928 kW, 19.5 m^3/s, N 9.6 air changes/h, total Q 402 l/s where 121 l/s enters from the hot duct at 23°C and 281 l/s is from the cold duct at 13°C, g_s 0.008282 kg/kg, fresh air 6%. Winter t_s 23°C d.b., t_m 19°C d.b., total supply Q 402 l/s comprising 268 l/s, 25°C from the hot duct plus 134 l/s at 19°C from the cold duct.

6 Ductwork design

5. 1.183 kg/m^3.
6. 46.6°C d.b.
7. 245 Pa, 5 kPa, 1226 Pa, 2942 Pa, 25 kPa.
8. 3.82 m/s.
9. 1 m^3/s, 2.26 m^3/s, 6.28 m^3/s, 25.13 m^3/s.
10. 1.145 kg/m^3, 28 Pa, −413 Pa.
11. 1.181 kg/m^3, 10.2 m/s, 361 Pa.
12.

Node	*P_t, Pa*	*P_v, Pa*	*P_s, Pa*
1	600	22	578
2	595	22	574
3	591	62	529
4	573	62	511

13.

Node	P_t, *Pa*	P_v, *Pa*	P_s, *Pa*
1	250	29	221
2	240	29	211
3	237	146	91
4	42	146	−104

14.

Node	P_t, *Pa*	P_v, *Pa*	P_s, *Pa*
1	400	221	179
2	320	221	99
3	144	38	106
4	126	38	89

Static regain *SR*, $(P_{S3} - P_{S2}) = 6$ Pa.

15.

Node	P_t, *Pa*	P_v, *Pa*	P_s, *Pa*
1	200	36	164
2	165	36	129
3	145	2	143
4	145	2	143

Static regain *SR*, $(P_{S3} - P_{S2}) = 13$ Pa.

16.

Node	P_t, *Pa*	P_v, *Pa*	P_s, *Pa*
1	400	21	379
2	391	21	370
3	389	30	360
4	379	30	349
5	375	23	352
6	365	23	342

Static regain *SR*, $(P_{S3} - P_{S2}) = -10$ Pa.

17.

Node	P_t, *Pa*	P_v, *Pa*	P_s, *Pa*
1	200	36	164
2	183	36	147
3	183	16	167
4	175	16	159
5	170	38	132
6	166	38	128

Static regain *SR*, $(P_{S3} - P_{S2}) = 20$ Pa.

18. v_1 15.3 m/s, p_{v1} 136 Pa, v_2 12.5 m/s, p_{v2} 91 Pa, p_{s1} −883 Pa, p_{t1} −747 Pa, p_{s2} 1162 Pa, p_{t2} 1253 Pa, *FTP* 2000 Pa, *FVP* 91 Pa, *FSP* 1909 Pa.
19. There is more than one correct solution to this question. Duct sizes can be:

    Section 1–2; 1200 mm × 1500 mm;
    Section 3–4; 600 mm × 600 mm;
    Section 5–6; 1200 mm × 700 mm;
    Section 6–7; 800 mm × 400 mm;
    Section 6–8; 1200 mm × 500 mm.

Section	*Length, l m*	$\frac{\Delta p}{l}, \frac{Pa}{m}$	$v, \frac{m}{s}$	*p_v, Pa*	*k*	*Fitting, Pa*	*Duct, Pa*	*Total, Pa*	
1–2	0	0.05	2.3	3	3	10	0	10	*
2–3	3	0.07	2.3	3	0.07	105	0	105	*
3–4	0	2.16	11.7	82	0.04	3	0	3	*
4–5	0	0.24	5.1	15	0.36	29	0	29	*
5–6	16	0.24	5.1	15	0.35	5	4	9	*
6–7	0	0.42	5	15	1.25	49	0	49	*
6–8	15	0.31	4.5	12	0.09	31	5	36	
Index route *				Total Δp for index route				206 Pa	

20. There is more than one correct solution to this question. Duct sizes can be:

    Section 1–2; 2000 mm × 1600 mm;
    Section 2–3; 1350 mm × 1600 mm;
    Section 3–4; 650 mm × 650 mm;
    Section 5–6; 1000 mm × 800 mm;
    Section 6–7; 800 mm × 600 mm;
    Section 6–8; 800 mm × 450 mm.

Section	*Length, l m*	$\frac{\Delta p}{l}, \frac{Pa}{m}$	$V, \frac{m}{s}$	*p_v, Pa*	*k, Pa*	*Fitting, Pa*	*Duct, Pa*	*Total, Pa*	
1–2	0	0.02	2	2	2.1	5	0	5	*
2–3	9	0.06	2.9	5	0.07	225	1	226	*
3–4	0	2.9	14.9	133	0.45	60	0	60	*
4–5	0	0.6	7.9	37	0.04	5	0	5	*
5–6	48	0.6	7.9	37	0.35	13	29	42	*
6–7	0	0.81	7.7	35	1.25	114	0	78	
6–8	45	1	7.4	33	0.09	73	45	118	*
Index route *				Total Δp for index route				456 Pa	

7 Controls

4. 6.25 volt, 1250 ohm.
7. 500 ohms.
23. Vapour compression reciprocating, scroll, screw and centrifugal. Absorption, gas-fired, waste heat supplied.

8 Commissioning and maintenance

1. Allowable leakage 169 l/s, test leakage 161 l/s, duct passes test.
2. Assume duct air p_S is 300 Pa to estimate ρ 1.234 kg/s, Δp 2256 Pa, test orifice flow Q 100 l/s, for a duct air leakage of 100 l/s, p_S is 290 Pa, watch for $X^{1/Y}$ function, h is 29.6 mm H_2O.
3. Allowable leakage 151 l/s, leakage flow on test 185 l/s, duct fails test.

9 Fans

30. F_2 6.3 kg.
31. *FVP* 34 Pa.
32. N_1 20 Hz, N_2 16 Hz, Q_2 6667 l/s, Δp_2 267 Pa, P_2 5267.5 W.
33. At the design air flow, power 4.286 kVA, at the closed damper air flow, power 0.857 kVA, phase current 1.19 ampere.
34. Q 4.9 m^3/s, *FTP* 90 Pa, 1.2 kVA, 1.7 amp.
35. Q 1200 l/s, *FTP* 680 Pa, 2.3 kVA, 3.2 amp.
36. Q 4200 l/s, *FTP* 550 Pa, 5.6 kVA, 7.8 amp.
37. Annual energy costs (a) £979.87, (b) £822.35, (c) £620.30, (d) £426.53. Annual cost savings (b) £157.52, (c) £359.57, (d) £553.34.

10 Fluid flow

2. 1.014 kg/s.
3. 15 mm
4. 15 mm 0.107 kg/s, 22 mm 0.311 kg/s, 28 mm 0.624 kg/s, 35 mm 1.12 kg/s, 42 mm 1.881 kg/s 54 mm 3.801 kg/s.
5. 1.264 kg/s.
6. 0.49 l/s.
7. 22 mm 1.13 l/s, 28 mm 2.3 l/s, 35 mm 4.19 l/s, 42 mm 7.11 l/s, 54 mm 14.56 l/s.
8. 0.507 m^3/s.
9. Q 0.782 m^3/s, v 4.92 m/s.
10.

$Q, \frac{m^3}{s}$	*d, mm*
0.12	200
0.3	250
0.6	350
1	450
2	600

11. For LTHW and 3.5 kW, M 0.07 kg/s; for LTHW and 420 kW, M 8.345 kg/s; for HTHW and 3.5 kW, M 0.027 kg/s; for HTHW and 420 kW, M 3.296 kg/s.
12. M 0.497 kg/s, 28 mm pipe, $\Delta p/l$ 330 Pa/m, v 0.941 m/s.
13. M 4.316 kg/s, 54 mm, $\Delta p/l$ 630 Pa/m, v 2.11 m/s.
14. For Table X copper pipes, LTHW M 1.987 kg/s, d 54 mm, $\Delta p/l$ 150 Pa/m, v 0.96 m/s. For galvanised metal duct, M 8.235 kg/s, Q 7.331 m^3/s, d 1.1 m, $\Delta p/l$ 0.475 Pa/m, v 7.72 m/s.

Distribute the air handling plant around the building to minimise air duct runs and supply each plant with LTHW pipework.

15. For Table X copper pipe, LTHW, M 1.49 kg/s, d 35 mm, $\Delta p/l$ 840 Pa/m, v 1.83 m/s, pump head H 157.5 kPa, pump power consumption 370.3 W.

    For galvanised steel air duct, Q 5.498 m^3/s, d 0.9 m, $\Delta p/l$ 0.75 Pa/m, v 8.64 m/s, fan pressure rise H 196.875 Pa, fan power consumption 1.665 kW.
17. Stored energy 35.2 kWh, heater input power 4.4 kW.
18. Storage capacity 864 MJ, heater power 30 kW, tank needs to be 2.907 m × 2.907 m × 1 m high.
19. Cooling load 115 kW, compressor runs for 30 minutes per hour for four runs of 7.5 minutes each, chilled water storage tank size is 4.44 m × 4.44 m × 1 m.

11 Air duct acoustics

5. 87 dBA.
6. 91 dBA.
10. 86, 83, 80, 78, 74, 70, 65, 58, 48 dB.
11. 83, 82, 81, 79, 77, 75, 72, 69, 65.
22. 72 dB.
23. 82 dB.
24. *NR* 35, yes.
25. *NR* 55, not suitable, more attenuation required or use a different fan.
26. *NR* 20, highly suitable.
27. Maximum *NR* allowed in a library is *NR* 35. 1 m of lined duct reduces higher frequencies to below *NR* 30.
28. 1 m of lined duct reduces the critical 500 Hz *SPL* to *NR* 35 suitable for a lounge.
29. Without duct attenuation the restaurant experiences just over *NR* 60 and this is unacceptable; *NR* 40 is needed. A serious problem is the harshness of the room, every surface is hard, bouncing sound and creating multiple echoes with a reverberation time of 2.5 s at 1000 Hz, the most noticeable for listening. Hanging 250 m^2 of acoustic panels from the ceiling and 100 m^2 of drapes to hide some of the stone wall halves reverberation time but does not reduce room sound below *NR* 60; additionally lining 3 m of duct lowers the room to *NR* 35; this would greatly improve the aural environment without permanently losing the interior heritage style.

12 Air conditioning system cost

No model answers.

13 Question bank

1.3	2.5	3.5	4.5	5.5	6.3	7.5	8.5	9.5	10.4	11.5	12.5	13.3
14.3	15.3	16.4	17.5	18.4	19.3	20.3	21.5	22.4	23.3	24.5	25.1	26.5
27.4	28.3	29.4	30.5	31.5	32.2	33.4	34.3	35.5	36.2	37.3	38.3	39.4
40.4	41.5	42.5	43.3	44.5	45.3	46.3	47.1	48.3	49.3	50.5	51.5	52.3
53.4	54.3	55.3	56.4	57.4	58.5	59.1	60.5	61.5	62.5	63.2	64.5	65.2
66.3	67.5	68.4	69.5	70.4	71.4	72.4	73.5	74.4	75.5	76.5		

14 Understanding units

1.4	2.4	3.2	4.3	5.5	6.3	7.4	8.5	9.4	10.5	11.4	12.3	13.4
14.5	15.5	16.2	17.5	18.4	19.5	20.4	21.5	22.1	23.1	24.4	25.3	26.2
27.1	28.2	29.5	30.2	31.3	32.3	33.5	34.2	35.5	36.4	37.3	38.3	39.3
40.5	41.1	42.5	43.3	44.3	45.1	46.4	47.1	48.4	49.3	50.5		

References and further reading

AIRAH Technical Handbook, edition 5, 2013, The Australian Institute of Refrigeration, Air Conditioning and Heating, http://www.airah.org.au

AS 1460-1983 Acoustics-Methods for the determination of noise rating numbers, http://infostore.saiglobal.com/store/

BSRIA (2010) Soft Landings Framework and CIBSE Building Log Book Toolkit Bundle, Chartered Institution of Building Services Engineers, http://www.cibse.org/

Building Services, The CIBSE Journal, Chartered Institution of Building Services Engineers, http://www.cibsejournal.com/

Cement and Concrete Association of Australia (2002) The Labyrinth under Federation Square, Melbourne, *Mix*, 10, July.

Chadderton, David V. (1997) *Building Services Engineering Spreadsheets*, E & F N Spon, London.

Chadderton, David V. (2013) *Building Services Engineering*, sixth edition, Routledge, Abingdon.

CIBSE (1997) *The Quest for Comfort, Centenary 1897–1997*, Centenary of the CIBSE Heritage Group.

CIBSE Commissioning Code A (2004) Air Distribution, Chartered Institution of Building Services Engineers, http://www.cibse.org/

CIBSE Commissioning Code B (2002) Boilers, Chartered Institution of Building Services Engineers, http://www.cibse.org/

CIBSE Commissioning Code C (2009) Automatic Controls, Chartered Institution of Building Services Engineers, http://www.cibse.org/

CIBSE Commissioning Code R (2002) Refrigerating Systems, Chartered Institution of Building Services Engineers, http://www.cibse.org/

CIBSE Commissioning Code W (2010) Water Distribution Systems, Chartered Institution of Building Services Engineers, http://www.cibse.org/

CIBSE Guide A (2006) Environmental Design, Chartered Institution of Building Services Engineers, http://www.cibse.org/

CIBSE Guide B (2005) Heating, Ventilation, Air Conditioning and Refrigeration, Chartered Institution of Building Services Engineers, http://www.cibse.org/

CIBSE Guide C (2007) Reference Data, Chartered Institution of Building Services Engineers, http://www.cibse.org/

CIBSE Guide H (2001) Building Control Systems, Chartered Institution of Building Services Engineers, http://www.cibse.org/

CIBSE Guide M (2008) Maintenance Engineering and Management, Chartered Institution of Building Services Engineers, http://www.cibse.org/

CIBSE KS02 (2005) Managing Your Building Services (CIBSE Knowledge Series KS2), Chartered Institution of Building Services Engineers, http://www.cibse.org/

CIBSE KS04 (2005) Understanding Controls (CIBSE Knowledge Series KS4), Chartered Institution of Building Services Engineers, http://www.cibse.org/

CIBSE PROBE reports available for members to download from: https://www.cibse.org/membersservices/downloads/listings.asp?pid=373

CIBSE TM26 (2000) Hygienic Maintenance of Office Ventilation Ductwork, Chartered Institution of Building Services Engineers, http://www.cibse.org/

CIBSE TM31 (2006) Building Log Book Toolkit (includes templates on CD-ROM), Chartered Institution of Building Services Engineers, http://www.cibse.org/

CIBSE TM52 (2013) The Limits of Thermal Comfort: Avoiding Overheating in European Buildings, Chartered Institution of Building Services Engineers, http://www.cibse.org/

DUALL Project (1996) Queen's Building, De Montfort University, Leicester, PROBE 4, *Building Services Journal*, April, http://duall.iesd.dmu.ac.uk/1010buildings/

HVCA (1995) Heating and Ventilating Contractors Association, HVCA Green Building of the Year Award, http://www.hvca.org.uk/index.php

HM Government Carbon Plan (2011) The Department of Energy and Climate Change, DECC, 3 Whitehall Place, London, SW1A 2AW, http://www.decc.gov.uk

Honeywell (1997) Engineering Manual of Automatic Control for Commercial Buildings, https://customer.honeywell.com

Johnson Controls (2008) Building Automation System over IP (BAS/IP), Design and Implementation Guide, http://www.cisco.com

Kyoto Protocol (1997) http://unfccc.int/kyoto_protocol/items/2830.php

Low Energy Cooling (n.d.) Good Practice Guide 5, http://www.islington.gov.uk

Moss, B. (1991) A New Approach to Airports, *The CIBSE Journal*, 13(10), October, p.22, http://www.cibsejournal.com/

Siemens AG Download Center App (n.d.) http://www.siemens.com

Spon's Mechanical and Electrical Services Price Book 2013, 44th edition, Davis Langdon, Routledge, http://www.pricebooks.co.uk

TermoDeck International Ltd (n.d.) http://www.termodeck.com/

The Australian Institute of Refrigeration, Air-Conditioning and Heating (1997) Application Manual DA20, Humid Tropical Air Conditioning, http://www.airah.org.au

The Government's Energy Efficiency Best Practice Programme (1997) New Practice Case Study 102, 1997.

Index